Metagenomics of the Human Body

Karen E. Nelson
Editor

Metagenomics of the Human Body

Foreword by Jane L. Peterson and Susan Garges

Springer

Editor
Karen E. Nelson
J. Craig Venter Institute
9704 Medical Center Drive
Rockville, MD 20850, USA
kenelson@jcvi.org

ISBN 978-1-4419-7088-6 e-ISBN 978-1-4419-7089-3
DOI 10.1007/978-1-4419-7089-3
Springer New York Dordrecht Heidelberg London

Library of Congress Control Number: 2010937845

Cover illustration: © vege - Fotolia.com

Printed on acid-free paper

Springer is part of Springer Science+Business Media (www.springer.com)

Karen E. Nelson would like to thank her colleagues with whom she has worked for several years, for their support and involvement in this publication. They have allowed her to appreciate the wonders of genomics. Special thanks are due to J. Craig Venter and Hamilton Smith.

Foreword
The Human Genome and the Human Microbiome

The first decade of the twenty-first century has seen, as a result of the completion of the Human Genome Project (HGP), the launch of an extraordinary new capability to make unprecedented progress in our understanding of human genetics and its role in disease. Now, at the end of that decade, scientific exploration is poised to add another new, perhaps just as revolutionary, capacity to our study of human biology and disease, the capability to study the human microbiome. The human microbiome, the complete set of microbes that live in and on the human body, is thought to play a major role in human health and disease, either directly through the expression of the microbial genes, or indirectly through interaction with human physiology, particularly with the immune system.

As was the case for the HGP, the call for a human microbiome project came from the scientific community: Julian Davies noted in 2001 that, although completing the human genome sequence was a "crowning achievement" in biology, it would be an incomplete tool for understanding human biology until the synergistic activities between humans and the microbes living in and on them are understood (Davies, 2001). Relman and Falkow (2001), at about the same time, called for a "second human genome project" that "would entail a comprehensive inventory of microbial genes and genomes at the four major sites of microbial colonization in the human body: mouth, gut, vagina, and skin."

In thinking about human biology in terms of both the human genome and the human microbiome, comparisons between the two are inevitable. But how should the two be compared and, in particular, how could one get a sense of the potential impact of the human microbiome on human health from such a comparison? One possible way to start would be to compare the number of human cells and microbial cells. The frequently stated number of microbial cells per human body is 10^{14}, which exceeds the number of human cells in the body by 10-fold. Of course, bacterial cells are considerably smaller in volume than human cells and the DNA content of bacterial cells is also much less. The genome sizes of microorganisms vary considerably and evolve continually due to lateral transfer of DNA between them. Overall, the human genome is probably 1,000 times bigger than a typical microbial genome (note that while bacteria make up the majority of the microbiome in humans, Archaea, viruses and eukaryotic microbes are also significant components); so, numerically, most of the DNA in the human organism is contained in its human cells. But looking at the relative number of human and microbial genes tells a different story. The best current estimate is that the human genome contains about 20,000 genes. But the number of unique genes in the microbiome is estimated to be at least 100 times greater, owing, of course, to the tremendous diversity of species and strains of microbes present. Thus, even though our microbial passengers are "invisible," as a community, they account for a very large number of genes whose expression could potentially have an important impact on our health. Understanding the resident microbial organisms, their genes and how they affect human health is the primary goal of the human microbiome projects around the world today.

The very nature of the genomes being studied by the HGP and a collective "human microbiome project" dictates that the course of the two projects will be different. The human genome is a relatively stable genome of 3 billion base pairs that varies, on average, only 0.1% between individuals. Thus the goals for sequencing the human genome could be clearly defined, because they were bounded by its size and limited variation (further description of the full set of variations across the human population, and interpretation of the human genome sequence are, of course, much more complicated and much less bounded endeavors). On the other hand, a human microbiome project appears to be effectively unbounded and likely to be ongoing for many years for many reasons, chief of which is the great amount of variation in the composition of the microbial populations at any one site on/in a human, at different sites on/in a single human, at different times in the life of a single human, and between individual humans.

Yet there actually may be more similarities than differences between the two projects, starting with the challenges faced by each. Both projects were proposed and subsequently launched at a time when it was not yet clear that available technology would be capable of producing the desired outcome. At the beginning of the HGP in 1990, it was not possible to sequence 1 Mb, nonetheless a genome! Similarly, in the early days of research on the human microbiome, it is not apparent how well our current technology and tools will allow for deep and accurate metagenomic analysis of the human microbiome. Both projects faced even more substantial bioinformatics challenges than were unappreciated at the beginning of each. Finally, they both committed to open, rapid data release, which was not the norm in either the human genetics or microbial research communities at the time. Most importantly, there is the underlying biology. Evidence indicates that the human genome and the human microbiome have evolved together, so that microbes that provide a benefit to the human (such as those that provide needed enzymes for digestion) thrive in the human, but when detrimental changes occur in the microbiome (such as in inflammatory bowel disease), human disease may arise.

For several years now, research agencies around the world have funded individual investigators to study the human microbiome. More recently, these agencies recognized that there are probably aspects of microbiome studies that would benefit from being carried out by large consortia that could generate, in a much more rapid and cost-effective manner, the answers to some very basic questions about the feasibility and value of microbiome studies and some very high-quality data resources that individual researchers could then use to study particular research questions as they investigate the complexity and function of the human microbiome in multiple body sites. This is another parallel with the HGP, which has inevitably led to the concept of a "Human Microbiome Project" (HMP).

But, what is a "Human Microbiome Project?" There are many different approaches that could (and should) be taken in studying the content and function of the human microbiome. Are systematic studies of the microbiome possible or is the inherent variation too great to allow noise to be distinguished from signal? Is there a core microbiome and can it be identified? If so, what is so important about those core microbes that their presence is conserved across environment, age, and time? How does the microbiome change under different environmental conditions (of which there are thousands)? How does it change during human development? Can a causal relationship between the composition of the microbiome and important biological phenomena (e.g., good health or specific disease) be demonstrated? Can manipulation of the human microbiome affect human health?

Different international research programs have taken different approaches to these large issues. In some cases (the European Union's effort and the effort in China), a specific biological system (the human gut) was chosen as the focus. By contrast, the National Institute of Health (NIH) decided to identify a limited number of factors that could be studied over a 5-year time period, to ensure that its project would be focused enough to answer a few important questions. The NIH HMP's goal is to demonstrate the feasibility and value of microbiome studies as an approach to human health by demonstrating enough about the correlation and, it is to be hoped, function, of the human microbiome to encourage more investment in studying the microbiome in humans. Then, in one more parallel with the HGP, an international consortium of researchers interested in common goals has been established. In 2008, the International Human Microbiome Consortium (IHMC) was established. At the time of this writing, 10 funded international groups studying the human microbiome at various body sites and in different diseases are working together to share data, eliminate redundancy and ensure common policies.

Together, these two major founding projects and those described in this volume are expected to create a new field of investigation that will allow us to understand the connection between the human microbiome and human biology, to both confirm what we now postulate to be the microbiome's important role in human health and use that knowledge to manipulate the human microbiome to maintain or restore individuals' health. The articles in this volume represent many of the current efforts aimed at accomplishing these critical goals.

Bethesda, Maryland

Jane L. Peterson
Susan Garges

Contents

Contributors

Paul J. Albert Weill Cornell Medical College, 525 East 68th Street, New York, NY 10065, USA, paa2013@med.cornell.edu

Elaine Allan Division of Microbial Diseases, UCL Eastman Dental Institute, 256 Gray's Inn Road, London WC1X 8LD, UK, e.allan@eastman.ucl.ac.uk

Jonathan H. Badger Microbial and Environmental Genomics, J. Craig Venter Institute, 10355 Science Center Drive, San Diego, CA, USA, jbadger@jcvi.org

Hong Wei Chu Department of Medicine, Integrated Department of Immunology, National Jewish Health and the University of Colorado, 1400 Jackson St., Denver, CO 80206, USA, chuhw@njhealth.org

Marie-France de La Cochetière INSERM, UFR Médecine, Université Nantes, Thérapeutiques Cliniques et Expérimentales des Infections, EA 3826, 1 Rue Gaston Veil, Nantes, France, mfdlc@nantes.inserm.fr, marie-france.de.la-cochetiere@inserm.fr

Willem M. deVos Laboratory of Microbiology, Wageningen University, Dreijenplein 10, 6703 HB Wageningen, The Netherlands, willem.devos@wur.nl

Appolinaire Djikeng Biosciences Eastern and Central Africa (BecA), International Livestock Research Institute (ILRI), BecA-ILRI hub, Nairobi, Kenya, adjikeng@gmail.com

S. Dusko Ehrlich Microbiology and the Food Chain Division, INRA, 78350 Jouy en Josas, France, dusko.ehrlich@jouy.inra.fr

Sue Garges Laboratory of Molecular Biology, NCI, National Institutes of Health, 37 Convent Dr. MSC 4255, Bethesda, MD 20892, USA, sgarges@helix.nih.gov

Elodie Ghedin Center for Vaccine Research, Department of Computational and Systems Biology, University of Pittsburgh School of Medicine, 3501 5th Avenue, Biomedical Science Tower 3, Pittsburgh, PA 15261, USA, elg21@pitt.edu

Matthew Haynes Department of Biology, San Diego State University, 5500 Campanile Dr., San Diego, CA 92182, USA, mhaynes@projects.sdsu.edu

Sarah Highlander Department of Molecular Virology and Microbiology, Human Genome Sequencing Center, Baylor College of Medicine, One Baylor Plaza, Houston, TX 77030, USA, sarahh@bcm.edu

Mengfei Ho Department of Microbiology, University of Illinois at Urbana-Champaign, 601 East John St., Champaign, IL 61820–5711, USA, mengho@life.illinois.edu

Eline S. Klaassens CSIRO Livestock Industries, Queensland Bioscience Precinct, 306 Carmody Road, St Lucia, QLD 4067, Australia, eline.klaassens@csiro.au

Michiel Kleerebezem Laboratory of Microbiology, Wageningen University, Dreijenplein 10, 6703 HB Wageningen, The Netherlands, michiel.kleerebezem@nizo.nl

Huiying Li Crump Institute for Molecular Imaging, California NanoSystems Institute, 570 Westwood Plaza, Los Angeles, CA 90095-1770, USA, huiying@mednet.ucla.edu

Lanjuan Li Dean of State Key Laboratory for Infectious Diseases Diagnosis and Treatment, The First Affiliated Hospital, College of Medicine, Zhejiang University, 79 QingChun Road, Hangzhou, Zhejiang 310003, China, ljli@zju.edu.cn

Zongxin Ling State Key Laboratory for Diagnosis and Treatment of Infectious Diseases, The First Affiliated Hospital, School of Medicine, Zhejiang University, Hangzhou, China 310003, lingzongxin_lzx@163.com

Liliana Losada J. Craig Venter Institute, 9704 Medical Center Drive, Rockville, MD 20850, USA, llosada@jcvi.org

Trevor G. Marshall Murdoch University, 90 South Street, Murdoch, WA 6150, Australia, trevor.m@trevormarshall.com

Emmanuel Montassier Hôpital Hôtel Dieu Nantes, UFR Médecine, Université Nantes, Thérapeutiques Cliniques et Expérimentales des Infections, EA 3826, 1 Rue Gaston Veil, Nantes, France, emmanuelmontassier@hotmail.com

Alison Morris University of Pittsburgh, NW628 Montefiore Hospital, 3459 Fifth Ave, Pittsburgh, PA 15213, USA, morrisa@upmc.edu

Mark Morrison CSIRO Livestock Industries, Queensland Bioscience Precinct, 306 Carmody Road, St Lucia, QLD 4067, Australia, mark.morrison@csiro.au

Peter Mullany Division of Microbial Diseases, UCL Eastman Dental Institute, 256 Gray's Inn Road, London WC1X 8LD, UK, p.mullany@eastman.ucl.ac.uk

Barbara Jones Nelson KentHill Biosciences, 29 Carriage Walk Court, Gaithersburg, MD 20879, USA, barabaranelson184@hotmail.com

Karen E. Nelson Department of Human Genomic Medicine, The J. Craig Venter Institute, 9704 Medical Center Drive, Rockville, MD 20850, USA, kenelson@jcvi.org

Pauline C. Ng Genome Institute of Singapore, 60 Biopolis St., Singapore, S138672, png@jcvi.org

William C. Nierman J. Craig Venter Institute, 9704 Medical Center Drive, Rockville, MD 20850, USA, wnierman@jcvi.org

Mark J. Pallen School of Biosciences, University of Birmingham, West Midlands, Birmingham, B15 2TT, UK, m.pallen@bham.ac.uk

Jane L. Peterson Division of Extramural Research, Large-scale Sequencing National Human Genome Research Institute National Institutes of Health, 5635 Fishers Lane, Bethesda, MD 20892-9305, USA, jane_peterson@nih.gov

Amy D. Proal Murdoch University, 90 South Street, Murdoch, WA 6150, Australia, amy.proal@gmail.com

Forest Rohwer Department of Biology, San Diego State University, 5500 Campanile Dr., San Diego, CA 92182, USA, forest@sunstroke.sdsu.edu

Bernd Schnabl Department of Medicine, University of California San Diego, MC0702, 9500 Gilman Drive, La Jolla, CA 92093, USA, beschnabl@ucsd.edu

Susan M. Thomas University of Illinois at Urbana-Champaign, 601 East John Street, Champaign, IL 61820-5711, USA, susant@life.illinois.edu

Sebastian Tims Laboratory of Microbiology, Wageningen University, Dreijenplein 10, 6703 HB Wageningen, The Netherlands, sebastian.tims@wur.nl

J. Craig Venter J. Craig Venter Institute, 9704 Medical Center Drive, Rockville, MD 20850, USA, jcventer@jcvi.org

Philip Warburton Division of Microbial Diseases, UCL Eastman Dental Institute, 256 Gray's Inn Road, London WC1X 8LD, UK, p.warburton@eastman.ucl.ac.uk

Brenda A. Wilson Host-Microbe Systems Theme, Department of Microbiology, Institute for Genomic Biology, University of Illinois at Urbana-Champaign, 601 South Goodwin Ave., Urbana, IL 61801, USA, bawilson@life.illinois.edu

Charlie Xiang State Key Laboratory for Diagnosis and Treatment of Infectious Diseases, The First Affiliated Hospital, School of Medicine, Zhejiang University, Hangzhou, China 310003, cxiang@zju.edu.cn

Erwin G. Zoetendal Laboratory of Microbiology, Wageningen University, Dreijenplein 10, 6703 HB Wageningen, The Netherlands, erwin.zoetendal@wur.nl

About the Editor

Dr. Karen E. Nelson is the director of the Rockville Campus of the J. Craig Venter Institute (JCVI), where she has been for the past 14 years. She was formerly the director of human microbiology and metagenomics in the Department of Human Genomic Medicine at the JCVI. She received her undergraduate degree from the University of the West Indies, and her Ph.D. from Cornell University in Ithaca, New York. She has authored or co-authored over 100 publications and is currently editor-in-chief of the Springer journal *Microbial Ecology*. She is also a standing member of the NRC Committee on Biodefense, a member of the American Society for Microbiology (ASM) Communications Committee and a Fellow of the ASM.

Chapter 1
The Human Genome, Microbiomes, and Disease

Jonathan H. Badger, Pauline C. Ng, and J. Craig Venter

Disease may stem from the environment, genetics, and the human microbiome. In this chapter, we discuss what is currently known about the genetic and microbial contribution to human disease. Because this is a large area, we provide a high-level review of a few diseases to reveal underlying themes of the contribution of genetics and metagenomics to human disease. As changing a microbiome is a far easier task than changing one's genome, manipulating it may prove an effective way to both treat and prevent disease where causal disease relationships can be established. Elucidating the interplay of the contributions of the genome and microbiome will be necessary to help us better understand disease.

Introduction: Variation in the Genome and Microbiome

The past decade in human genetic disease has been dominated by genome-wide association studies and, more recently, by sequencing human genomes. The goal is to discover disease genes by identifying variants that cause or increase the risk of disease. In genome-wide association studies, patients with disease are compared with unaffected controls. DNA is collected from both groups, and approximately a million genetic variants are surveyed in their genomes. Variants that are at a higher frequency in patients than in controls are associated with an increased risk of disease. More than a thousand disease-associated variants have been identified from over 400 genome-wide association studies, but much is still unknown about the genetic heritability of disease. For example, genome-wide association studies have identified 40 markers associated with height, but these markers only account for ~5% of height's heritability (Maher, 2008). Despite studying thousands of individuals, <10% of heritability is explained for most diseases (Manolio et al., 2009).

J.H. Badger (✉)
J. Craig Venter Institute, 10355 Science Center Drive, San Diego, CA, USA
e-mail: jbadger@jcvi.org

Jonathan H. Badger and Pauline C. Ng contributed equally.

K.E. Nelson (ed.), *Metagenomics of the Human Body*,
DOI 10.1007/978-1-4419-7089-3_1,

Individual human genomes have been sequenced, and there are approximately 3 million to 4 million variations with respect to the reference genome (Frazer et al., 2009). It is thought that some of these variants will cause phenotypic differences that can lead to disease or apparent physical traits. It is estimated that 3–8% of the human genome is functional (Siepel et al., 2005), so it is unlikely that all the variation in the 3 Gb human genome will lead to phenotypic differences. Rather, functional variants may be localized to the 90–240 Mb of human genome that contains transcribed coding genes, regulatory elements, RNA genes, and other functional elements.

By contrast, the human microbiome has extensive diversity. Each location (skin, mouth, intestine, etc) has its own metagenome. Recent studies have suggested that healthy individuals have up to 15,000 species-level phylotypes in their gastrointestinal tracts as determined by 16S rRNA sequencing (>97% identity) (Peterson et al., 2008) (Fig. 1.1 of this chapter), and that the two major phylogenetic groups present are the Firmicutes and Bacteriodetes. The average genome size of sequenced organisms from these groups is 3.4 Mb (Liolios et al., 2008), and the percentage of these genomes that codes for protein-coding genes is approximately 92%. Therefore, the functional part of the gastrointestinal microbiome can be estimated to be approximately 47,000 Mb (15,000 × 3.4 × 0.92), which is more than two orders of magnitude greater than the above-mentioned estimate of the functional part of the human genome.

Because human metagenomics is a nascent field, fewer individuals have been studied relative to the human genetics field. In a similar manner, the microbiome compositions observed in patients differ from those of the controls for some diseases. Like inherited genetic variation, this is not a complete predictor: species that

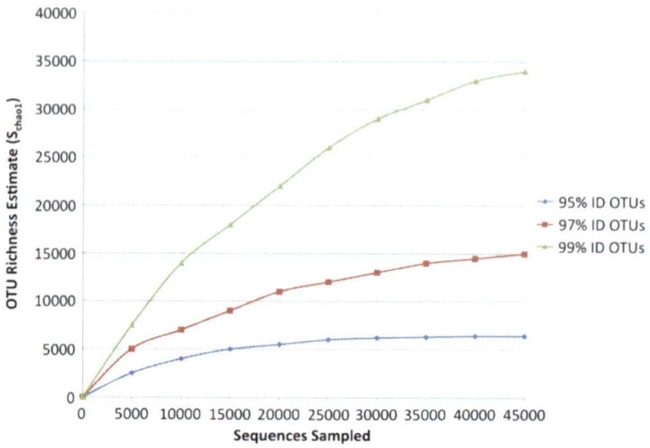

Fig. 1.1 OTU richness vs. sequences sampled (adapted from Peterson et al. (2008)). 16S rRNA collector's curves showing the number of observed operational taxonomic units (OTUs) given the number of sequences collected and a pairwise sequence identity (ID%) cutoff. A 99% ID level correlates with accepted strain classifications, likewise 97% with species and 95% with genus. When a collector's curve begins to plateau, more sequencing is unlikely to return new OTUs at that percent identity level

are predominant in the disease will also be observed in controls, and vice versa. In most cases, the microbiomes' constitution is associated rather than causative, meaning that while a correlation between the disease state and the microbiome has been established, there is no evidence that the microbial constitution caused the disease.

The microbiome is an ecosystem in which the various members maintain equilibrium. So far, it appears that many diseases are associated with an abnormal proportion of the same taxonomic groups that are present in healthy individuals. For example, patients with Crohn's disease show a lower than normal frequency of bacteria from the phylum Bacteriodetes in their gastrointestinal tract (e.g., Gophna et al., 2006), whereas patients suffering from active celiac disease have a higher than normal frequency of Bacteriodetes (e.g., Nadal et al., 2007). Biodiversity also plays a role in the microbiome. Patients with bacterial vaginosis have a less diverse vaginal microbiome than healthy individuals, suggesting that the biodiversity normally residing in a healthy vagina may serve as a resilient buffering against disease.

Another common theme is that diseases with a microbial component are often disorders of the immune system. Psoriasis of the skin and Crohn's disease in the gut are some examples. Seeing as the immune system modulates the interaction between the microbiome and the native human body, it is not surprising that the immune system plays a huge role, but this connection may also hint at possible therapies.

Genome-wide association studies typically examine 10^6 variants and focus mainly on common variation. But recent technological advances have enabled sequencing whole genomes of human individuals, which can identify rare mutations that cause disease (Altshuler et al., 2008). Similarly, high-throughput sequencing will allow us to sequence metagenomes at greater depth to obtain finer resolution of the various species in a microbial sample.

Steps to Treatment

A human's DNA is not easily amenable to change. The genome in one cell is static and identical (for the most part) to the genomes in the individual's other cells. Thus, knowing disease-associated genetic variation does not mean you can directly treat the disease. The genome-wide association discoveries are meant to be followed up by functional studies to investigate biological pathways involved in disease, and then see which drugs target these pathways. This can take many years to translate genetics research into a clinical treatment. There are exceptions; age related macular degeneration is one disease where treatment was rapidly implemented after genome-wide discoveries (Gehrs et al., 2006).

Microbiomes may be easier to treat. Antibiotics erase the current microbiome state. The use of probiotics, that is, ingesting bacteria to improve the intestinal microbial balance, is gaining popularity for multiple gastrointestinal diseases (Preidis and Versalovic, 2009). Probiotics (from the Greek "pro bios," meaning "for, or promoting, life") are foods or supplements containing live bacteria thought to promote health. While yogurt and other fermented products have traditionally been considered “health food,” studies of the human microbiome promise to understand what effect they have on the composition of the microbiome and ultimately health. Studies have shown that probiotics can manipulate the microbiome, as well as affect the intestinal barrier (Zyrek et al., 2007) and the immune system (Fitzpatrick et al., 2007).

Prebiotics are oligosaccharides or complex saccharides that stimulate the growth or activity of the beneficial commensal bacteria that already are present in the host. Although traditionally pro- and prebiotics have been defined in the context of the gastrointestinal tract, recent work has explored the use of probiotics (Lee et al., 2008) and prebiotics (Bockmuhl et al., 2007) on skin, and probiotics in the vagina (Mastromarino et al., 2009).

Factors in Experimental Design

Human studies can be complicated by population differences, because different ethnicities can have different patterns of linkage disequilibrium, and different allele frequencies can bias genetics studies. Furthermore, genome-wide association studies test 10^6 markers. As a consequence, a large number of individuals must be studied to get a significant p-value after correction for multiple testing.

The study of the microbiome has numerous complications as well. One of the most important ones is the choice of percent sequence identity to define an operational taxonomic unit (OTU). This cutoff can greatly affect the estimated sample diversity (see Fig. 1.1). A low value will group highly diverse organisms together, underestimating biodiversity, but a high value may overestimate biodiversity, particularly in the case of next-generation sequencing methods, as they often have a higher rate of sequencing error than the traditional Sanger method (Gomez-Alvarez et al., 2009).

Another choice that needs to be made is what to sequence in order to measure biodiversity. The cheapest option is to sequence one or more variable regions of the 16S rRNA gene (Huber et al., 2007). This has the advantage of not requiring assembly as such regions can fit on a single Roche 454™ read. The next cheapest option is to sequence full-length (or nearly full-length) 16S rRNA sequences, typically by sequencing a forward and reverse read from a PCR-amplified product, followed by assembly. A complication in both forms of 16S rRNA diversity studies is the selection of PCR primers and annealing temperature because poor selections can greatly bias the results (Sipos et al., 2007). The most expensive, and yet potentially most informative method of studying microbial biodiversity is random environmental sequencing, known as metagenomics. More diversity is seen in the protein-coding parts of genomes from environmental sequencing than what is seen from 16S rRNA sequencing.

Finally, there are many factors that need to be considered in the study of the microbiome. Although many of them may be the same factors that influence genome-wide association studies such as age, sex, and ethnicity, there are also unique factors that need to be considered. As the microbiome is not static, time of sampling is likely to be a factor. Time and composition of the last meal may play a major role in the gastrointestinal tract microbiome. Sanitary habits and use of cosmetics may influence the skin microbiome. Sexual activity and the current state of the menstrual cycle may influence the vaginal microbiome.

Given the diversity of the human microbiome relative to the human genome, there is a huge effort to understand the human microbiome. The Human Microbiome Project (HMP) aims to study the microbiomes of the nasal, oral, skin, gastrointestinal, and urogenital environments. By establishing reference microbiomes from healthy normal individuals, these can be compared to disease states to find potential causes and associations. Much like sequencing the human genome led to a better understanding of human genetic variation (Lander et al., 2001; Venter et al., 2001), the HMP is set up to do the same.

The Gastrointestinal Microbiome

Perhaps the body site that comes first to mind when the "human microbiome" is mentioned is the gastrointestinal system. The importance of the intestinal microbiota to health was first suggested a hundred years ago by the Russian Nobel-laureate microbiologist Ilya Ilyich Mechnikov, who hypothesized that the supposed health benefits of *kefir* (a traditional fermented milk beverage in Eastern Europe), derived from the ability of the live bacteria in the drink to colonize the intestine, displacing "unhealthy" bacteria and improving digestion (Metchnikov, 1908). Only recently have molecular methods allowed detailed studies of the relationships between disease and microbiota, particularly in relation to three disorders: Crohn's disease, celiac disease, and obesity.

Crohn's Disease

Crohn's disease is a common inflammatory bowel disease with high heritability. In Crohn's disease, it appears that the immune system misidentifies benign members of the commensal microbiota as pathogens and mounts an immune response against them (Strober et al., 2007). An analysis of three complete genome-wide scans looking at over 3,000 cases and 4,000 controls identified 31 loci involved in Crohn's disease and many were in or near immune-related genes (*IL23R*, *CCR6*, *IL12B*, *STAT3*, *ICOSLG*, *PTPN2*, *PTPN22*, *ITLN1*). The *ITLN1* gene is notable because it encodes a lectin that recognizes galactofuranosyl residues found in the cell walls of microorganisms but not present in mammals. Its physical contact with microbes may demonstrate a direct interaction between the microbiome and human genome. These 31 loci in humans explain about 20% of the heritability of Crohn's disease (Barrett et al., 2008). Either as a cause or a result of this immune response, Crohn's sufferers tend to have lower than normal levels of Firmicutes, Bacteriodetes (Frank et al., 2007; Gophna et al., 2006), and *Lachnospiraceae* (group IV and XIVa Clostridia; Peterson et al., 2008) in their gastrointestinal microbiota. As Peterson et al. mention (Peterson et al., 2008), this suggests a testable hypothesis; as the difference between the diseased and normal microbiomes may either be causal or be reactive, restoring a normal microbiome to Crohn's sufferers would have no therapeutic effect if the microbiome is reactive to the disease, but would work as a treatment if it is causal. Peterson and colleagues propose an extensive long-term research program (Peterson et al., 2008) for understanding the nature of Crohn's and similar bowel diseases. This proposal comprises monozygotic twins studies, microbiota transplantation between healthy and diseased individuals, and characterization of the B- and T-cell repertoires of patients with differing microbiomes.

Celiac Disease

Celiac disease is an autoimmune disease in which the lining of the small intestine is damaged from eating protein gluten (found in bread, pasta, and other foods that contain wheat, barley, or rye). If an immune response is triggered due to inappropriate diet, damage can occur in the small intestine such that there is decreased nutrient absorption, and consequently vitamin deficiencies that can lead to stunted growth. Much of the known heritability (~35%) is attributable to HLA haplotypes, which was discovered from a linkage study on 60 families (Liu et al., 2002). Genome-wide association studies on 2,000 individuals have subsequently identified an additional eight regions that account for ~3–4% additional heritability (Hunt et al., 2008; van Heel et al., 2007). Seven of the eight celiac disease regions identify immune genes, involved in T or B cell function, indicating the importance of the immune system.

Initial studies of the microbiota of patients with active celiac disease (Nadal et al., 2007) have revealed that their duodenal microbiota differ from healthy individuals as well as from controlled celiac patients (those who have been on a strict gluten-free diet for over a year). In particular, the microbiota of the patients with active celiac disease had significantly higher percentage of bacteria from the *Bacteroides–Prevotella* group as well as a higher percentage of *Escherichia coli* as compared to the other two groups. Although members of these groups constitute a large fraction of the commensal microbiota in healthy individuals, some *Bacteroides* species (such as *Bacteroides vulgatus*) have been shown to cause immune-system-induced intestinal inflammation in animal models (Setoyama et al., 2003). Both the patients with active and controlled celiac disease displayed reduced levels of *Lactobacillus* and *Bifidobacterium* as compared to healthy individuals.

Obesity

Family and twin studies have shown that genetic factors account for 40–70% of the population variation in BMI (Frazer et al., 2009; Siepel et al., 2005). A meta-analysis of genome-wide association studies on more than 32,000 individuals by the GIANT identified 20 loci associated with obesity. Most of the obesity markers discovered so far have minor effects on obesity (odds ratios <2). The gene FTO, the marker with the strongest association with obesity, only accounts for ~1% of the heritability (Walley et al., 2009).

Studies comparing the microbiomes of 31 identical twin pairs (Turnbaugh et al., 2009) revealed that obese twins tended to have a higher proportion of Firmicutes and a lower proportion of Bacteriodetes in their gastrointestinal microbiomes than their normal or underweight twin. In addition, the microbiomes of obese individuals were less diverse phylogenetically than in the normal case. Interestingly, similar trends were seen in 23 pairs of fraternal twins (Turnbaugh et al., 2009), who are not closer genetically to each other than other siblings. A related study (Turnbaugh et al., 2008) involving mice also saw an increase in Firmicutes and a decrease in Bacteriodetes when the mice were fed a diet high in fat and sugar (resembling a typical "Western" diet) but not when fed a low-fat diet. This suggests a link (perhaps causal) between a high-fat, high-sugar diet and obesity that is mediated by the microbiome.

Colorectal Cancer

Colorectal cancer is the third most common cancer and approximately 650,000 people die from colorectal cancer annually. Mutations in the FAP and HNPCC genes account for approximately 5% of colorectal cancer cases (http://www.genome.gov/10000466). Whole exome sequencing of colorectal cancer tissues have identified an additional 140 genes that were enriched with somatic cancer mutations in the cancer samples (Wood et al., 2007). These genes were shown to be mutated in >10% of colorectal cancers, and in other cancers.

A diet high in red meat and fat is associated with high risk of colorectal cancer. This can possibly be traced back to the colon microbiome. Many strains of Bacteroides species can convert bile to fecapenentaenes, which may cause cancer (see Fig. 1.3) (Moore and Moore, 1995). A comparison of colorectal cancer patients, high-risk individuals with Western diets, and low-risk individuals with non-Western diets showed that the low-risk individuals had a lower percentage of *B. vulgatus* (Fig. 1.4).

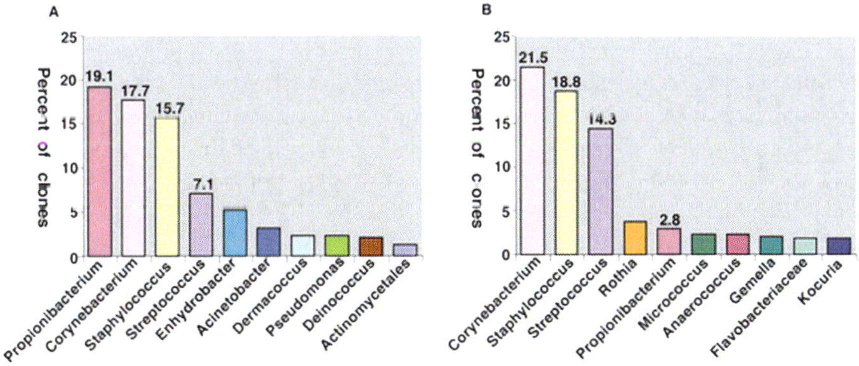

Fig. 1.2 Bacterial genera from healthy skin (**a**) and psoriatic lesions (**b**). Taken from Gao et al. (2008) under the Creative Commons Attribution License

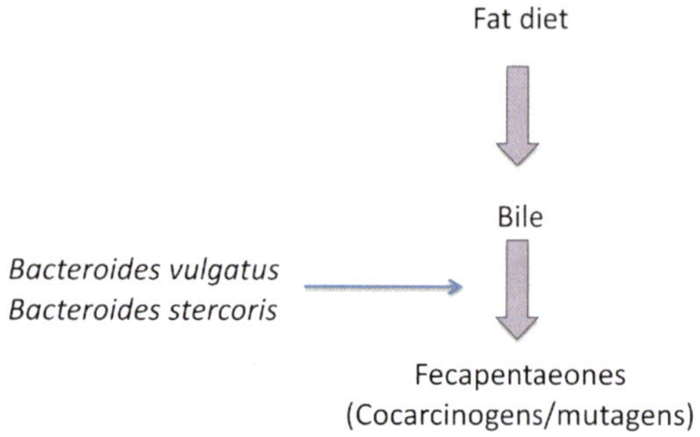

Fig. 1.3 A fatty diet can increase the risk of colorectal cancer

Vaginal Bacterial Biota

The vagina has different types of environments depending on age, hormonal fluctuations, sexual activity, sanitary habits such as douching, etc. In healthy women, the bacterial biota is dominated by lactobacilli with more diversity observed in African-American woman than in Caucasian women (Zhou et al., 2007).

Bacterial vaginosis is a common disease in women whose symptoms include abnormal vaginal discharge and increased susceptibility to HIV. The vaginal microbiome of women with bacterial vaginosis is more diverse than that of healthy women. Healthy women's vaginas were initially thought to be dominated by *Lactobacillus* species. *Lactobacillus* may provide a protective environment against pathogens, and disturbance of the ecology may eventually lead to bacterial vaginosis. It is thought that *Lactobacillus* metabolizes the glycogen to lactic acid in the vaginal epithelium to produce a low pH environment that prevents the growth of pathogenic organisms (Fig. 1.4) (Zhou et al., 2007).

However, based on 16S rRNA sequencing, *Lactobacillus* is not always found on healthy human vaginal epitheliums (Hyman et al., 2005; Zhou et al., 2007). *Gardnerella vaginalis* and *Atopobium vaginae* can also be detected in healthy women (Hyman et al., 2005), even though they are associated with bacterial vaginosis (Srinivasan and Fredricks, 2008). It has been suggested that the co-occurrence of *G. vaginalis* and *A. vaginae* may play a role in bacterial vaginosis (Bradshaw et al., 2006).

Human genetic variation also plays a role in HIV susceptibility. The CCR5 is the major HIV co-receptor expressed in the female genital tract (Patterson et al., 1998). In Caucasians, an allelic variant of the CCR5 with a 32-bp deletion that results in a frameshift and a nonfunctional receptor protects against HIV infection (Samson et al., 1996). CCR5 mRNA is significantly increased in a woman with STDs and inflammatory conditions, so the bacterial environment in conjunction with a woman's genetics could determine her susceptibility to HIV.

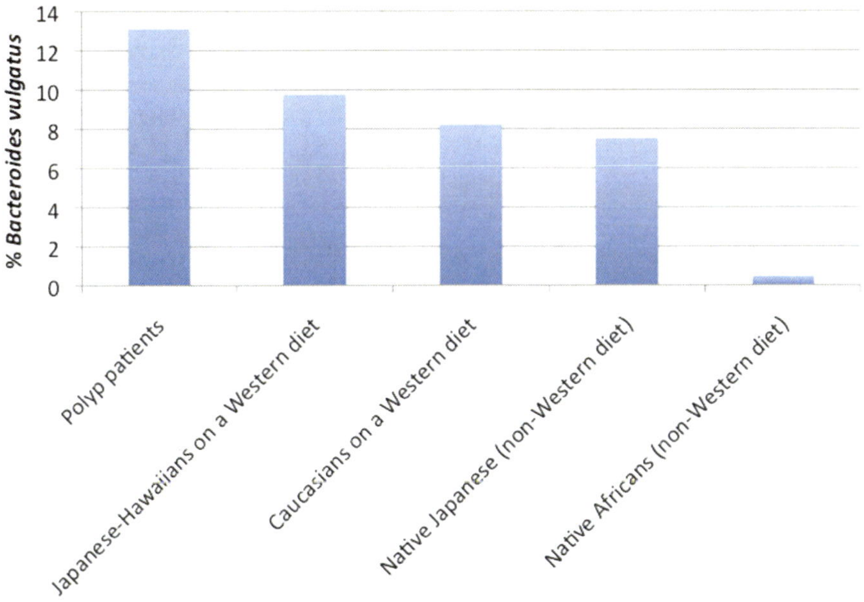

Fig. 1.4 Intestinal microflorae differ between individuals with low and high risk for colorectal cancer. From percentages in Moore and Moore, 1995

Oral

The oral environment is challenging because it undergoes constant changes in pH, redox potential, atmospheric conditions, salinity, and water activity from saliva (Avila et al., 2009). People who are diabetic, HIV positive, pregnant, lactating, or who have recently taken antibiotics can also have altered oral bacterial compositions (Lepp et al., 2004). Within the oral cavity, different environments (e.g., soft tissues and plaques) are dominated by different species (Aas et al., 2005). More than 700 species reside in the oral cavity, and some of these can cause disease. *Porphyromonas gingivalis*, *Tannerella forsythia*, and *Treponema denticola* have been found to be associated with periodontal disease; *Streptococcus mutans*, *Lactobacillus* spp., *Bifidobacterium* spp., and *Atopobium* spp. have been found to be associated with dental caries.

Periodontal disease (gingivitis and periodontitis) is a chronic bacterial infection. Besides the pathogenic microflora that causes periodontitis, genetics can also contribute to disease risk. Genetic studies have shown that approximately half of the population variance in chronic periodontitis is attributable to genetic factors. A genome-wide association study on aggressive periodontitis identified an intronic variant in the gene *GLT6D1*, which is expressed in gingiva (Schaefer et al., 2010). The G allele of the variant is found in 12% more of the periodontitis cases than in controls, and the variant appears to lie in a GATA-3 transcription factor binding site. Furthermore, the authors show that the disease-associated G allele has reduced binding affinity for GATA-3. GATA-3 regulates Th2 cells and is involved in immunity against extracellular parasites. Further work is needed to elucidate possible interactions between GLT6D1 and the oral microbiome.

It is interesting to observe that the oral microbiome from children with severe cavities is much less diverse than their healthy counterparts (Li et al., 2007). Similar to Crohn's disease in the gastrointestinal tract and bacterial vaginosis in the vagina, the affected individuals have less diverse microflora than healthy individuals.

Skin Microbiome

The skin is the largest organ in the body, and within the skin, there are microenvironments such as sebaceous (e.g., upper chest and back), moist (e.g., armpit and inside of nostril), and dry (e.g., forearm and buttock) (Grice et al., 2009). Based on 16S rRNA sequencing, the majority of microbes are Proteobacteria (Grice et al., 2008). Skin also shows a low level of interpersonal variation in contrast to the gastrointestinal tract (Grice et al., 2008).

Psoriasis

Psoriasis is a common skin condition that causes skin redness and irritation. It is a chronic autoimmune disease that afflicts ~2% of the European population. The number of species per skin sample is not significantly different between healthy individuals and those with psoriasis, but the composition differs (Fig. 1.5). *Streptococcus* spp. is found more frequently in psoriatic lesions than in normal skin from the same patients. Meanwhile, *Propionibacterium* is observed more frequently on normal skin (21% from healthy, 12% from normal skin on psoriatic

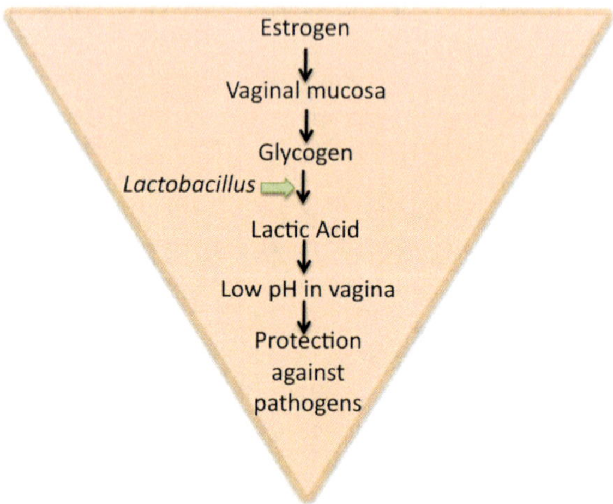

Fig. 1.5 The role of microbiota in a healthy vagina

Table 1.1 Genes with nearby variants that are strongly associated with psoriasis (Nair et al., 2009)

Notable nearby genes	Location of associated variant	Role in immune response
MHC		
HLA-C	13 kb upstream	Major histocompatibility complex class I heavy chain receptor which presents small peptides to the immune system, so that the immune system can recognize foreign antigens
IL-23 signaling		
IL12B	24 kb downstream	IL12B encodes subunit of IL-23
IL23A	3.7 kb downstream	encodes a subunit of IL-23
IL23R	Intron	subunit of IL-23 receptor
NF-κB Pathway		
TNFAIP3	Intron	TNFα-induced protein 3 inhibits NF-κB activation and terminates NF-κB-mediated responses
TNIP1	1 kb upstream	TNFAIP3-interacting protein 1
Other		
IL13	Nonsynonymous Coding	Modulate humoral immune response mediated by Th2 cells
STAT2	Intron	Responds to interferon. STAT2 interacts with P300/CBP, which is thought to be involved in the process of blocking IFN-α response by adenovirus

patients) than on psoriatic lesions (2.9%). It is not known if *Propionibacterium* protects the host, and removal of *Propionibacterium* therefore plays a causative role. Alternatively, *Propionibacterium* could simply be correlated to healthy skin, and a decreased number simply indicates its displacement from more aggressive microbes (Gao et al., 2008).

A genome-wide association study revealed that many of the variants that increase the risk of psoriasis are near, or in genes, that play a role in the human immune system (Table 1.1) (Nair et al., 2009). For example, variants in the IL-23 pathway (the IL12B, IL23A, IL23R genes) are associated with psoriasis. IL-23 signaling promotes cellular immune responses by promoting the survival and expansion of a subset of T cells that protects epithelia against microbial pathogens (Altshuler et al., 2008).

Acne

Acne vulgaris is the most common skin disorder and drives a multi-billion dollar industry. There are two stages of the disorder: (1) blackheads and whiteheads are formed in a noninflammatory phase and (2) the blackheads and whiteheads develop into inflamed lesions in an inflammatory phase (Leeming et al., 1985).

In the first stage, *Pityrosporum* spp., Stapylococci, and Propionibacteria were found more frequently at blackheads and whiteheads than at normal follicles, but were not found in all of these sites (Leeming et al., 1985). Mature blackheads are more frequently colonized, and it may be that microbial colonization is not necessary for whitehead/blackhead formation, but that colonization occurs as the whitehead/blackhead enlarges.

Propionibacterium acnes, which is present in both normal and acne-afflicted follicles, could be responsible for instigating inflammation in the second stage of acne (Burkhart et al., 1999). *Propionibacterium acnes* directly secretes factors that attract immune cells, and acne patients have an immune response to *P. acnes*. Thus, controlling the skin microbiome by decreasing the *P. acnes* population and/or inhibiting its production of inflammatory factors could be effective in treating acne.

Conclusion

In this chapter, we review some diseases known to involve both human genetics and the human microbiome. Undoubtedly, more relationships between microbiome and human genetics will be elucidated with further research. However, from this chapter we can summarize some general trends across diseases that are evident today. First, the disease state is often associated with a reduced level of diversity in the microbiome (Turnbaugh et al., 2009; Zhou et al., 2007; Li et al., 2007) as compared to the healthy state. This may mean that the disease state represents a breakdown in the complex ecosystem that is the microbiome. Second, many diseases associated with the microbiome are associated with problems in the immune system, particularly disorders that involve misidentifying commensals as pathogens (e.g., Crohn's disease, celiac disease, acne, and psoriasis).

The future of microbiome studies will likely include attempts at curing diseases through the modification of the microbiomes of diseased individuals. This should not be done in a vacuum but should be combined with genome-wide association studies and studies of the immune system in order to make the best use of the results. The synergy of both metagenomics and human genetics holds the promise to improve human health for many common diseases.

References

Aas JA, Paster BJ, Stokes LN, Olsen I, Dewhirst FE (2005) Defining the normal bacterial flora of the oral cavity. J Clin Microbiol 43:5721–5732

Altshuler D, Daly MJ, Lander ES (2008) Genetic mapping in human disease. Science 322:881–888

Avila M, Ojcius DM, Yilmaz O (2009) The oral microbiota: living with a permanent guest. DNA Cell Biol 28:405–411

Barrett JC et al (2008) Genome-wide association defines more than 30 distinct susceptibility loci for Crohn's disease. Nat Genet 40:955–962

Bockmuhl D et al (2007) Prebiotic cosmetics: an alternative to antibacterial products. Int J Cosmet Sci 29:63–64

Burkhart CG, Burkhart CN, Lehmann PF (1999) Acne: a review of immunologic and microbiologic factors. Postgrad Med J 75:328–331

Bradshaw CS et al (2006) The association of *Atopobium vaginae* and *Gardnerella vaginalis* with bacterial vaginosis and recurrence after oral metronidazole therapy. J Infect Dis 194:828–836

Fitzpatrick LR et al (2007) Effects of the probiotic formulation VSL#3 on colitis in weanling rats. J Pediatr Gastroenterol Nutr 44:561–570

Frank DN et al (2007) Molecular-phylogenetic characterization of microbial community imbalances in human inflammatory bowel diseases. Proc Natl Acad Sci USA 104:13780–13785

Frazer KA, Murray SS, Schork NJ, Topol EJ (2009) Human genetic variation and its contribution to complex traits. Nat Rev Genet 10:241–251

Gao Z, Tseng CH, Strober BE, Pei Z, Blaser MJ (2008) Substantial alterations of the cutaneous bacterial biota in psoriatic lesions. PLoS One 3:e2719

Gehrs KM, Anderson DH, Johnson LV, Hageman GS (2006) Age-related macular degeneration – emerging pathogenetic and therapeutic concepts. Ann Med 38:450–471

Gomez-Alvarez V, Teal TK, Schmidt TM (2009) Systematic artifacts in metagenomes from complex microbial communities. ISME J 3:1314–1317

Gophna U, Sommerfeld K, Gophna S, Doolittle WF, Veldhuyzen van Zanten SJ (2006) Differences between tissue-associated intestinal microfloras of patients with Crohn's disease and ulcerative colitis. J Clin Microbiol 44:4136–4141

Grice EA et al (2008) A diversity profile of the human skin microbiota. Genome Res 18:1043–1050

Grice EA et al (2009) Topographical and temporal diversity of the human skin microbiome. Science 324:1190–1192

Huber JA et al (2007) Microbial population structures in the deep marine biosphere. Science 318:97–100

Hunt KA et al (2008) Newly identified genetic risk variants for celiac disease related to the immune response. Nat Genet 40:395–402

Hyman RW et al (2005) Microbes on the human vaginal epithelium. Proc Natl Acad Sci USA 102:7952–7957

Lander ES et al (2001) Initial sequencing and analysis of the human genome. Nature 409:860–921

Lee J, Seto D, Bielory L (2008) Meta-analysis of clinical trials of probiotics for prevention and treatment of pediatric atopic dermatitis. J Allergy Clin Immunol 121:116–121 e111

Leeming JP, Holland KT, Cunliffe WJ (1985) The pathological and ecological significance of microorganisms colonizing acne vulgaris comedones. J Med Microbiol 20:11–16

Lepp PW et al (2004) Methanogenic Archaea and human periodontal disease. Proc Natl Acad Sci USA 101:6176–6181

Liolios K, Mavromatis K, Tavernarakis N, Kyrpides NC (2008) The Genomes On Line Database (GOLD) in 2007: status of genomic and metagenomic projects and their associated metadata. Nucleic Acids Res 36:D475–479

Liu J et al (2002) Genomewide linkage analysis of celiac disease in Finnish families. Am J Hum Genet 70:51–59

Li Y, Ge Y, Saxena D, Caufield PW (2007) Genetic profiling of the oral microbiota associated with severe early-childhood caries. J Clin Microbiol 45:81–87

Maher B (2008) Personal genomes: the case of the missing heritability. Nature 456:18–21

Manolio TA et al (2009) Finding the missing heritability of complex diseases. Nature 461:747–753

Mastromarino P et al (2009) Effectiveness of lactobacillus-containing vaginal tablets in the treatment of symptomatic bacterial vaginosis. Clin Microbiol Infect 15:67–74

Metchnikov I (1908) The prolongation of life: optimistic studies. G.P. Putnam's Sons, New York & London

Moore WE, Moore LH (1995) Intestinal floras of populations that have a high risk of colon cancer. Appl Environ Microbiol 61:3202–3207

Nadal I, Donat E, Ribes-Koninckx C, Calabuig M, Sanz Y (2007) Imbalance in the composition of the duodenal microbiota of children with coeliac disease. J Med Microbiol 56: 1669–1674

Nair RP et al (2009) Genome-wide scan reveals association of psoriasis with IL-23 and NF-kappaB pathways. Nat Genet 41:199–204

Patterson BK et al (1998) Repertoire of chemokine receptor expression in the female genital tract: implications for human immunodeficiency virus transmission. Am J Pathol 153:481–490

Peterson DA, Frank DN, Pace NR, Gordon JI (2008) Metagenomic approaches for defining the pathogenesis of inflammatory bowel diseases. Cell Host Microbe 3:417–427

Preidis GA, Versalovic J (2009) Targeting the human microbiome with antibiotics, probiotics, and prebiotics: gastroenterology enters the metagenomics era. Gastroenterology 136:2015–2031

Samson M et al (1996) Resistance to HIV-1 infection in caucasian individuals bearing mutant alleles of the CCR-5 chemokine receptor gene. Nature 382:722–725

Schaefer AS et al (2010) A genome-wide association study identifies GLT6D1 as a susceptibility locus for periodontitis. Hum Mol Genet 19:553–562

Setoyama H, Imaoka A, Ishikawa H, Umesaki Y (2003) Prevention of gut inflammation by bifidobacterium in dextran sulfate-treated gnotobiotic mice associated with bacteroides strains isolated from ulcerative colitis patients. Microbes Infect 5:115–122

Siepel A et al (2005). Evolutionarily conserved elements in vertebrate, insect, worm, and yeast genomes. Genome Res 15:1034–1050

Sipos R et al (2007) Effect of primer mismatch, annealing temperature and PCR cycle number on 16S rRNA gene-targetting bacterial community analysis. FEMS Microbiol Ecol 60:341–350

Srinivasan S, Fredricks DN (2008) The human vaginal bacterial biota and bacterial vaginosis. Interdiscip Perspect Infect Dis 2008:750479

Strober W, Fuss I, Mannon P (2007) The fundamental basis of inflammatory bowel disease. J Clin Invest 117:514–521

Turnbaugh PJ et al (2009) A core gut microbiome in obese and lean twins. Nature 457:480–484

Turnbaugh PJ, Backhed F, Fulton L, Gordon JI (2008) Diet-induced obesity is linked to marked but reversible alterations in the mouse distal gut microbiome. Cell Host Microbe 3:213–223

van Heel DA et al (2007) A genome-wide association study for celiac disease identifies risk variants in the region harboring IL2 and IL21. Nat Genet 39:827–829

Venter JC et al (2001) The sequence of the human genome. Science 291:1304–1351

Walley AJ, Asher JE, Froguel P (2009) The genetic contribution to non-syndromic human obesity. Nat Rev Genet 10:431–442

Wood LD et al (2007) The genomic landscapes of human breast and colorectal cancers. Science 318:1108–1113

Zhou X et al (2007) Differences in the composition of vaginal microbial communities found in healthy Caucasian and black women. ISME J 1:121–133

Zyrek AA et al (2007) Molecular mechanisms underlying the probiotic effects of *Escherichia coli* Nissle 1917 involve ZO-2 and PKCzeta redistribution resulting in tight junction and epithelial barrier repair. Cell Microbiol 9:804–816

Chapter 2
Host Genotype and the Effect on Microbial Communities

Sebastian Tims, Erwin G. Zoetendal, Willem M. de Vos, and Michiel Kleerebezem

Introduction

Microbial communities inhabit a variety of body surfaces of mammalian hosts. In the human mouth, 10^2–10^3 different species from nine bacterial and a single archaeal phyla have been found (Aas et al., 2005). Teeth, cheek, and tongue all have their own specific communities with anaerobic bacteria present at the gum-line and between the teeth. For each of these sites, the selective trait is based on the surface adherence capabilities of the microbes, typically resulting in multispecies-biofilm formation (Aas et al., 2005). Microbial diversity and abundance normally decrease further down the gastrointestinal (GI) tract until the stomach. In the esophagus approximately 100 species from six phyla are found, most of which are similar to the species found in the mouth (Pei et al., 2004). The stomach is generally regarded derelict for any microbial species except for *Helicobacter pylori* (Finegold, 1983). Nevertheless, 16S rRNA surveys have reported up to 128 species from eight phyla in the stomach; however, it seems likely that these findings represent remnants from ingested strains rather than true residents (Pavoine et al., 2004). After the stomach, bacterial populations increase again in the small intestine, ranging from 10^4–10^5 g^{-1} in the duodenum and jejunum to 10^7 g^{-1} in the terminal ileum. In this region, intestinal transit slows down and the microbiota composition changes, favoring the more anaerobic species (Finegold, 1983; Hayashi et al., 2005). Next, along the GI tract is the colon. In the ascending colon, polysaccharide hydrolysis and carbohydrate fermentation support rapid microbial growth, whereas in the transverse and descending colon, amino acid and host-derived glycans (mucin) fermentation occurs, coinciding with a reduction of bacterial growth rate (Cummings and Macfarlane, 1991). The fermentations along the entire colon cause a microbial population increase up to 10^{11}–10^{12} cells g^{-1} in feces accompanied by a strong proportional decrease of facultative anaerobes (Marteau et al., 2001). In the colon, the

---

S. Tims (✉)
Laboratory of Microbiology, Wageningen University, Dreijenplein 10, 6703 HB, Wageningen, The Netherlands
e-mail: sebastian.tims@wur.nl

K.E. Nelson (ed.), *Metagenomics of the Human Body*,
DOI 10.1007/978-1-4419-7089-3_2,

most numerous species are obligate anaerobes belonging to the phyla *Bacteroidetes* and *Firmicutes* (Backhed et al., 2005; Eckburg et al., 2005).

No other body site attains the high bacterial abundance as observed in the colon, although the mouth harbors a taxonomic richness approximating that of the colon. Furthermore, the recent expansion of the human skin ribosomal operon database indicates a diversity level close to that of the GI tract as well (Gao et al., 2008). Unfortunately, interactions between host and skin microbiota are for the most part unexplored. Next to skin and GI tract, the urinary tract is being studied on a microbial level as well. The vaginal microbiota is generally assumed to have low diversity and to be dominated by *Lactobacilli*. However, *Lactobacilli* are not dominant vaginal inhabitants in all healthy women (Fredricks et al., 2005). One can conclude that the GI microbiota is the most densely populated microbial ecosystem of the mammalian host and has been subjected to many studies. Hence the remainder of this chapter will mainly focus on the knowledge obtained from GI ecosystems.

A variety of functional metabolic activities are generally thought to derive from most, if not all, of the resident communities. Several general processes necessitate the presence of microbial inhabitants in order to function properly, such as the maturation of the immune system, resistance to pathogens, digestion of nutritional components, and the production of essential nutrients. Especially important to the host are microbiota-derived nutrient conversions and contributions that cannot be executed by the host itself, including the degradation of complex polymers in the GI tract that cannot be (completely) digested by the host's enzyme machinery (nondigestible carbohydrates, proteins, and lipids) (Cummings and Macfarlane, 1997). Besides the improvement of our nutritional access to complex nutrients, microorganisms in the GI tract provide a significant supply of essential amino acids (Metges, 2000) and other important compounds such as various vitamins, including vitamin K and several B vitamins (Ramotar et al., 1984; Albert et al., 1980). Conversion of several bio-active molecules, such as sex steroid hormones in order to promote their circulation, is another health-related function humans receive from their GI inhabitants (Begley et al., 2005; Adlercreutz et al., 1984). These types of microbial functions complement the metabolic potential of the host, as the host itself does not encode for the required proteins. In addition to the beneficial contributions, the microbial communities can introduce metabolic activities that are detrimental for their host, such as the production of hydrogen sulfide (Attene-Ramos et al., 2006; Schicho et al., 2006) and the potentially tumor-promoting secondary bile acids (Summerton et al., 1985).

It is becoming a generally accepted view that multicellular organisms, especially mammals, should not be considered as autonomously living entities. "Super-organism" is a popular term that better describes mammals for what they really are: a cohort of host cells and microbial cells. This coalition of cells from the different domains of life is striving for the common mutual cause of survival. The microbiome is a commonly used term for the genetic composition of all microbial cells belonging to a super-organism. According to the super-organism concept, metagenomics can be defined as the mammalian host genes combined with the entire microbiome (Turnbaugh et al., 2006). Currently, there are no completed metagenomes available for any super-organism. Hence, it remains difficult to establish or estimate the importance of the host genome. It is obvious that many organisms, such as humans, contribute far less genes to their metagenome than their microbiome counterparts. In the human GI tract alone there are already ten times more microbial cells present than host cells in the entire human body (Savage, 1977). Humans are believed to contain approximately 23,000 genes in their genomes (Wei and Brent, 2006), whereas current estimations for the GI microbiota unique gene count are up to 9,000,000 (Yang et al., 2009). In other words, our human genes are outnumbered by several orders of magnitude by the GI microbiome alone! As more and more body sites are being sampled, this difference can only increase in favor of our microbiome. Such observations underline the limited human genetic input in the whole "super-organism" and pose the question: To what extent our genes matter during our life as a super-organism? Maybe, we as hosts have predominantly lost functions during evolution because our microbes provided them and could execute the corresponding functions more efficiently? Possibly, our evolutionary efficiency is increased by encoding a "limited" gene set, which could be specialized in molecular communication with microbes in order to recruit, nourish, and maintain a microbiota that is able to complement for the essential functions that are lacking in our own genomes? After all, we are still around despite our "limited" genotype.

Interactions in a Super-Organism

In mammals, a dynamic and complex relationship exists among diet, host phenotype, and the associated GI microbiota. All interactions are dependent on the host genotype, which can be seen as a matrix on which host phenotype and the resident microbiota are projected (Fig. 2.1). Diet and transient food organisms are external, yet important, components that complete the complex host–microbiota interactions.

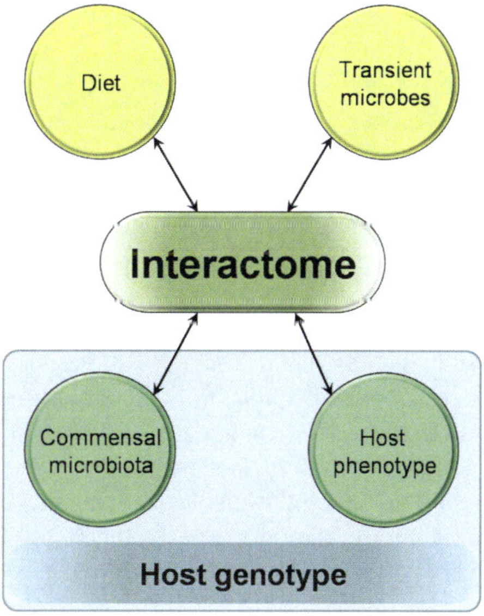

Fig. 2.1 Factors involved in host–microbiota–diet interrelations

Classical views on many disorder-associated phenotypes do not take into account all underlying factors. For example, the main focus usually lies on diet and host genotype in disorders such as obesity, diabetes, and many cardiovascular diseases. However, the GI microbiota should not be excluded in studies or treatment of these disorders, since changes in the gut communities have been associated with some of these disorders (Turnbaugh et al., 2006; Cani et al., 2007; Holmes and Nicholson, 2007; Wen et al., 2008; Zhang et al., 2010). Even though it is hard to determine the causality of observed microbiota deviations with respect to these diseases and disorders, several studies do suggest causal contributions from the gut communities. Microbiota transplants from obese mice to germ-free (GF) littermates induce the obese fat-storage phenotype (Turnbaugh et al., 2006). Furthermore, already before the onset of type-1 diabetes in genetically predisposed rat models, which are on the same diet, the gut microbiota was different in the rats that eventually developed diabetes compared to those that did not (Brugman et al., 2006). Moreover, antibiotic treatment of these rats significantly delayed and lowered the incidence of diabetes development (Brugman et al., 2006). These findings suggest a causal role of the gut microbiota in the development of diabetes.

Even in the extensively studied GI tract, the interactions between host and microbiota are not yet understood. Most studies to date are restricted to composition analyses. However, more and more insights are emerging. Especially the use of high-throughput technologies to study the diversity and functionality of the GI tract is greatly enhancing the current knowledge level. These technologies use a variety of approaches, such as the revival of culturing methods that are high-throughput (Zengler et al., 2005; Ingham et al., 2007), metabolite detection (Holmes and Nicholson, 2007), phylogenetic microarrays based on 16S rRNA sequences (Palmer et al., 2007; Rajilic-Stojanovic et al., 2009), and sequencing of 16S rRNA genes as well as sequencing of random microbial DNA (Turnbaugh et al., 2006, 2009a). Sequencing of the GI microorganisms has opened up the possibilities for functional metagenomics, which will allow further exploration of the microbial activity patterns.

It is generally accepted that the GI tract is sterile at birth and is swiftly colonized by microbes acquired from maternal and environmental sources. Recently, bacteria have been detected in the fluid in intact amniotic sacs of women in preterm labor (DiGiulio et al., 2008). This finding questions the broadly accepted view of postnatal GI tract colonization, since fetuses swallow and "inhale" amniotic fluid continuously, hence exposing their respiratory and GI tracts to everything that resides in it. However, bacteria were only found in 15% of the subjects ($n = 166$) (DiGiulio et al., 2008). Regardless of the precise moment of initial colonization, it remains without doubt that the human GI microbiota evolves over time. The development is especially drastic in the first 2 years of life, followed by stabilization of the GI community into a microbiota that resembles that of an adult (Conway, 1995). In adults the fecal microbiota is shown to be highly stable over time within one individual, as well as specific for its host (Zoetendal et al., 1998). Interestingly, despite a high variability in the GI species composition between individuals, functional capacity seems to be much more uniform between human adults (Kurokawa et al., 2007).

Moreover, a significant proportion of microbial phylotypes found in the gut are continuously present during a 10-year timeframe (Rajilic-Stojanovic et al., 2007). Therefore it seems likely that adults have, next to some transient guest organisms, a stable individual core of permanent GI tract colonizers (Rajilic-Stojanovic et al., 2007; Zoetendal et al., 2008). Tap et al. have reported 66 microbial phylotypes, which are present in more than 50% of the samples they investigated ($n = 17$) (Tap et al., 2009). Such findings suggest that besides an individual core, a limited number of microbial phylotypes are more prevalently found in people (>50% of the individuals) and appear to represent a common core of the human GI tract microbiota.

Host Genotype and Microbiota Selection

As noted before, the mechanism(s) behind GI tract colonization, succession within the community, and community structure itself are poorly understood. One hypothesis is that colonization at weaning is determined by the primary nutrient foundation supplied by the host (Hooper et al., 1999). This may be true, but there are several indications that colonization is influenced by the host genotype as well. Mouse studies have revealed that the composition of fecal microbiota is affected by the major histocompatibility complex (Toivanen et al., 2001). Furthermore, different mammalian host species develop a different make-up of their GI microbiota.

Additionally, studies with GF hosts that received inter-species GI microbiota transplants indicate that the hosts might be able to modulate their received microbial lineages toward a composition normally found in their non-GF, conventional status (Rawls et al., 2006). This can be attributed to obvious variables such as the type of food, the environment the host lives in, the nutritional requirements of the host, or more physiological aspects like host intestinal tract anatomy, body temperature, intestinal peristalsis, and residence time, etc. Intriguing results were obtained from the reciprocal GI microbiota-transplantation of zebra fish and mice, raised under GF conditions (Rawls et al., 2006). The GF hosts did not have any community legacy, and appeared to retain all intestinal species that were "given" to them during transplantation. Yet they reconstructed, in terms of relative abundance, the gut communities normally associated with conventionally raised animals (Rawls et al., 2006), indicating that a powerful and poorly understood host-mediated mechanism must be in place to coordinate microbial community composition. It seems that zebra fish and mouse host genetics play a prominent role in the natural selection of gut inhabitants, although there are many confounding factors in the differences between these hosts, such as differences in body and environment temperatures, host habitat and activity, bowel anatomy and dimensions, residence time, and dietary intake.

Last but not least, in adult humans, the extent of the variation of the dominant bacteria is associated with the degrees of relatedness between the subjects (Zoetendal et al., 2001; Van de Merwe et al., 1983). This family relatedness, especially with respect to twins, provides perhaps the most profound evidence of host genotypic influences on microbial communities, and will be discussed in more detail in the next sections. In conclusion, the GI microbiota composition must be dependent on the host genotype, but the exact degree of this dependence remains to be elucidated.

The effect of host genetics on the gut microbiota is most profoundly observed in studies conducted on related individuals. Especially revealing were studies conducted with samples obtained from identical (monozygotic) and/or fraternal (dizygotic) twins. Already in 1983, indications were found that monozygotic twins have more similar fecal microbiota than dizygotic twins (Van de Merwe et al., 1983). Although these findings were based on cultivation-dependent techniques, which gives an incomplete picture of the GI microbiota (Zoetendal et al., 2008), this study provided a clear indication of host genetics and its influence on the fecal communities. Two decades later, Zoetendal et al. confirmed the significantly higher bacterial profile similarity in monozygotic twins with a culture-independent technique (Zoetendal et al., 2001). This study was performed 10 years ago using denaturing gradient gel electrophoresis (DGGE) on fecal bacterial 16S rRNA gene amplicons to assess the bacterial composition similarity. Samples in this study originated from human adults with varying degrees of genetic relatedness (ranging from parents and children, non-twin siblings to twin siblings). This study revealed a positive relationship between the DGGE profile similarity and the genetic relatedness of the subjects (Fig. 2.2). Marital partners showed slightly higher similarities than unrelated individuals, but this was not found to be significant (Zoetendal et al., 2001). The latter is quite remarkable as marital partners essentially live in the same

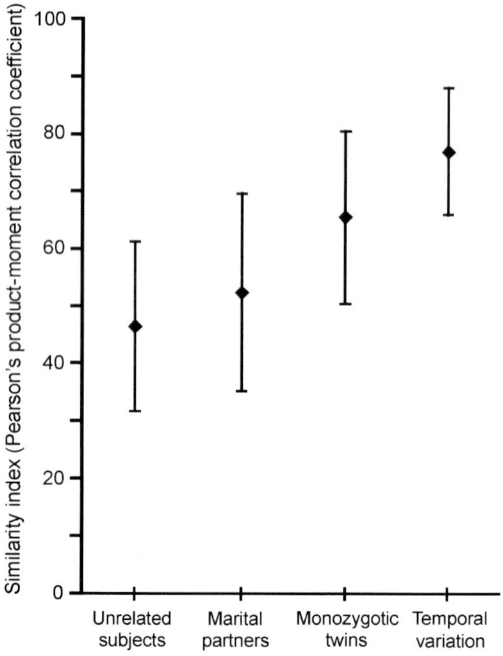

Fig. 2.2 Plot of the similarity indices (Pearson's product-moment correlation coefficient) from unrelated subjects, marital partners, monozygotic twins, and temporal variation comparisons. The mean (*diamonds*) and standard deviation (*black bars*) are plotted. DGGE profile of the total bacterial community was used to calculate the similarities. Adapted from Zoetendal et al. (2001)

environment and generally have similar dietary habits. Overall, these results indicate that host-genotype factors indeed have a strong impact on the bacterial community in the adult GI tract.

In a later study, a slightly different cultivation-independent technique, temporal temperature gradient gel electrophoresis (TTGE), was used to assess the influence of host genetics on the fecal microbiota composition in children (Stewart et al., 2005). TTGE profile similarity was again the lowest among unrelated children, higher between dizygotic twins, and clearly the highest between monozygotic twins (Stewart et al., 2005).

The high-throughput cultivation-independent Human Intestinal Tract Chip (HITChip), a phylogenetic microarray developed by Rajilic et al. (2009), was used to re-analyze the five monozygotic twin pair samples from Zoetendal et al. (2001). Average similarities between the twins was notably higher than the similarity between random unrelated individuals (Rajilic-Stojanovic et al., 2007). Even though only five twin pairs and five unrelated individuals were compared, the observed similarity difference was already borderline significant ($p = 0.067$) (Rajilic-Stojanovic et al., 2007). Thus the previous results obtained by DGGE were confirmed by phylogenetic microarray analysis. A recently conducted study on 40 monozygotic twin pairs showed that HITChip profiles were significantly ($p < 0.001$) more similar between the twins than between random unrelated subjects within this cohort (Fig. 2.3, Tims et al., unpublished observations). Palmer et al. also used

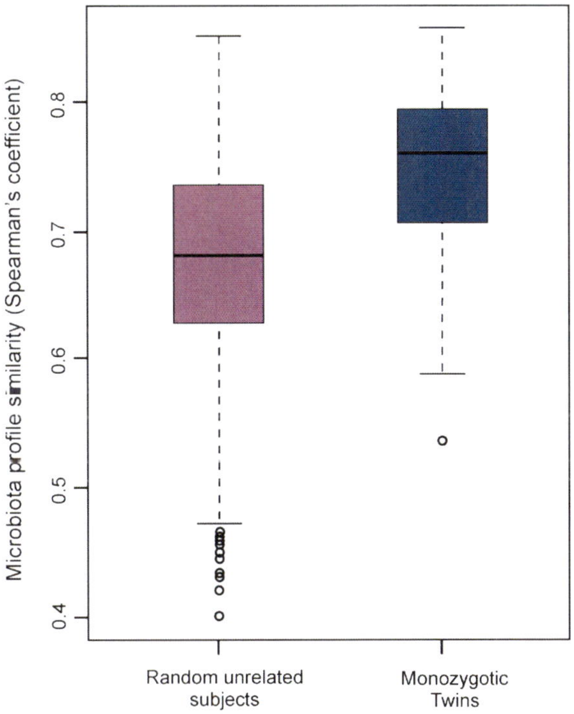

Fig. 2.3 Box-whisker plots of the similarity between the total microbiota profiles expressed as Spearman's correlation coefficient. *Purple box* represents similarities between random unrelated subjects and the *blue box* represents similarities between monozygotic twins. Similarities were calculated with total microbiota HITChip profiles (Tims et al., unpublished observations)

a phylogenetic microarray (different from the HITChip) to study GI microbiota development in human infants (Palmer et al., 2007). They included one dizygotic twin pair and this pair showed a more similar microbiota profile, at any stage of development, compared to the other 12 unrelated children (Palmer et al., 2007).

Recent developments in sequencing technologies and corresponding reductions of sequencing costs have been of great importance for GI microbiota research. Turnbaugh et al. performed pyrosequencing on variable regions of the 16S rDNA on fecal microbial DNA extracts in a cohort of 154 subjects (Turnbaugh et al., 2009a). On average, nearly 4,000 16S V2 region sequences were obtained for all subjects and additionally nearly 25,000 sequences per sample were acquired for 33 subjects. The studied group was composed of 31 monozygotic twin pairs, 23 dizygotic twin pairs, and the mothers of 46 of the twins, and included two samples per individual collected with a 57-day interval. Subjects included in this cohort were differentiated based on concordant leanness or obesity among the twin pairs. Most of the twins (71%) did not live together anymore. Pyrosequencing confirmed the previous observation that each individual has a unique GI microbiota composition and short-term changes are inferior to the inter-subject variations (Turnbaugh et al., 2009a; Zoetendal et al., 1998). Among all of the 154 individuals, no shared 16S rRNA gene-based phylotypes could be identified with an abundance of 0.5% or more. These results are in apparent contradiction with previous suggestions concerning the existence of a shared common core between humans, but different criteria were employed to define such a core community (>50% phylotype prevalence (Tap et al., 2009) versus 100% prevalence of phylotypes (Turnbaugh et al., 2009a)). Nevertheless, subjects from the same family had more similar microbial community structures and shared significantly more phylotypes (Turnbaugh et al., 2009a). Neither the obesity status per individual nor distance between the family members' homes confounded the observed higher similarities for the families (Turnbaugh et al., 2009a). In contrast to earlier findings (Stewart et al., 2005), the similarity between dizygotic twins were not lower than the similarities between the monozygotic twins (Turnbaugh et al., 2009a). Noteworthy is the fact that pyrosequencing still suffers from artifacts such as the formation of chimeric sequences during amplification and/or sequencing errors that are interpreted as distinct phylotypes. When such errors are not effectively removed they obviously influence the results obtained from sequencing-based studies. Perhaps future improvement of sequencing technology (in terms of reduced sequence error rates, advanced data analysis software suites, and/or extended sequence length) combined with increased sequencing depths may resolve the apparent contradictions in the current conclusions. Recently, Claesson et al. showed that pyrosequencing at the deepest sequencing-depth currently feasible, i.e., at approximately 400,000 sequences per sample, still does not capture the full microbial richness of GI tract samples (Claesson et al., 2009).

A common approach in the GI tract studies is inferring the possible microbiota functionalities from the microbial lineages present as detected by 16S rRNA gene sequences. Current functional metagenomic study designs are moving toward random sequencing of as much microbial DNA as possible (Zoetendal et al., 2008). Through these random sequences, a more direct view on the functional repertoire of the microbiota can be obtained. Therefore, in addition to the 16S rRNA gene sequencing, Turnbaugh et al. also analyzed the samples of six families ($n = 18$) by random shotgun pyrosequencing (Turnbaugh et al., 2009a). In line with the 16S rRNA-based study, the profiles of the functional categories present in the gut communities were more similar between relatives. Interestingly, GI microbiota seemed to have even more similar functional profiles among all subjects despite their sometimes highly distinct microbiota composition profiles (Turnbaugh et al., 2009a). This raises the hypothesis that a core microbiome exists at a functional (and metabolic) level rather than at the level of microbial composition. In line with this hypothesis are results of earlier studies that determined in situ concentrations of short-chain fatty acids (SCFAs) in fecal material of different individuals. Different population survey data show the same fecal composition with respect to the ratios of the three main SCFAs acetate, propionate, and butyrate (Topping et al., 2001). Dietary changes have been shown to modulate SCFA production and absorption somewhat, but the SCFA ratios are not drastically altered (Cummings, 1981). As an individual has its own unique GI microbiota composition, the fairly constant SCFA ratios are quite remarkable and indicate that the gut microbes perform similar overall functions. Even though a core GI microbiota is likely to exist more on a functional level it must be noted that the species composition, although hypervariable and seemingly chaotic, is not random. Grouping of mammals based on their gut community composition is associated with their dietary needs, i.e., whether a mammal is a herbivore, carnivore, or omnivore (Ley et al., 2008). This could indicate that when roughly similar digestive tasks are required, the GI community tends to be more similar at higher taxonomic levels (Ley et al., 2008). However, the latter statement needs to be verified as the GI tract anatomy differs between individual mammalian species belonging to the carnivores, herbivores, and omnivores.

More comprehensive studies following both monozygotic and dizygotic twins from birth to adulthood can provide valuable information on interactions between host genotype and its inhabiting microbiota. In such studies diet should be taken into account, due to its drastic influences.

Dietary Influences

Diet is a strongly confounding factor in the ambition to obtain insight in the exact roles of the host genotype and GI microbiome in a super-organism. Not only the GI community is influenced by the dietary influx of microbes and nutrients but also the host phenotype itself, mainly through regulation of gene expression and physiological adaptation. Several genes from a group of nuclear hormone receptors called peroxisome proliferator-activated receptors (PPARs) can be used to illustrate host adaptation to its diet on a transcriptional level. Unsaturated fatty acids are, among others, activating ligands for PPAR-γ (Kliewer et al., 1997), and thereby PPAR-γ activities are directly influenced by the diet. PPAR-γ acts as a regulator of a variety of relevant physiological processes, including transcription control of many genes involved in fat-cell differentiation, insulin sensitivity, and lipid homeostasis

(Debril et al., 2001). The host genotype, in terms of PPAR-γ gene variants, affects the extent to which the host reacts to its diet. For example, a PPAR-γ2 gene polymorphism (Yen et al., 1997) correlates with the inter-individual variability in serum triacyl–glycerol levels after administration of $n-3$ fatty acids (Lindi et al., 2003). This example is just one of the many available in literature, which indicates that although the host contributes with a marginal amount of genes to the metagenome, variations in the host genotype can still have drastic effects for the super-organism. Furthermore, this example indicates the major role of the host's diet as well. This raises the question of how important the diet is exactly.

Obesity in Animal Models

Obesity has gained popularity with respect to study of the importance of dietary influences on the gut microbiota and the host. The prime cause of obesity is an excessive caloric intake. Such a surplus intake disturbs the normal balance between the amount of energy harvested from the diet and the amount used by the host. This balance, or energy homeostasis, is at least partly defined by the GI microbiota. Studies with GF mice clearly show the microbial impact on host energy homeostasis. GF mice are resistant to obesity development when fed a "Western" (high-fat/high-sugar) diet (Backhed et al., 2007), but GI tract colonization was stimulating weight gain in these mice (Samuel et al., 2008). Colonization of GF mice leads to an increased release of monosaccharides and SCFAs from complex dietary polysaccharides, enhanced conversion rates of fatty acids toward complex lipids in the liver, and by regulating host genes involved in storage of the converted lipids into adipocytes (Backhed et al., 2004).

However, next to the GI microbiota, host-dependent factors are also required to develop obesity. For instance, mice lacking a functional copy of the Gpr41 gene, a G-protein-coupled receptor that binds SCFAs, do not readily develop obesity (Samuel et al., 2008). Without Gpr41, the mice displayed an increased intestinal motility and decreased SCFA absorption (Samuel et al., 2008). On a side note, another G-protein-coupled receptor called Gpr43 displays a different role in mice (Maslowski et al., 2009). Mouse models of colitis, arthritis, and asthma required stimulation of Gpr43 by SCFAs to counteract their inflammation (Maslowski et al., 2009). Thus, besides metabolic regulation, immune and inflammatory responses are affected by the bacterial SCFA production as well.

Simply having a gut microbiota is not the only microbial factor affecting the energy balance in mice. The composition of the microbial community determines to what extent the microbiota improves energy harvest from food. Both GF mice inoculated with distal gut microbiota from conventionalized obese and lean animals resulted in an increase of bodyweight and total body fat (Turnbaugh et al., 2006). Yet this increase was found to be significantly larger in the GF mice that had received the microbiota from the obese animals (Turnbaugh et al., 2006). Correlations between phenotypic variations and attributes of the microbial gut community were reported after studies conducted with lean and genetically obese (ob/ob) mice (Turnbaugh et al., 2006; Ley et al., 2005) as well as with lean and genetically obese rats (Waldram et al., 2009). Variations of the members belonging to the bacterial phyla Bacteroidetes, Firmicutes, and Acintobacteria appeared to be associated with leanness or obesity (Turnbaugh et al., 2006; Ley et al., 2005; Waldram et al., 2009). Especially the Bacteroidetes to Firmicutes ratio was found to be lower in the obese rodents.

Most animal models used to study obesity consist of specific gene knockout mutants, such as those focusing on the role of the *ob* gene in mice that predisposes them to develop obesity. The *ob* gene encodes for the protein hormone leptin, which regulates body weight, metabolism, and reproduction in mammals (Friedman et al., 1998). Both inactivating mutations in the leptin (*ob*) gene and in its receptor (*db*) gene lead to genetically obese mice (Friedman et al., 1998). Therefore, the host genotype is important in the development of obesity as well. Another study involving wild-type and *Apoa*-I knockout mice indicates that all three aspects are implicated in the development of obesity and metabolic syndrome (MS) (Zhang et al., 2010). *Apoa*-I knockout mice have an impaired glucose tolerance and high body-fat levels. Groups of wild-type and of *Apoa*-I knockouts were fed a high-fat diet and normal chow diet for 25 weeks. Diet as well as genetic mutation could explain 57 and 12% of the observed variation found in the GI microbiota communities, respectively (Zhang et al., 2010). The results of this study indicate a stronger, possibly dominating, role of the diet compared with the genetic variations of the host. Nevertheless, the influence of host genotype is not negligible and should therefore be taken into account in MS studies.

Obesity in Humans

Extrapolation of the relations among diet, host genotype, and GI microbiota discovered in animal studies to human obesity seems sensible but has proven to be difficult. Nevertheless, deviating leptin concentrations are associated with obesity in humans as well (Considine et al., 1996). Leptin normally suppresses hunger and increases metabolism, and it has been suggested that obese humans are insensitive to leptin. However, although several cases have been described (Clement et al., 1998), mutations in the human *db* gene are only rarely seen in obese people. However, to accurately assess the importance of *db* gene variations with respect to obesity development risks, the *db* mutation frequencies in lean individuals should also be investigated.

In contrast to mice studies, research with human subjects provide conflicting results regarding the association of obesity with relative abundances or abundance ratios of specific bacterial phyla (Ley et al., 2006; Schwiertz et al., 2009; Duncan et al., 2008). However, human studies reported to date employed different molecular techniques and targeted populations of different geographic origin. Duncan et al. found no differences in bacterial phyla abundances or abundance ratios, but identified a significantly higher proportion of butyrate producers in the obese subjects (Duncan et al., 2008). This observation is in agreement with the finding of increased butyrate concentrations in the cecum of *ob/ob* mice as compared to their lean littermates (Turnbaugh et al., 2006). The SCFA butyrate is mainly produced during carbohydrate fermentation in the colonic lumen, mainly by members of the *Firmicutes* phylum, especially those belonging to *Clostridium* cluster IV. Luminal butyrate is quickly absorbed by the colon mucosa where it serves as the main energy source for the colonocytes (colonic epithelial cells) (Cummings, 1981; Roediger, 1980). However, the exact physiological effects of butyrate are not fully understood. Different, but related cell-line models can yield direct opposite results regarding the role of butyrate in the modulation of cell proliferation, differentiation, and apoptosis (Hague et al., 1997; Medina et al., 1998). These conflicting results, commonly referred to as the "butyrate paradox," are extensively reviewed elsewhere (Hamer et al., 2008). In short, in vivo human data are insufficient but most studies support beneficial roles for butyrate, including the restraining of inflammation and carcinogenesis, reinforcement of various components of the mucosal barrier, lowering of colonic oxidative stress, and promotion of satiation (Hamer et al., 2008). Overall, it is obvious that the production of butyrate by the GI microbiota has a major influence on colonic mucosa. Thereby the differential abundance levels of butyrate-producing microbes, as reported by Duncan et al., seem relevant with respect to human health and may be associated with energy homeostasis and obesity risk.

Diet: Transient or Permanent Effects?

One may conclude from the animal experiments and the observations in human volunteers that the interactive factors, constituted by dietary intake and gut microbial ecology, are of major importance for the host's well-being. This raises the question whether dietary effects should be seen as transient or can also generate permanent effects. The microbiota transplant approach in mice revealed that efficient energy-harvesting traits are transferable by the GI microbiota (Turnbaugh et al., 2006). In continuation of this approach, C57BL/6 J mice were conventionalized in such a way that all animals inherited similar gut microbiota (Turnbaugh et al., 2008). Also in these mice, the change from a chow diet (low-fat/high-fiber) to the "Western" diet (high-fat/high-sugar) resulted in an increased weight gain (Turnbaugh et al., 2008). In the diet-induced obese mice the relative abundances of the *Firmicutes* were higher, whereas those of the *Bacteroidetes* were lower compared with their lean status (Turnbaugh et al., 2008). These findings are in agreement with previous results obtained with genetically obese (*ob/ob*) mice (Ley et al., 2005). However, the changes in the Firmicutes phylum were not division wide but appeared to be mainly restricted to an increased abundance of the Mollicutes class (Turnbaugh et al., 2008). Apparently, these diet-induced changes invoked an adaptation of the microbiota to the quality and quantity of the available nutrients, and these diet-induced microbiota adaptations are apparently reversible (Turnbaugh et al., 2008). Follow-up experiments in mice receiving human gut microbiota transplants confirmed that the diet-induced changes on the GI communities in these so-called humanized mouse models are reversible as well (Turnbaugh et al., 2009b). Interestingly, obese and lean associated microbial communities could be maintained by diet alone. Maintenance of community structure and diversity was even achieved across several generations of mice following initial transplantation (Turnbaugh et al., 2009b). This again illustrates the prominent influence of the diet. Whether the diet is able to overrule the host genotype by permanent alterations of the GI community remains to be explored. In rats and humans, dietary "metabolic imprinting" through epigenetic modifications on the host genotype seems likely (Bateson et al., 2004; Godfrey and Barker, 2006; Lillycrop et al., 2005). However, the reversibility of the gut ecology in the inter-host as well as intra-host species microbiota transplants described above seem to rule out the possibility of imprinting through the GI community (Turnbaugh et al., 2008, 2009b). Although not permanent, it remains a fact that diet has a major and rapid impact on the microbiota.

Host–Microbiota Co-evolution: Selective Geographic Pressure?

Environment can be easily overlooked while studying GI tract microbiota, but is quite important as it determines physiological as well as microbial influences, e.g., through availability of food, food consumption habits, temperature, and humidity. Many environmental aspects are geographically confined, which raises the question if geography and its associated factors, such as climate, availability of food, and composition of the diet, could be an important aspect in host–microbe interactions. Several studies indicate an intimate co-evolution of humans and the gastric pathogen *H. pylori* (Linz et al., 2007; Moodley et al., 2009). *Helicobacter pylori* appears to have spread from east Africa 58,000 years ago along with its human host. This finding implies that geography can also influence the microbial community, although modern commuting could blur the extent of such geographical impact. Naturally, the exact course and speed of this "blurring" depends on the magnitude of the evolutionary developed differences between the ethnic groups involved.

Many studies try to minimize the drastic effects of diet on the microbiota. Dietary habits usually depend on the geographical location of the subjects. Hence, not many studies have been performed on GI microbiota composition across country boundaries. One of the first studies across several countries was performed by Lay et al. and conducted on 91 subjects from five Northern European countries (France, Denmark, Germany, the Netherlands, and the United Kingdom) who consumed a nonrestricted Western European diet (Lay et al., 2005). However, the identified bacterial proportion of the gut microbiota did not significantly differ in composition when grouping the samples according origin, gender, or age (Lay et al., 2005). Mueller et al. conducted a study that included as many as 230 healthy subjects from the more distant European countries such as France, Germany, Italy, and Sweden (Mueller et al., 2006). Several differences were found regarding country, age, gender, and combinations thereof (Mueller et al., 2006). Without much effort, dietary justifications can be found for most of the observed phylogenetic differences. An interesting example is the relative abundance of *Faecalibacterium prausnitzii*. Strict vegetarians appear to have no detectable amounts of *F. prausnitzii* (Hayashi et al., 2002). The authors suggest that the highest levels of *F. prausnitzii* and related species was in the Swedish subjects and may be related to a high level of fish and meat consumption, which is a known dietary tradition of the Swedish population (Mueller et al., 2006). However, at the time of this study the Swedes, Italians and French consumed the same amount of animal products (World Resources Institute - EarthTrends Environmental, http://earthtrends.wri.org/), hence other dietary influences are probably implicated as well.

The European studies seem to indicate that differences in GI microbiota composition increase with the distance between the geographic origins of the subjects. At a larger intercontinental distance, comparable studies were performed by American (Eckburg et al., 2005; Gill et al., 2006) and Chinese (Li et al., 2008) research groups. Although at phylum level, the Chinese and American subjects exhibited comparable phylogenetic GI tract compositions, principal coordinate analysis showed clear differences at species-level composition (Fig. 2.4) (Li et al., 2008). Importantly, these findings reflect the differences in nuclear magnetic resonance based metabolic urine phenotypes found between large groups of Chinese and American subjects. Many of the differential urine metabolites do not have a mammalian origin but are derived from microbial sources (Dumas et al., 2006). Thus, it can be hypothesized that there is a co-variation between gut microbiota structure and the host metabolic

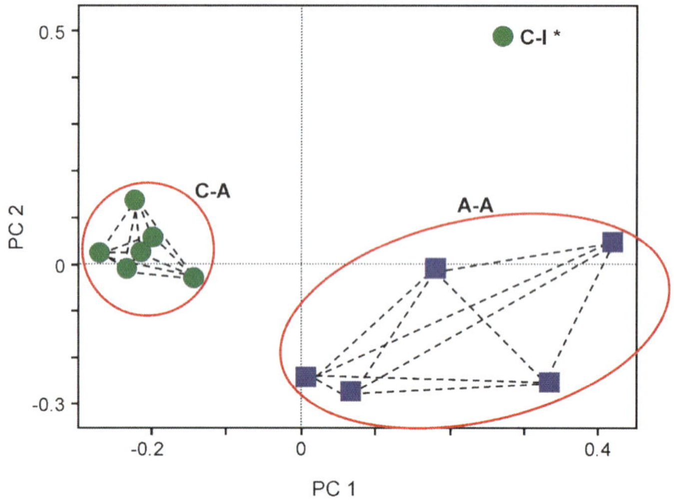

Fig. 2.4 Principal component (PC) analysis on the species-level composition of the gut microbiota of Chinese family members (*green circles*) and American volunteers (*blue squares*). UniFrac metrics were used to generate the principal coordinate scores plot. The percentage of variation described by PC 1 is 19.8% and by PC 2 is 13.5%. *Red circles* are drawn around the group of adult Chinese (C-A) and around the American subjects (A-A). Intersample differences within the adult groups are visualized with dashed lines. C-I* indicates the Chinese baby 1.5 years of age, which was left out the C-A group due to its still developing gut microbiota. Figure adapted from Li et al. (2008) (American microbiota data: (Eckburg et al., 2005; Gill et al., 2006))

phenotype. Urine metabolite profiles associated with microbial products are able to discriminate Japanese living in Japan and Japanese living in America (Holmes et al., 2008). Therefore, the differences found between the Americans and Chinese could be more dependent on diet and/or geography than on host genotype. Interestingly, the Chinese volunteers (Fig. 2.4) all belonged to the same family and had more similar GI communities when compared among each other than the unrelated subjects within the American cohort. Hence, this observation does associate genetic relationship with GI microbiota composition and therefore points at a role for the host genotype in gut community development.

Whether geographic pressure has left its mark on the co-evolution of men and GI microbes is impossible to tell from the studies described in the previous paragraphs, without the knowledge on the influence of the diet and host genetics. Geography could be important with respect to the availability or prevalence of environmental or dietary microbial lineages for host colonization. Although the host (genotype) is dependent on the microbes it receives, it might be able to put selective pressure on sub-populations (Rawls et al., 2006). However, currently, no colonization restrictions have been observed in people who migrated from their traditional home countries into other geographic regions.

Intriguingly, the variation in colon cancer prevalence among Afro-American and native African populations appears to be in agreement with co-evolutionary relationships between human and GI microbiota caused by geographic and environmental conditions. Americans of African origin have the highest risk of all American subpopulations to develop colon cancer, whereas native Africans rarely suffer from this type of cancer (O′Keefe et al., 2007). The biggest difference between these two populations is probably their dietary habits, which prominently influences their GI microbiota communities (Ley et al., 2006; Turnbaugh et al., 2008, 2009b). Initial cultivation-dependent analyses confirmed differences in GI communities between native Africans and Afro-Americans (O'Keefe et al., 2007), although these methods provide an incomplete impression of the microbiota. Newborn Afro-Americans are not exposed to the (relative) abundant levels of many microorganisms as they would in the native African region. Thus, no major colonization occurs of microbes that normally dominantly reside among the child's kin group, which essentially is a settled population established through generations-long consistent geographic and lifestyle factors. This means that the emigration process that commonly coincides with changes in dietary habits seems to disrupt the co-evolved mutualism between the host genotype and its GI microbiota, resulting in the increased colon cancer risk in African Americans. In conclusion, the functioning of the human super organism appears to be affected by currently fading barriers in human and environmental ecology caused by urbanization, global traveling, and emigration. These fading barriers probably coincide with a reduction in the number of microbe encounters, which could lead in humans to an underdeveloped immune system according to the hygiene hypothesis (Guarner et al., 2006). Such disruptions of long-term co-evolved interactions between man and microbe might partially explain the observed increase of chronic and degenerative disease frequencies in industrialized countries (O′Keefe et al., 2007; Guarner et al., 2006).

Host-Genotype Polymorphisms

The genetic make-up of humans is very similar, and small genetic polymorphisms form an important aspect of genetic variation among human beings. Evidence is accumulating that these genetic variations are important determinants for the interactions with host-associated microbiota. Intuitively, the highly polymorphic immune-related genes are eminent candidates to define the interaction between host and microbe. The mucosal epithelial barrier has traditionally been considered to prevent contact between microbiota and underlying cells, including immune cells. Any contact was thought to provoke immune reactions that would even eradicate the commensal organisms from the host. However, current knowledge clearly establishes frequent and essential communications between immune system and microbiota (Rakoff-Nahoum et al., 2004; Macdonald and Monteleone, 2005). Immune tolerance is promoted by local immune modulations that only inhibit further host tissue penetration of the commensal microbes (Macdonald and Monteleone, 2005; Cebra, 1999; Macpherson, 2005). This "peaceful situation" is normally maintained despite the massive presence of bacterial molecules that are capable of activating the host's bacterial molecular pattern recognition receptor and their cognate immune regulation cascades (Rakoff-Nahoum et al., 2004). Polymorphisms in any of the immune-related genes involved in immune tolerance are prominent candidate determinants of the bacterial selection by the host.

Host–Microbiota Communication: Innate Immune System

Important bacterial molecular pattern recognition receptors capable of initiating innate immune responses are the toll-like receptors. Toll-like receptor-4 (TLR-4) can recognize lipopolysaccharide (LPS), a cell wall component of Gram-negative bacteria (Hoshino et al., 1999; Poltorak et al., 1998). Mutations of the TLR-4 gene result in a weakened immune response to LPS in mice (Poltorak et al., 1998). Moreover, TLR-4 sequence variants in humans correlated with a reduced response to inhaled endotoxins (Arbour et al., 2000). For Gram-positive-produced lipoproteins and lipoteichoic acids TLR-2 seems to be the main mammalian receptor (Lorenz et al., 2000). Moreover, TLR-2 either alone or in heterodimer form with other TLRs can recognize more than a single ligand (Opitz et al., 2009). Naturally occurring mutations in the human TLR-2 gene were shown to have diminished response to lipoproteins harvested from the Gram-positive bacteria *Borrelia burdorferi* and *Treponema pallidum* (Lorenz et al., 2000). Although the polymorphisms in TLR-4 and TLR-2 are medically relevant, no studies have been reported, which determine their impact on human (or other mammalian organisms) associated microbiota. However, it is reasonable to suspect gene variants of important components of the innate immune system to be involved in the host–microbiota interactions. From mice studies, it is known that a genetically disabled innate immune system is associated with a collitogenic murine gut community (Garrett et al., 2007). Unfortunately, in these types of studies it is unclear whether the colitis is caused by a change in community structure or by the defective defense system of the murine hosts.

Other immune system components have been shown to be involved with the host–microbiota crosstalk. Besides TLRs, the innate immune system also depends heavily on the nucleotide-binding oligomerization domain (NOD) receptors (Inohara et al., 2001). An overview on the current knowledge of TLRs, NOD receptors, and other innate immune system receptors is given elsewhere (Opitz et al., 2009). NOD2 variants clearly show a relation between host genotype and gut microbiota composition, which is associated with increased risk for Crohn's disease (CD) (Hugot et al., 2001; Ogura, Y., et al., 2001). Normally, NOD2 binds the muramyl dipeptides of bacterial peptidoglycan (Girardin et al., 2003), but certain polymorphisms in NOD2 can result in failure of muramyl dipeptide detection. Such a failure results in a lack of tolerance development for commensal bacteria and dietary antigens and consequently leads to "inappropriate" immune response against them (Landers et al., 2002). Considering these results, it is not surprising that CD patients were found to have lower diversity and diminished levels of normally abundant bacteria (Manichanh et al., 2006). Especially the phylogenetic group *Clostridium leptum* in the Firmicutes phylum seems to be reduced in patients with CD. Several molecular techniques have indicated that *F. prausnitzii*, which is a member of the *C. leptum* group, is depleted in the mucosa-associated communities (Manichanh et al., 2006; Frank et al., 2007; Martinez-Medina et al., 2006). Importantly, secreted metabolites of *F. prausnitzii* have been shown to exert anti-inflammatory effects in vitro (Sokol et al., 2008). In addition, *F. prausnitzii* is associated with an in vivo reduction of pro-inflammatory cytokine synthesis and increase of anti-inflammatory cytokine production in the colon. The observations concerning *F. prausnitzii* combined with the association of CD and NOD2 mutations illustrate a potential link among host genotype, phenotype, and microbiota composition. NOD2 mutant genotypes are likely to promote a negative selection of *F. prausnitzii* in the colon, which in turn can lead to the development of CD. Nevertheless, it is possible that other variations in the host genotype can allow the selection of other microbes similar to anti-inflammatory influences as *F. prausnitzii*.

Host–microbe interactions are not restricted to direct interactions between bacterial ligands and receptors of the innate immune system. Other proteins that should be considered are, for instance, further down the signaling pathway of TLRs or are involved in other cellular processes supporting immune responses, such as cell movement and restructuring. The gene MEFV encodes the protein pyrin, which is involved in innate immune response regulation, but has currently no definite function assignment (Ting et al., 2006). This gene has no direct contact with microbes as it has been found in the cytoskeleton, but mutations of MEFV can lead to Mediterranean fever, a hereditary auto-inflammatory disorder (Khachatryan et al., 2008). In addition, patients with this disease have lower bacterial diversity and prominent population shifts in the Bacteroidetes, Firmicutes and Proteobacterium phyla during periods of active disease, whereas when the disease is in remission the bacterial gut community is more similar to normal microbial composition but still atypical (Khachatryan et al., 2008). Therefore, even polymorphisms in genes encoding proteins that are not supposed to be in direct contact with the gut microbes can influence host–microbe interactions. Recently, in healthy human mucosa, gene expression patterns have been found, which correlate with the development of immune tolerance for the organism *Lactobacillus plantarum* (van Baarlen et al., 2009). Even for one organism, these patterns involve many genes and therefore may be affected by many potential polymorphisms, which again exemplifies the complex nature of host–microbe interactions.

Host–Microbiota Communication: Non-immune-Related Mechanisms

Variations in genes not directly involved in immune system pathways can exert influence on the microbiota composition and functioning as well. Changing the conditions for the GI microbiota through modulation of host-derived resources or available attachment site in the mucosal layer seems of considerable importance. A nice illustration can be given by the abundant commensal species in the human and murine gut *Bacteroides thetaiotaomicron*. This organism matches its demand for fucose, a growth substrate, by upregulating fucosylated glycan production of epithelial cells in mice whenever pentose sugars are scarce (Hooper et al., 1999). Other bacteria might (ab)use this epithelial fucose synthesis regulation of *B. thetaiotaomicron*, as such this mechanism is important for multiple species in the GI microbiota (Hooper and Gordon, 2001a). Extrapolating this finding to the human situation, one can imagine the involvement of the fucosyltransferase enzyme polymorphisms, which are determinant for human blood types (Becker and Lowe, 2003). This type of gene has many variants since human glycoproteins are likely to be continuously evolving through natural selection provided by commensal and pathogenic microorganisms (Hooper and Gordon, 2001b). Notably, some correlations between blood type and gut community variations indeed have been reported in the past (Van de Merwe et al., 1983; Hoskins, 1993).

Another example of potential host-genotype-dependent interactions with gut microbes is mucin encoding (MUC) genes. These MUCs encode protein backbones for mature mucin molecules, which are heavily glycsylated by O-linked oligosaccharides on the threonine, proline, and/or serine repeats (Herrmann et al., 1999). The terminal oligosaccharides of mucin molecules contain sulfate and/or O-acetyl-substituted sialic acids. Mucins can both promote and prevent bacterial cell adhesion, depending on the exact structures of their O-glycan chains (Hollingsworth and Swanson, 2004). Changes in mucin composition have been associated with inflammatory bowel diseases (Morita et al., 1993). Next to microbial mucin degradation, these changes could be due to genotype-related issues, such as polymorphisms in MUC genes, variations in MUC mRNA or protein levels, and variable post-translational modification changes (i.e., the extent of glycosylation and sulfation) (Morita et al., 1993).

Many more host genes can be found that are involved in the functioning of the GI tract and thereby polymorphisms in such genes could have an impact on the microbial communities. Mutants in host enzymes responsible for nutrient breakdown and/or absorption could potentially influence the microbiota by altered nutrient composition and availability in the different regions of the intestine. For example, enterocytes in the small intestine can absorb glucose through active and passive glucose transporters (Leonie Los et al., 2007). Variants of these host-encoded transporters could cause alterations in the rate of carbohydrate absorption and thereby modulate the carbon source availability for the resident microbes, which may favor different microbial communities. More complex host–microbiota interactions are mediated by bile acids, which, next to their digestive functions (i.e., solubilization of lipids and lipid-soluble vitamins to enhance their absorption), have a role in maintaining the intestinal barrier (Martin et al., 2007). Mammalian-microbiota co-metabolism result in the so-called secondary bile acids, which exert biologically important effects on both host and microbiota constituents (Martin et al., 2007). Variations in level or composition of bile may have prominent effects on microbial communities and may correspond with specific consequences in mucosal cell biology. Primary bile acids are synthesized in the liver by a cascade of enzymes, providing many possibilities for gene variants that influence bile composition and corresponding host–microbe interactions.

Direct Interactions of Host Genome and Microbiome

Of all the possible interactions taking place in or around a host (Fig. 2.1), modulation of the host genotype by the microbiota seems extraordinary. Nevertheless, when 223 genes in the rough draft of the human genome were found to potentially have a bacterial origin, horizontal gene transfer (HGT) from bacteria to humans has been suggested (Lander et al., 2001). This could indicate that bacteria can manipulate their host, likely for their own benefit. However, HGT between human and bacteria is a difficult process because the genes should be stably integrated into the host DNA of germ line cells, to which bacteria normally do not have physical access. Furthermore, in 2001, Salzberg et al. carefully reexamined protein sequences of human, four other eukaryotes, and all completed prokaryote genomes at that time (Salzberg et al., 2001). They only found about 40 human genes to be possible candidates for HGT from bacteria to humans. Therefore, HGT between bacteria and humans remains doubtful since alternative, more plausible biological and technical explanations may be responsible for the few shared genes that are observed. One such biological factor is the high probability that the analyzed species had lost several genes from the common eukaryotic ancestor gene pool. Furthermore, nucleotide substitution rates can vary between genes within one genome as well as between similar genes in different organisms (i.e., evolutionary rate variation (Li, 1997)). Therefore, in HGT analyses, evolutionary relatedness cannot be based on sequence similarity alone, indicating an important technical limitation in the currently available studies on this topic (Eisen, 1998). Furthermore, only five eukaryotic genomes were available at the time of analysis, three of which belong to the animal lineage (*Caenorhabditis elegans*, *Drosophila melanogaster*, and *Homo sapiens*) (Lander et al., 2001; Salzberg et al., 2001). Hence, the total eukaryotic diversity was poorly represented. By contrast, the available prokaryotic genomes at that time embody a much broader evolutionary diversity (Nelson et al., 2000). This limited sample size of eukaryotic genomes is yet another technical problem confounding the HGT from bacteria to human. Concluding, it seems unlikely that bacteria have permanently manipulated their human hosts through HGT.

It seems more feasible that microbes sometimes pick genes up from their hosts. For example, the possible transfer of genes encoding serpins, which are protease inhibitors involved in the regulation of many physiological processes (Ivanov et al., 2006). Although serpins are found in all three domains of life, which indicate that they could originate from a common ancient, serpins are found in relatively few prokaryotes. The latter would imply that serpins are not essential for survival or that they may have been acquired by prokaryotes as the result of HGT. Even though the serpin of *Bifidobacterium longum* is distantly related to eukaryotic serpins, Ivanov et al. showed that it exerts inhibitory functions through an identical mechanism. Another clearer example is the presence of *nptA* gene, a sodium/phosphate co-transporter, in *Vibrio cholerae*. This gene is also present in animals but seems to be absent in any other bacterial species (Lebens et al., 2002). Furthermore, *V. cholerae* has been shown to exhibit activity similar to that of its animal homologs (Lebens et al., 2002). It is likely that this transporter facilitates *V. cholerae* in the GI tract and consequently could be involved in the pathogenicity of this microbe. However, in both the serpin and the *nptA* case, it remains difficult to prove that the genes are not derived from an ancestral gene instead of being transferred from a (mammalian) host species.

Mitochondria provide off course evidence that indeed bacterial DNA resides in the mammalian genotype. Although these eukaryotic organelles originate from the endosymbiosis of an alpha-*Proteobacterium* ancestor, they show no indications that they were introduced in order to manipulate the host for their own benefit. By contrast, recent findings show that it was the eukaryotic host that took control and manipulated the bacterial endosymbiont for its own benefit (Gabaldón and Huynen, 2007). During the transformation of bacterium to organelle, many bacterial genes not involved in energy conversion were lost or replaced by genes originating from the eukaryotic host (Gabaldón and Huynen, 2007).

Conclusions

Humans and other mammalian hosts provide only a minor quantity of genes to their super-organism metagenome. Yet these genes are essential and decisive in defining the final host–microbiota interactions. Disturbances in the host genotype can lead to malfunctioning of the super-organism, i.e., all kinds of metabolic disorders, immune diseases, and other disorders.

Human genotypes consist of a huge amount of variables, many of which could be of importance for host–microbiota interactions. Not only genes directly related to the immune system should be considered in future studies. For instance, *Escherichia coli* has been implicated to, via quorum sensing, cross-communicate with the host epinephrine signaling pathway (Sperandio et al., 2003). Although this is coming from a pathogenic species, one cannot exclude this type of non-immune-system-related communication between the host and its commensals. Another form of communication is through metabolites, such as SCFAs or (secondary) bile acids (Maslowski et al., 2009; Martin et al., 2007). Hence, many different mechanisms constitute the overall host–microbe interactions pallet. Basically, the hierarchy in importance of human genotypes in relation to host–microbe interactions is unknown. Currently, the majority of predictions hint at immune-related factors, i.e., gene polymorphisms. However, genes involved in metabolic functions and their control or those involved in biosynthetic pathways, such as mucus production and modification, or bile metabolism are likely to be important modulators as well.

New human genotyping efforts using extensive volunteer cohorts, combined with in-depth microbiota profiling, provide a possibility to mine for all factors underlying the relation between human host and its microbial communities. Furthermore, comprehensive studies following both monozygotic and dizygotic twins from birth to adulthood will provide vital information to assess the relative contribution of host genotype to the GI microbiota composition. These studies will be most successful when they acquire additional metadata, such as dietary habits, and actual short-term nutrient intake, and preferably also include intergenerational analysis of the subjects' families. Such multivariant analyses will be essential to dissect influences of dietary, environmental, and host-genotype factors.

Nevertheless, diet will probably always be an obscuring factor due to its dramatic, but apparently reversible effects on the microbiota. Future studies could benefit from the consistent use of family members, different ethnic groups, or both. Difficult studies, from an ethical point of view, in which the subjects are isolated for longer periods of time and under strict dietary regimes, might provide better insight in the human–diet–microbiota relationship. Regardless of the chosen study types, it will be a "life-changing" experience to finally fully understand both dietary and host genotype influences involved in shaping and interacting with the intestinal microbiota. Such knowledge may enable the definition of dietary regimes that provide prophylactic and therapeutic possibilities for a variety of disorders and/or diseases, provided that a causal relationship underlies the observed diet and microbiota correlation with these disorders or diseases. Specific dietary design might be attempted to correct deviating microbiota compositions and/or activity associated with specific diseases toward a more "healthy microbiota." The rapidly developing field of (functional) metagenomics may allow us in the near future to actually come to the accurate description of what can be considered a "healthy microbiota," which could then be employed as a biomarker in diagnosis and treatment of diseases and/or disorders.

> "He who does not know food, how can he understand the diseases of men?" – Hippocrates (460–357 B.C.)

Acknowledgments Authors ST and EGZ received funding from the European Community's Seventh Framework Programme (FP7/2007-2013): MetaHIT, grant agreement HEALTH-F4-2007-201052, coordinated by S. Dusko Ehrlich (Institut National de la Recherche Agronomique, France).

References

Aas JA et al (2005) Defining the normal bacterial flora of the oral cavity. J Clin Microbiol 43(11):5721–5732

Albert MJ, Mathan VI, Baker, SJ (1980) Vitamin B12 synthesis by human small intestinal bacteria. Nature 283(5749):781–782

Adlercreutz H et al (1984) Studies on the role of intestinal bacteria in metabolism of synthetic and natural steroid hormones. J Steroid Biochem 20(1):217–229

Attene-Ramos MS et al (2006) Evidence that hydrogen sulfide is a genotoxic agent. Mol Cancer Res 4(1):9–14

Arbour NC et al (2000) TLR4 mutations are associated with endotoxin hyporesponsiveness in humans. Nat Genet 25(2):187–191

Backhed F et al (2004) The gut microbiota as an environmental factor that regulates fat storage. Proc Natl Acad Sci USA 101(44):15718–15723

Backhed F et al (2005) Host-bacterial mutualism in the human intestine. Science 307(5717):1915–1920

Backhed F et al (2007) Mechanisms underlying the resistance to diet-induced obesity in germ-free mice. Proc Natl Acad Sci USA 104(3):979–984

Bateson P et al (2004) Developmental plasticity and human health. Nature 430(6998):419–421

Becker DJ, Lowe JB (2003) Fucose: biosynthesis and biological function in mammals. Glycobiology 13(7):41R–53R

Begley M, Gahan CG, Hill C (2005) The interaction between bacteria and bile. FEMS Microbiol Rev 29(4):625–651

Brugman S et al (2006) Antibiotic treatment partially protects against type 1 diabetes in the bio-breeding diabetes-prone rat. Is the gut flora involved in the development of type 1 diabetes? Diabetologia 49(9):2105–2108

Cani PD et al (2007) Selective increases of bifidobacteria in gut microflora improve high-fat-diet-induced diabetes in mice through a mechanism associated with endotoxaemia. Diabetologia 50(11):2374–2383

Cebra JJ (1999) Influences of microbiota on intestinal immune system development. Am J Clin Nutr 69(5):1046S–1051S

Claesson MJ et al (2009) Comparative analysis of pyrosequencing and a phylogenetic microarray for exploring microbial community structures in the human distal intestine. PLoS One 4(8):e6669

Clement K et al (1998) A mutation in the human leptin receptor gene causes obesity and pituitary dysfunction. Nature 392(6674):398–401

Considine RV et al (1996) Serum immunoreactive-leptin concentrations in normal-weight and obese humans. N Engl J Med 334(5):292–295

Conway PL (1995) Microbial ecology of the human large intestine, in human colonic bacteria. In: Gibson GR, Macfarlane GT (eds) Role in nutrition, physiology, and pathology. CRC Press: Boca Raton, FL, pp 1–24

Cummings JH (1981) Short chain fatty acids in the human colon. Gut 22(9):763–779

Cummings JH, Macfarlane GT (1991) The control and consequences of bacterial fermentation in the human colon. J Appl Bacteriol 70(6):443–459

Cummings JH, Macfarlane GT (1997) Colonic microflora: nutrition and health. Nutrition 13(5):476–478

Debril M-B et al (2001) The pleiotropic functions of peroxisome proliferator-activated receptor? J Mol Med 79(1):30–47

DiGiulio DB et al (2008) Microbial prevalence, diversity and abundance in amniotic fluid during preterm labor: a molecular and culture-based investigation. PLoS One 3(8):e3056

Dumas ME et al (2006) Assessment of analytical reproducibility of 1H NMR spectroscopy based metabonomics for large-scale epidemiological research: the INTERMAP Study. Anal Chem 78(7):2199–2208

Duncan SH et al (2008) Human colonic microbiota associated with diet, obesity and weight loss. Int J Obes (Lond) 32(11):1720–1724
Eckburg PB et al (2005) Diversity of the human intestinal microbial flora. Science 308(5728):1635–1638
Eisen JA (1998) Phylogenomics: improving functional predictions for uncharacterized genes by evolutionary analysis. Genome Res 8(3):163–167
Finegold SM (1983) Normal indigenous intestinal flora. In: Hentges DJ (ed) Human intestinal microflora in health and disease. Academic, New York, NY, pp 3–31
Frank DN et al (2007) Molecular-phylogenetic characterization of microbial community imbalances in human inflammatory bowel diseases. Proc Natl Acad Sci USA 104(34): 13780–13785
Fredricks DN, Fiedler TL, Marrazzo JM (2005) Molecular identification of bacteria associated with bacterial vaginosis. N Engl J Med 353(18):1899–1911
Friedman JM, Halaas JL (1998) Leptin and the regulation of body weight in mammals. Nature 395(6704):763–770
Gabaldón T, Huynen MA (2007) From endosymbiont to host-controlled organelle: the hijacking of mitochondrial protein synthesis and metabolism. PLoS Comput Biol 3(11):e219
Gao Z et al (2008) Substantial alterations of the cutaneous bacterial biota in psoriatic lesions. PLoS One 3(7):e2719
Garrett WS et al (2007) Communicable ulcerative colitis induced by T-bet deficiency in the innate immune system. Cell 131(1):33–45
Gill SR et al (2006) Metagenomic analysis of the human distal gut microbiome. Science 312(5778):1355–1359
Girardin SE et al (2003) Nod2 is a general sensor of peptidoglycan through muramyl dipeptide (MDP) detection. J Biol Chem 278(11):8869–8872
Godfrey KM, Barker DJ (2001) Fetal programming and adult health. Public Health Nutr 4(2B):611–624
Guarner F et al (2006) Mechanisms of disease: the hygiene hypothesis revisited. Nat Clin Pract Gastroenterol Hepatol 3(5):275–284
Hague A, Singh B, Paraskeva C (1997) Butyrate acts as a survival factor for colonic epithelial cells: further fuel for the in vivo versus in vitro debate. Gastroenterology 112(3): 1036–1040
Hamer HM et al (2008) Review article: the role of butyrate on colonic function. Aliment Pharmacol Ther 27(2):104–119
Hayashi H et al (2005) Molecular analysis of jejunal, ileal, caecal and recto-sigmoidal human colonic microbiota using 16S rRNA gene libraries and terminal restriction fragment length polymorphism. J Med Microbiol 54(Pt 11):1093–1101
Hayashi H, Sakamoto M, Benno Y (2002) Fecal microbial diversity in a strict vegetarian as determined by molecular analysis and cultivation. Microbiol Immunol 46(12):819–831
Herrmann A et al (1999) Studies on the "insoluble" glycoprotein complex from human colon. Identification of reduction-insensitive MUC2 oligomers and C-terminal cleavage. J Biol Chem 274(22):15828–15836
Hollingsworth MA, Swanson BJ (2004) Mucins in cancer: protection and control of the cell surface. Nat Rev Cancer 4(1):45–60
Holmes E, Nicholson JK (2007) Human metabolic phenotyping and metabolome wide association studies. Ernst Schering Found Symp Proc 4:227–249
Holmes E et al (2008) Human metabolic phenotype diversity and its association with diet and blood pressure. Nature 453(7193):396–400
Hooper LV, Gordon JI (2001a) Commensal host–bacterial relationships in the gut. Science 292(5519):1115–1118
Hooper LV, Gordon JI (2001b) Glycans as legislators of host–microbial interactions: spanning the spectrum from symbiosis to pathogenicity. Glycobiology 11(2):1R–10R
Hooper LV et al (1999) A molecular sensor that allows a gut commensal to control its nutrient foundation in a competitive ecosystem. Proc Natl Acad Sci USA 96(17):9833–9838

Hoshino K et al (1999) Cutting edge: toll-like receptor 4 (TLR4)-deficient mice are hyporesponsive to lipopolysaccharide: evidence for TLR4 as the Lps gene product. J Immunol 162(7): 3749–3752
Hoskins LC (1993) Mucin degradation in the human gastrointestinal tract and its significance to enteric microbial ecology. Eur J Gastroenterol Hepatol 5(4):205–213
Hugot JP et al (2001) Association of NOD2 leucine-rich repeat variants with susceptibility to Crohn's disease. Nature 411(6837):599–603
Ingham CJ et al (2007) The micro-petri dish, a million-well growth chip for the culture and high-throughput screening of microorganisms. Proc Natl Acad Sci USA 104(46): 18217–18222
Inohara N et al (2001) Human Nod1 confers responsiveness to bacterial lipopolysaccharides. J Biol Chem 276(4):2551–2554
Ivanov D et al (2006) A serpin from the gut bacterium *Bifidobacterium longum* inhibits eukaryotic elastase-like serine proteases. J Biol Chem 281(25):17246–17252
Khachatryan ZA et al (2008) Predominant role of host genetics in controlling the composition of gut microbiota. PLoS One 3(8):e3064
Kliewer SA et al (1997) Fatty acids and eicosanoids regulate gene expression through direct interactions with peroxisome proliferator-activated receptors α and γ. Proc Natl Acad Sci USA 94(9):4318–4323
Kurokawa K et al (2007) Comparative metagenomics revealed commonly enriched gene sets in human gut microbiomes. DNA Res 14(4):169–181
Lander E et al (2001) Initial sequencing and analysis of the human genome. Nature 409(6822): 860–921
Landers CJ et al (2002) Selected loss of tolerance evidenced by Crohn's disease-associated immune responses to auto- and microbial antigens. Gastroenterology 123(3):689–699
Lay C et al (2005) Colonic microbiota signatures across five northern European countries. Appl Environ Microbiol 71(7):4153–4155
Lebens M et al (2002) The nptA gene of Vibrio cholerae encodes a functional sodium-dependent phosphate cotransporter homologous to the type II cotransporters of eukaryotes. J Bacteriol 184(16):4466–4474
Leonie Los E et al (2007) Intestinal capacity to digest and absorb carbohydrates is maintained in a rat model of cholestasis. Am J Physiol Gastrointest Liver Physiol 293(3):G615–622
Ley RE et al (2005) Obesity alters gut microbial ecology. Proc Natl Acad Sci USA 102(31): 11070–11075
Ley RE et al (2006) Microbial ecology: human gut microbes associated with obesity. Nature 444(7122):1022–1023
Ley RE et al (2008) Evolution of mammals and their gut microbes. Science 320(5883):1647–1651
Li W-H (1997) Molecular evolution. Sinauer Associates, Sunderland, MA
Li M et al (2008) Symbiotic gut microbes modulate human metabolic phenotypes. Proc Natl Acad Sci USA 105(6):2117–2122
Lillycrop KA et al (2005) Dietary protein restriction of pregnant rats induces and folic acid supplementation prevents epigenetic modification of hepatic gene expression in the offspring. J Nutr 135(6):1382–1386
Lindi V et al (2003) Impact of the Pro12Ala polymorphism of the PPAR-gamma2 gene on serum triacylglycerol response to n–3 fatty acid supplementation. Mol Genet Metab 79(1):52–60
Linz B et al (2007) An African origin for the intimate association between humans and *Helicobacter pylori*. Nature 445(7130):915–918
Lorenz E et al (2000) A novel polymorphism in the toll-like receptor 2 gene and its potential association with staphylococcal infection. Infect Immun 68(11):6398–6401
Macdonald TT, Monteleone G (2005) Immunity, inflammation, and allergy in the gut. Science 307(5717):1920–1925
Macpherson AJ, Geuking MB, McCoy KD (2005) Immune responses that adapt the intestinal mucosa to commensal intestinal bacteria. Immunology 115(2):153–162

Manichanh C et al (2006) Reduced diversity of faecal microbiota in Crohn's disease revealed by a metagenomic approach. Gut 55(2):205–211
Marteau P et al (2001) Comparative study of bacterial groups within the human cecal and fecal microbiota. Appl Environ Microbiol 67(10):4939–4942
Martin F-PJ et al (2007) A top-down systems biology view of microbiome–mammalian metabolic interactions in a mouse model. Mol Syst Biol 3:112
Martinez-Medina M et al (2006) Abnormal microbiota composition in the ileocolonic mucosa of Crohn's disease patients as revealed by polymerase chain reaction-denaturing gradient gel electrophoresis. Inflamm Bowel Dis 12(12):1136–1145
Maslowski KM et al (2009) Regulation of inflammatory responses by gut microbiota and chemoattractant receptor GPR43. Nature 461(7268):1282–1286
Medina V et al (1998) Sodium butyrate inhibits carcinoma development in a 1,2-dimethylhydrazine-induced rat colon cancer. JPEN J Parenter Enteral Nutr 22(1):14–17
Metges CC (2000) Contribution of microbial amino acids to amino acid homeostasis of the host. J Nutr 130(7):1857S–1864S
Moodley Y et al (2009) The peopling of the Pacific from a bacterial perspective. Science 323(5913):527–530
Morita H et al (1993) Glycosylation and sulphation of colonic mucus glycoproteins in patients with ulcerative colitis and in healthy subjects. Gut 34(7):926–932
Mueller S et al (2006) Differences in fecal microbiota in different European study populations in relation to age, gender, and country: a cross-sectional study. Appl Environ Microbiol 72(2):1027–1033
Nelson KE et al (2000) Status of genome projects for nonpathogenic bacteria and archaea. Nat Biotechnol 18(10):1049–1054
Ogura Y et al (2001) A frameshift mutation in NOD2 associated with susceptibility to Crohn's disease. Nature 411(6837):603–606
O'Keefe SJ et al (2007) Why do African Americans get more colon cancer than Native Africans? J Nutr 137(1 Suppl):175S–182S
Opitz B et al (2009) Role of Toll-like receptors, NOD-like receptors and RIG-I-like receptors in endothelial cells and systemic infections. Thromb Haemost 102(6):1103–1109
Palmer C et al (2007) Development of the human infant intestinal microbiota. PLoS Biol 5(7):e177
Pavoine S, Dufour AB, Chessel D (2004) From dissimilarities among species to dissimilarities among communities: a double principal coordinate analysis. J Theor Biol 228(4):523–537
Pei Z et al (2004) Bacterial biota in the human distal esophagus. Proc Natl Acad Sci USA 101(12):4250–4255
Poltorak A et al (1998) Defective LPS signaling in C3H/HeJ and C57BL/10ScCr mice: mutations in Tlr4 gene. Science 282(5396):2085–2088
Rajilic-Stojanovic M et al (2007) Dynamics of the adult gastrointestinal microbiota, in diversity of the human gastrointestinal microbiota – novel perspectives from high throughput analysis. PhD thesis, Wageningen University and Research
Rajilic-Stojanovic M et al (2009) Development and application of the human intestinal tract chip, a phylogenetic microarray: analysis of universally conserved phylotypes in the abundant microbiota of young and elderly adults. Environ Microbiol 11(7):1736–1751
Rakoff-Nahoum S et al (2004) Recognition of commensal microflora by toll-like receptors is required for intestinal homeostasis. Cell 118(2):229–241
Ramotar K et al (1984) Production of menaquinones by intestinal anaerobes. J Infect Dis 150(2):213–218
Rawls JF et al (2006) Reciprocal gut microbiota transplants from zebrafish and mice to germ-free recipients reveal host habitat selection. Cell 127(2):423–433
Roediger WE (1980) Role of anaerobic bacteria in the metabolic welfare of the colonic mucosa in man. Gut 21(9):793–798
Salzberg SL et al (2001) Microbial genes in the human genome: lateral transfer or gene loss? Science 292(5523):1903–1906

Samuel BS et al (2008) Effects of the gut microbiota on host adiposity are modulated by the short-chain fatty-acid-binding G-protein-coupled receptor, Gpr41. Proc Natl Acad Sci USA 105(43):16767–16772

Savage DC (1977) Microbial ecology of the gastrointestinal tract. Annu Rev Microbiol 31: 107–133

Schicho R et al (2006) Hydrogen sulfide is a novel prosecretory neuromodulator in the Guinea-pig and human colon. Gastroenterology 131(5):1542–1552

Schwiertz A et al (2009) Microbiota and SCFA in lean and overweight healthy subjects. Obesity (Silver Spring) 18(1):190–195

Sokol H et al (2008) Faecalibacterium prausnitzii is an anti-inflammatory commensal bacterium identified by gut microbiota analysis of Crohn's disease patients. Proc Natl Acad Sci USA 105(43):16731–16736

Sperandio V et al (2003) Bacteria–host communication: the language of hormones. Proc Natl Acad Sci USA 100(15):8951–8956

Stewart JA, Chadwick VS, Murray A (2005) Investigations into the influence of host genetics on the predominant eubacteria in the faecal microflora of children. J Med Microbiol 54(12): 1239–1242

Summerton J et al (1985) Effect of deoxycholic acid on the tumour incidence, distribution, and receptor status of colorectal cancer in the rat model. Digestion 31(2–3):77–81

Tap J et al (2009) Towards the human intestinal microbiota phylogenetic core. Environ Microbiol 11(10):2574–2584

Ting JP, Kastner DL, Hoffman HM (2006) CATERPILLERs, pyrin and hereditary immunological disorders. Nat Rev Immunol 6(3):183–195

Toivanen P, Vaahtovuo J, Eerola E (2001) Influence of major histocompatibility complex on bacterial composition of fecal flora. Infect Immun 69(4):2372–2377

Topping DL, Clifton PM (2001) Short-chain fatty acids and human colonic function: roles of resistant starch and nonstarch polysaccharides. Physiol Rev 81(3):1031–1064

Turnbaugh PJ et al (2006) An obesity-associated gut microbiome with increased capacity for energy harvest. Nature 444(7122):1027–1031

Turnbaugh PJ et al (2008) Diet-induced obesity is linked to marked but reversible alterations in the mouse distal gut microbiome. Cell Host Microbe 3(4):213–223

Turnbaugh PJ et al (2009a) A core gut microbiome in obese and lean twins. Nature 457(7228): 480–484

Turnbaugh PJ et al (2009b) The effect of diet on the human gut microbiome: a metagenomic analysis in humanized gnotobiotic mice. Sci Transl Med 1(6):6ra14–6ra14

van Baarlen P et al (2009) Differential NF-kappaB pathways induction by Lactobacillus plantarum in the duodenum of healthy humans correlating with immune tolerance. Proc Natl Acad Sci USA 106(7):2371–2376

Van de Merwe JP, Stegeman JH, Hazenberg MP (1983) The resident faecal flora is determined by genetic characteristics of the host. Implications for Crohn's disease? Antonie Van Leeuwenhoek 49(2):119–124

Waldram A et al (2009) Top-down systems biology modeling of host metabotype–microbiome associations in obese rodents. J Proteome Res 8(5):2361–2375

Wei C, Brent MR (2006) Using ESTs to improve the accuracy of de novo gene prediction. BMC Bioinformatics 7:327

Wen L et al (2008) Innate immunity and intestinal microbiota in the development of Type 1 diabetes. Nature 455(7216):1109–1113

Yang X et al (2009) More than 9,000,000 unique genes in human gut bacterial community: estimating gene numbers inside a human body. PLoS One 4(6):e6074

Yen C-J et al (1997) Molecular scanning of the human peroxisome proliferator activated receptor [gamma] (hPPAR[gamma]) gene in diabetic caucasians: identification of a Pro12Ala PPAR[gamma]2 missense mutation. Biochem Biophys Res Commun 241(2): 270–274

Zengler K et al (2005) High-throughput cultivation of microorganisms using microcapsules. Methods Enzymol 397:124–130

Zhang C et al (2010) Interactions between gut microbiota, host genetics and diet relevant to development of metabolic syndromes in mice. Isme J 4(2):232–241. Epub 2009 Oct 29.

Zoetendal EG, Akkermans AD, De Vos WM (1998) Temperature gradient gel electrophoresis analysis of 16S rRNA from human fecal samples reveals stable and host-specific communities of active bacteria. Appl Environ Microbiol 64(10):3854–3859

Zoetendal EG, Rajilic-Stojanovic M, de Vos WM (2008) High-throughput diversity and functionality analysis of the gastrointestinal tract microbiota. Gut 57(11):1605–1615

Zoetendal EG et al (2001) The host genotype affects the bacterial community in the human gastronintestinal tract. Microb Ecol Health Dis 13(3):129–134

Chapter 3
The Human Microbiome and Host–Pathogen Interactions

Mark J. Pallen

Microbial Pathogenesis: Looking Beyond Koch

In the late nineteenth century, Robert Koch and his associates devised what are now known as Koch's postulates (Koch, 1884; Koch, 1892) to establish a causal relationship between a specific disease and a single pathogenic microorganism, capable of causing disease after growth in pure culture. A century or so later, Falkow formulated "molecular Koch's postulates" (Falkow, 1988) to provide a set of experimental criteria to identify virulence factors in bacterial pathogens: (a) a gene is hypothesised to contribute to a virulence-related phenotype to the bacterium under study, (b) inactivation of the gene abolishes the phenotype, (c) reintroduction of the gene restores the wild-type phenotype to the mutant.

At the heart of both sets of postulates lies a simplifying assumption of a one-to-one relation: between a single pathogen and a single disease or between a single gene, ripped out of its evolutionary and ecological context, and a single virulence-related phenotype. These simple models of pathogens and pathogenesis are now under threat from a range of conceptual and experimental advances in our understanding of host–pathogen interactions, which are examined in this chapter (although, with such a broad scope, references will often be restricted to recent reviews). These advances include the growing recognition of the importance of the human microbiota in health and disease; methodological progress in our ability to catalogue the germs and genes within the human microbiome; a new awareness of the importance of polymicrobial infection; the blurred boundaries between pathogen and commensal; the sophisticated crosstalk between the resident microbiota and its host and the origins of virulence in the bacterial "struggle for existence" with microscopic adversaries. We are thus moving towards a new paradigm, in which pathogen–host interactions are evaluated against the backdrop of the resident complex communities of microbes and genes (Fig. 3.1).

M.J. Pallen (✉)
School of Biosciences, University of Birmingham, Birmingham, West Midlands, B15 2TT, UK
e-mail: m.pallen@bham.ac.uk

K.E. Nelson (ed.), *Metagenomics of the Human Body*,
DOI 10.1007/978-1-4419-7089-3_3,

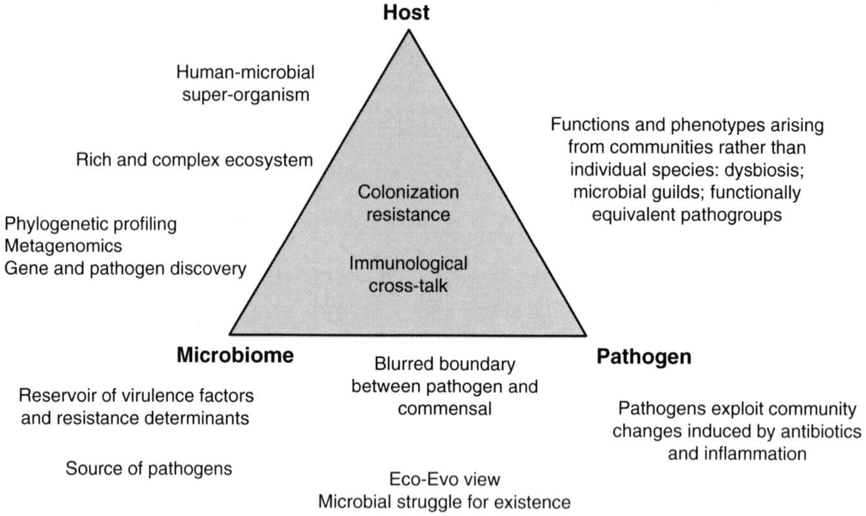

Fig. 3.1 Host–pathogen–microbiome interactions

The Human Microbiota in Health and Disease

The human body is home to a microbiota as rich and stunning in complexity as the flora and fauna of a rain forest or coral reef. The superlative metaphors used to describe this complex microbial community veer towards cliché: a community of bacterial cells ten-fold more numerous than the eukaryotic cells of its human host; an additional multi-cellular organ, as complex as the liver, encoded by a microbiome or "third genome," with >100 times more genes than the human nuclear genome; humans viewed as "superorganisms," built from human and microbial cells (Shanahan, 2002; Gill et al., 2006; Foxman et al., 2008; Carroll et al., 2009) or seen as "large, highly complex microbial communities attached to some relatively uninteresting organic matter" (Davies, 2009).

Distinct microbial communities inhabit different host environments on what can be seen topologically as the body's surface: the skin, the oral cavity, the upper respiratory tract, the genital tract and the stomach (Cogen et al., 2008; Srinivasan and Fredricks, 2008; Avila et al., 2009; Nasidze et al., 2009b; Parahitiyawa et al., 2009). However, the richest and most abundant community exists in the large bowel, with up to 10^{12} bacterial cells per gram of luminal contents (Frank and Pace, 2008; Kinross et al., 2008; Tuohy et al., 2009).

Although often dismissed merely as commensals, many components of the human microbiota engage in a mutualistic interaction with the host. Metchnikoff was the first to highlight the importance of the microbiota (particularly lactic acid bacteria) to human health in his 1907 work *The Prolongation of Life* (Metchnikoff and Mitchell, 1908) – although his view that "the large intestine ... is certainly useless in the case of man" might now be the subject of ridicule!

The mammalian microbiota is now known to exert decisive effects on normal development and tissue homeostasis, particularly in the gut, where comparisons of germ-free and colonized laboratory animals have shown that the microbiota has effects on epithelial and enteroendocrine cell differentiation and on the production and composition of mucin (Falk et al., 1998; Shanahan, 2002; Mason et al., 2008; Collins and Bercik, 2009). In addition, germ-free mice are known to have abnormally long intestinal villi and experience cecal enlargement, twinned with altered gastrointestinal motility (Collins and Bercik, 2009).

The intestinal microbiota also has an important influence on the development and maintenance of the mucosal immune system (for a review, see (Falk et al., 1998)). For example, in germ-free mice, secondary lymphoid structures fail to form and inflammatory cells remain sparse in the intestinal mucosa, and the indigenous microbiota has been shown to play a crucial role in the expansion and maintenance of viral-specific CD8 memory T cells within the lungs of mice infected with murine cytomegalovirus (Tanaka et al., 2007; Collins and Bercik, 2009).

The gut microbiota also plays a role in the assimilation and production of nutrients. In fact, the colonic microbiome encompasses a complex anaerobic food web, fed by nutrients that have not been absorbed in the upper bowel (see review by Stecher and Hardt, 2008). In humans, these microbes synthesize folic acid, biotin and vitamin K, which can supplement dietary intake of these vitamins. In addition, complex polysaccharides are broken down into sugars by catabolic enzymes produced by the most abundant taxa, such as *Bacteroides*. Primary bacterial fermentation of these sugars produces short-chain fatty acids, which provide up to 10% of the host's dietary energy intake and which are thought to play beneficial roles in protection against cancer (Stecher and Hardt, 2008). Secondary fermenters, such as propionibacteria, sulphate reducers and methanogens, exploit the products of primary fermentation.

Interactions between the host and resident microbiota have now been implicated in many human diseases that are not normally thought of as infections. The "hygiene hypothesis" suggests that interrupting exposure to environmental or even commensal microbes early in development may lead to allergic disease (Vassallo and Walker, 2008; Bjorksten, 2009). Abnormal responses to bowel organisms are thought to underlie the pathogenesis of inflammatory bowel disease (see recent reviews (Alverdy and Chang, 2008; Salzman and Bevins, 2008; Reiff and Kelly, 2009)). Variation in the gut microbiome has been linked to bowel cancer and may even play a role in carcinogenesis (Lampe, 2008). The gut microbiome may also play a role in the pathogenesis of irritable bowel syndrome and diabetes (Vaarala et al., 2008; Preidis and Versalovic, 2009).

Perhaps most surprisingly of all, recent studies have implicated the gut microbiota in obesity. Ley et al. (2006) reported that levels of the bacteria from the phylum Firmicutes (relative to the Bacteroidetes) were higher in obese than in lean humans and observed that weight loss was accompanied by a reduction in the relative number of Firmicutes. In a subsequent study, Turnbaugh et al. transferred the microbiota from obese to lean mice and demonstrated that the lean mice then gained excess weight (Turnbaugh et al., 2006).

From Microbiota to Microbiome: Surveying the Unknown

The first studies on the microbial communities associated with the human body relied on culture-based approaches. In the late 1970s, Savage (1977) reviewed what was known of the microbial ecology of the gastrointestinal tract just before the dawn of the molecular era. By then, culture-based studies had established that obligate anaerobes, particularly the *Bacteroidetes*, outnumbered facultative organisms such as *Escherichia coli* by $\geq$1000 to one. Among the facultative anaerobes, Gram-positive bacteria, such as bifidobacteria and lactobacilli, held pride of place. However, these studies typically enumerated a few dozen or at most a few hundred microbial species, with most bacterial species in the human microbiota resisting culture in the laboratory. Nonetheless, Savage was able to apply key ecological concepts to the human microbiota such as succession (the changes in community structure during human development) and the idea of a robust microbial "climax community."

In the 1980s, Woese pioneered the use of 16S (also called small subunit or SSU) ribosomal RNA (rRNA) to study bacterial phylogenetics and evolution (Woese, 1987). Subsequently application of the polymerase chain reaction (PCR) to DNA sequences encoding 16S rRNA genes (16S rDNA sequences), twinned with electrophoretic or sequencing approaches, found widespread use in cataloguing uncharacterised and uncultured organisms in mixed microbial populations—an approach sometimes called "phylogenetic profiling" (Theron and Cloete, 2000). The Ribosomal Database Project now holds >400,000 16S rRNA gene sequences, illustrating the rich microbial diversity of our planet and hinting at what remains to be discovered.

Alongside the sequencing of 16S rDNA genes, fluorescent in situ hybridisation (FISH), molecular fingerprinting and use of microarrays also provided insights into the composition of the human microbiota (Frank and Pace, 2008). Crucially, such studies showed that most of the bacterial 16S rDNA sequences from the human microbiota, particularly gut microbes, represent previously unknown bacterial species (Streit and Schmitz, 2004; Frank and Pace, 2008).

It is now clear that the human microbiota is predominantly bacterial, with very few archaea or eukaryotes. In the first large-scale culture-independent survey of the lower bowel microbiota, Eckburg et al. (2005) surveyed over 13,000 small-subunit rDNA amplicons from the colonic mucosal and faeces of healthy adult humans. Sequences from just two phyla predominated: the Firmicutes (75%; mostly clostridia) and Bacteroidetes (16%). However, they identified nearly 400 bacterial phylotypes, of which ~60% were novel.

At least two similar large-scale studies of the human intestinal biota have been reported (Hold et al., 2002; Frank et al., 2007), in addition to the study by Eckberg et al. Drawing on all three studies, Frank et al. (2007) estimated that the bacterial component of the human gut microbiome consists of $\geq$1,800 genera and an astonishing 15,000–36,000 species, depending on whether species are classified conservatively or liberally! At the species level, extreme person-to-person variation might cast doubt on how far a "core" human microbiome actually exists. Instead it has been suggested that a conserved cluster of functions is delivered by a core set of "microbial guilds" – that is, groups of microbial species sharing a common ecological niche. On this view, different combinations of species could fulfil the same functional roles in different hosts (Tschop et al., 2009).

Eckburg et al. (2005) reported just one archaeal species in the human gastrointestinal tract: *Methanobrevibacter smithii*. Other archaeal taxa such as *Methanosphaera stadtmanae*, *Methanobrevibacter oralis*, members of the Methanosarcinale and one putative new order of methanogens have been detected occasionally in the human gut microbiota (Mihajlovski et al., 2008; Scanlan et al., 2008).

In most surveys, eukaryotes are also relatively rare in the human intestinal microbiome. However, healthy Westerners predominate in these surveys and one wonders whether the picture would be different if samples originated in the developing world, where intestinal colonisation with protozoal and metazoan parasites is commonplace. Fungi also occur in the human microbiome – one recent survey reported nearly 50 different fungal phylogroups in the human intestine, including *Candida* spp., *Penicillium* spp. and *Saccharomyces* spp. (Ott et al., 2008).

Several studies have exploited high-throughput sequencing of 16S DNA amplicons to document the extensive variation in the human microbiome at various sites outside the large bowel sites. Such surveys have targeted the microbial inhabitants of the mouth, the skin, the stomach and the vagina (Grice et al., 2009; Nasidze et al., 2009b; Schellenberg et al., 2009).

In a phylogenetic profile of the salivary microbiome, Nasidze et al. reported over 14,000 partial 16S rRNA sequences from saliva samples from 120 healthy individuals. These could be assigned to 101 known bacterial genera (over a third of them not seen before in the human oral cavity) and to an additional 64 unknown genera (Nasidze et al., 2009a).

Grice and colleagues performed phylogenetic profiling of the skin, generating 112,283 near-full-length bacterial 16S rDNA sequences from 20 skin sites on 10 healthy humans (Grice et al., 2009). They detected just over 200 genera, belonging to 19 bacterial phyla. Most sequences were assigned to what are traditionally viewed as Gram-positive groups: about half to the Actinobacteria and a quarter to the Firmicutes. They also confirmed that physiologically comparable sites harbour similar bacterial communities.

Bik and co-workers analysed 1,833 16S rDNA sequences PCR-amplified from 23 gastric biopsy samples (Bik et al., 2006). They identified 128 phylotypes – far more diversity at this site than previously expected. One in ten of the phylotypes was previously uncharacterised, including a *Deinococcus*-like organism, and statistical analyses revealed a considerable inter-subject variability in the gastric microbiome.

Although phylogenetic profiling provides insights into the composition of complex communities, it tells us little or nothing about their metabolic or other biological capabilities. The term "metagenome" was coined in the late 1990s to describe the collective genomes and the biosynthetic machinery of the soil microbiota (Handelsman et al., 1998) and soon generalised to cover the collective genome any complex microbial community.

A few years later, the term "metagenomics" was invented to describe wholesale shotgun sequencing of DNA extracted from mixed bacterial communities, in the hope of characterising the physiology and ecology of the mostly uncultured members of these communities (Streit and Schmitz, 2004; Mongodin et al., 2005; Tringe and Rubin, 2005). The random shotgun approach implicit in metagenomics may seem counter-intuitive as a means of uncovering gene function within complex microbial communities. However, as ≥80% of the average bacterial genome encodes protein sequence, most metagenomics sequences can, through homology searches with the encoded peptides, be linked to a putative function.

Around the turn of the century, Joshua Lederberg suggested the term "microbiome" to cover the rich diversity of cells, genomes and genes within microbial communities, which, in the human setting, rivalled that of the host (Hooper and Gordon, 2001). This proved a prescient analogy, as, a few years later, the human genome project reported fewer genes in the human genome than expected, whereas bacterial genome sequencing revealed unanticipated bacterial diversity at the genomic level (Fitzgerald and Musser, 2001; Dutta and Pan, 2002; International Human Genome Sequencing Consortium, 2004).

Metagenomics took a great leap forward in 2004 with a landmark study led by Venter, revealing a vast diversity of protein-coding genes in ocean-dwelling microbes (Venter et al., 2004). Metagenomics has since proven an informative approach for interrogating physiology and function in microbial communities. Over a hundred metagenomics studies have now extended such analysis to the microbial and viral components of numerous microbiomes.

In 2006, Gill et al. reported a metagenomics study of the distal human gut microbiome, with ~65,000–75,000 reads from DNA extracted from faecal samples from two healthy adults (Gill et al., 2006). When reads were compared with the genomes of two exemplar organisms, there was a uniformly good match to *Methanobrevibacter smithii* sequences, but considerable heterogeneity in alignments with *Bifidobacterium longum*, suggesting the presence of multiple closely related strains or phylotypes. A survey of 16S rDNA sequences in the library revealed an under-representation of the *Bacteroidetes*, probably due to bias in faecal lysis and DNA extraction. An analysis of the metabolic potential of the microbiome revealed genes engaged in the metabolism of glycans, amino acids and xenobiotics, as well as methanogenesis and the biosynthesis of vitamins and isoprenoids.

In a similar study published in 2007, Kurokawa et al. (2007) sampled 13 human intestinal microbiomes from adults, children and unweaned infants, reporting and obtained 479 Mb of sequence data. Comparisons with other metagenomic datasets revealed an enrichment of genes dedicated to carbohydrate metabolism in the human intestinal microbiome, but with polysaccharide-degrading enzymes predominant in the adult microbiome versus sugar transporters in the infant microbiome.

Also in 2007, the Human Microbiome Project (HMP) was launched, as an international collaboration with the aim of collecting and collating genomic information from many diverse human microbiomes (Peterson, 2009). Again, superlatives abound: for example, Davies has called the HMP "the largest international life-science project of all time" (Davies, 2009).

The superlatives may well be justified, as the HMP is being powered by a new enabling technology – high-throughput sequencing – that promises sequences 100 times cheaper and 100 times faster than before (Metzker, 2010). This new research platform avoids the need for plasmid libraries and also benefits from new sequencing chemistries. Crucially, high-throughput sequencing brings a fresh opportunity to obtain deep and definitive catalogues of bacterial phylotypes and bacterial genes in the human microbiome.

It is worth stressing that these new technologies are showing a steady Moore's-law-like improvement in performance, so it is not hard to envisage a time in the next few years when we can perform exhaustive phylogenetic profiling and comprehensive metagenomics of the human microbiome. Already, at the time of writing, more than 500 bacterial genomes have been pursued within the HMP, with most of them already in draft sequencing pipelines or already finished and deposited in GenBank (Peterson, 2009).

Blurring the Boundaries Between Pathogen and Commensal

Long before the advent of genomics or metagenomics, it was clear that there could be no fixed boundary between pathogens and commensals. In his 1977 review, Savage recognised that "Some components of the biota induce disease when given an opportunity to do so, for example when they are injected into normally sterile areas of the body during surgery or when host resistance mechanisms fail." In fact, it is clear that the human microbiome acts as a source of disease in numerous clinical settings. The bowel in particular acts as a reservoir for microorganisms that cause infections elsewhere – for example urinary tract infections, vaginal candidosis, peritonitis, biliary sepsis, Gram-negative pneumonia and septicaemia in critically ill patients. A strain of *E. coli* labelled "uropathogenic" in a young woman lives as a commensal in her male partner. Similarly, the agents of a fatal peritonitis following a burst appendix would count as mere commensals if confined to the gastrointestinal tract.

The ready availability of genomic and metagenomic data has further blurred the distinction between pathogen and non-pathogen, highlighting the presence of what might naively be called "virulence factors" (according to molecular Koch's postulates) in non-pathogenic members of the human microbiota. We now recognise the substantial overlap between "virulence factors" and the "colonisation factors" needed by non pathogens to survive in the relevant human microenvironment. It is clear that pathogens, commensals and symbionts all rely on similar strategies and molecular systems in their interactions with eukaryotic hosts (e.g. phase variation and type III secretion). Furthermore, our view of host–pathogen interactions has undergone a Copernican shift away from an anthropocentric focus on how bacteria damage humans to a broader "eco-evo" perspective that views interactions between "eukaryophilic" bacteria and eukaryotes in a wider ecological and evolutionary context (Pallen and Wren, 2007). We now recognise that many bacterial pathogens only incidentally infect our species and often carry virulence factors active against non-mammalian adversaries as diverse as plants, insects, grazing protozoa, nematodes, predatory bacteria and bacteriophages (Koval and Bayer, 1997; Waterfield et al., 2004; Pallen and Wren, 2007). Inherent in this view is the recognition that many so-called virulence factors have been shaped by evolutionary forces outside the context of the human host–pathogen interaction. For example, when considering the cell envelope, the evolution of both the components (e.g. capsules) and dynamics (e.g. antigenic diversity and phase variation) of this structure are likely to have been originally driven by interactions with non-human adversaries (bacteriophages, predatory bacteria or protozoa) (Coward et al., 2006; Stummeyer et al., 2006; Leiman et al., 2007; Pallen and Wren, 2007).

Enterohaemorrhagic *E. coli* O157:H7 (EHEC) provides an interesting test case for the eco-evo view. This organism is a commensal in the bovine gut, but can act as a devastating pathogen in humans. Recent studies have shown that a pilus adherence factor crucial to the virulence of this bacterium in humans is shared with human–commensal strains of *E. coli* (Rendon et al., 2007). This highlights an obvious deficit with molecular Koch's postulates – many factors required for virulence are in fact better characterised as "colonisation factors," given that pathogens and commensals alike have to colonise mucosal surfaces and exploit similar mechanisms to do so.

EHEC also illustrates the evolutionary origins of virulence. In a recent study, the Shiga-toxin phage was shown to increase bacterial survival in the presence of a grazing ciliate, *Tetrahymena pyriformis* (Steinberg and Levin, 2007). This supports the view that interactions with non-mammalian adversaries have driven the evolution of virulence in this bacterial lineage, rather than interactions with humans or other mammals. A survey of type III effector genes in this bacterium, co-ordinated by my own group, has also highlighted a key component of the human microbiome: the viral metagenome (Tobe et al., 2006). We found that bacteriophages encode the majority of type-III secretion effectors in the EHEC genome and link the evolution of bacterial virulence to a vast and dynamic phage metagenome.

In addition to providing a reservoir of potential colonisation and virulence determinants, it is clear that the human microbiome harbours a rich variety of potential resistance factors (sometimes called the "resistome"). A recent study documented >200 unique inserts derived from the gut-microbiome and encoding antibiotic resistance – furthermore, many of these determinants were quite distinct from known resistance genes (Sommer et al., 2009).

Immunological Crosstalk Between Microbiota and Host

Despite the huge number of bacterial cells within the human microbiome, these organisms do not usually induce a systemic antibody response or provoke local inflammatory pathology. This may be because the microbiota is anatomically and physiologically compartmentalised away from host tissues and the immune system – as evidenced by the fact that most immunocompromised humans are not overwhelmed by their own microbiota. However, it is now clear that the immune system and microbiota do indeed meet and engage in a sophisticated crosstalk that underpins physiological inflammation and immunological homeostasis.

In the intestine, Paneth cells directly sense the gut microbiota and maintain homeostasis through expression of multiple antimicrobial factors (Vaishnava et al., 2008). The resident bacterial microbiota produces cell envelope components (such as lipopolysaccharide, lipoteichoic acid, flagellin and peptidoglycan) that interact with the innate immune system, often acting as ligands for toll-like receptors. If unchecked, such interactions can trigger pathological inflammation – thus, in mice lacking the anti-inflammatory cytokine IL-10, inflammatory responses against commensal bacteria drive a chronic enterocolitis (Kuhn et al., 1993). Similarly, inappropriate immune responses to the commensal gut bacteria are thought to underlie the pathogenesis of inflammatory bowel disease (Alverdy and Chang, 2008; Salzman and Bevins, 2008; Reiff and Kelly, 2009).

Commensals can also evoke beneficial anti-inflammatory effects. The non-pathogenic gut commensal *Bacteroides thetaiotaomicron* can selectively suppress pro-inflammatory signalling by NF-κB (nuclear factor κ-light-chain-enhancer of activated B cells), via a pathway dependent on peroxisome proliferator activated receptor-gamma (PPAR-gamma) (Kelly et al., 2004). *Enterococcus faecalis* from neonatal humans also regulates endogenous PPAR-gamma activity and IL-10 levels in colonic epithelial cells (Are et al., 2008). In a recent study, the commensal *Bifidobacterium infantis* was found to suppress secretion of the pro-inflammatory chemokine CCL20 by human enterocytes in response to a variety of enteropathogens (Sibartie et al., 2009). Such effects are not limited to the gut microbiota – commensal bacteria can act as negative regulators of inflammation elsewhere. For example, lipoteichoic acid from staphylococci (itself a TLR2 ligand) selectively suppresses TLR3-mediated inflammation in the skin (Lai et al., 2009).

The Human Microbiome: A Shield Against Pathogens

As we have seen, the human microbiome can act as a source of disease in various clinical settings. However, this microbial community also acts as a barrier against infection, providing a "colonisation resistance" that prevents harmful pathogens from gaining a foothold (Stecher and Hardt, 2008). For example, the absence of a normal gut microbiota in germ-free or antibiotic-treated mice significantly increases vulnerability to enteric colonisation/infection with *Salmonella* spp., *Streptococcus mutans*, *Clostridium difficile* and *Shigella flexneri* (Stecher and Hardt, 2008).

This phenomenon of colonisation resistance is multi-factorial. Contributory mechanisms include:

1. Competition for space and mucosal adherence sites/receptors.
2. Competition for nutrients.
3. Production of bacteriocins – for example, a bacteriocin from *Lactobacillus salivarius* helps prevent invasive listeriosis in mice (Corr et al., 2007).

4. Production of anti-virulence compounds – for example, probiotic strains of *Lactobacillus acidophilus* and *L. reuteri* produce agents that down-regulate expression of virulence genes in EHEC (Medellin-Pena et al., 2007).
5. Stimulation of the host defences, including gut motility, secretion of antimicrobial peptides and enhanced mucin production (Kim et al., 2008).
6. Production of metabolites (e.g. short-chain fatty acids) that reduce the growth of pathogens, directly or indirectly – for example, butyrate can induce an antimicrobial peptide that protects against *Shigella* infection (Raqib et al., 2006).

Diet, Antibiotics, the Human Microbiome and Risks of Infection

It is generally agreed that a healthy, balanced colonic microbiota is primarily saccharolytic and is dominated by bifidobacteria and lactobacilli. Building on this view, "probiotic" strains from these genera have been shown to suppress infections by enteric pathogens such as *Salmonella* (Truusalu et al., 2008; Vieira et al., 2008; Wagner et al., 2009). Prebiotics, non-digestible foodstuffs that selectively stimulate the growth of bifidobacteria and lactobacilli, have also been advocated as health-promoting agents. Advocates of probiotics and prebiotics champion their anti-infective properties as well as preventative effects against non-infective conditions such as colon cancer, inflammatory bowel disease and cardiovascular disease (de Vrese and Schrezenmeir, 2008). However, the evidence from animal models for an effect of these agents on susceptibility to gastrointestinal infection is mixed.

On the one hand, in mice, oral administration of inulin or galacto-oligosaccharides (with or without bifidobacteria) helps protect against invasive salmonellosis (Buddington et al., 2002; Searle et al., 2009), whereas in rats, an inulin–oligofructose diet restricts proliferation of *Salmonella* in the large bowel (Kleessen and Blaut, 2005). On the other hand, mice fed diets containing fructo-oligosaccharide or xylo-oligosaccharide, and then challenged orally with *Salmonella*, had significantly higher numbers of the pathogen in liver, spleen and mesenteric lymph nodes than did mice fed a cornstarch-based control diet (Petersen et al., 2009).

Perturbation of normal microbial ecology can disrupt colonisation resistance and facilitate infection. The most obvious evidence for this in humans is provided by conditions such as the catastrophic colitis caused by *Clostridium difficile* (Bartlett, 2009; Honda and Dubberke, 2009; Raza et al., 2009) or life-threatening invasive fungal disease caused by *Candida albicans* (Lewis, 2009; Pennisi and Antonelli, 2009; Rueping et al., 2009). In addition, it is clear that antibiotic treatment can render mice more susceptible to enteric infection by *Salmonella, Shigella* and *E. coli* – in fact, a number of animal models of infection exploit this phenomenon to study human disease (Stecher and Hardt, 2008).

Several recent studies have highlighted the substantial perturbations in microbial ecology and physiology that result from antimicrobial chemotherapy. Administration of amoxicillin to suckling rats led to a reduction in lactobacilli within the bowel microbiome and to significant changes in intestinal gene expression (Schumann et al., 2005). Yap et al. showed that depletion of gut Firmicutes by administration of vancomycin to mice influences gastrointestinal carbohydrate metabolism (Yap et al., 2008). In another study, high-throughput DNA sequencing technology was used to document perturbations to the human microbiome induced by a fluoroquinolone (Dethlefsen et al., 2008). Surprisingly, despite general ecological resilience, several microbial taxa were still depleted 6 months after oral ciprofloxacin therapy.

Although disturbing the human microbiome with antimicrobial agents is generally seen as undesirable, there are some settings in which this is encouraged, in the name of disease prevention. Hospitalised patients tend to acquire a bowel and upper respiratory tract microbiota that harbours potentially pathogenic Gram-negative bacteria. A process known as "Selective Decontamination of the Digestive Tract" (or SDD) is widely used in an attempt to prevent life-threatening Gram-negative pneumonia in mechanically ventilated patients (Bonten and Krueger, 2006; de Smet and Bonten, 2008). In this approach, both topical and systemic antibiotics are administered to "decontaminate" the upper respiratory tract and gastrointestinal tract. The antimicrobial agents are chosen in the hope that they target potentially pathogenic Gram-negative bacteria, but leave the anaerobic microbiota (principally the Bacteroidetes) intact.

SDD has been implemented in numerous studies for more than two decades. However, controversy rages over its effect on disease outcomes and patient survival (Bonten and Krueger, 2006; de Smet and Bonten, 2008). Curiously, there appear to have been no large-scale phylogenetic profiling studies to determine what actually happens to the human microbiome when SDD is used.

Pathogen-Induced Inflammation Perturbs the Microbiome

Inflammation is another powerful influence on the human microbiome. Inflammation, particularly in the gastrointestinal tract, can perturb the microbiome, disrupt colonisation resistance and facilitate the growth of pathogens. Recent evidence suggests that some enteric pathogens exploit these effects by actively inducing inflammation. Thus, in a mouse model of salmonellosis, wild-type *S. typhimurium* induces intestinal inflammation, but an attenuated strain, defective in type-III secretion, does not (Stecher et al., 2007). In this model the attenuated strain was outgrown by the resident microbiota within 4 days of infection, whereas the wild-type strain grew well and triggered a dramatic change in the composition of the local microbiota within a similar timescale. Furthermore, the attenuated strain showed enhanced colonisation even when inflammation was induced by non-infective mechanisms, confirming the crucial contribution of inflammation to pathogenesis.

Citrobacter rodentium causes transient bacterial colitis in laboratory mice. Lupp et al. reported that pathogen abundance in the bowel is highest a week after infection, when perturbations in the gut microbiota are also at their high point. Interestingly, they suggest that the *Citrobacter*-induced changes they see in the mouse microbiome resemble those seen in humans and mice with colitis from other causes (Lupp et al., 2007). In a subsequent paper, Hoffmann et al. used high-throughput sequencing to paint a more detailed picture of the gut microbiome's response to infection with *C. rodentium* (Hoffmann et al., 2009). The mucosal and luminal microbiota showed striking compositional changes 9–14 days after infection, with a marked decrease in the abundance of lactobacilli.

Why gut inflammation disrupts colonisation resistance and enhances pathogen growth is unclear. In a recent opinion piece, Stecher and Hardt propose several possible mechanisms (Stecher and Hardt, 2008):

- *The differential killing hypothesis*: host antimicrobial factors unleashed by inflammation target all bacteria, but pathogens are more resistant to these effects; the normal microbiota suffers bystander collateral damage.
- *The food hypothesis*: inflammation releases nutrients and improves oxygenation, conditions which favour pathogens over commensals.
- *The commensal-network-disruption hypothesis*: inflammation causes disruption of the intestinal food web, which eliminates probiotic, anti-pathogen species.

In fact, more than one of these mechanisms may be at work. A recent study by Stecher et al. (2008) illustrates the complexity of the situation. They showed that, in a murine salmonellosis model, flagellar motility enabled the pathogen to benefit from localised high-energy nutrients (galactose-containing glyco-conjugates, mucin) released during the intestinal inflammatory response. Crucially, loss of flagellar function impaired the pathogen's fitness in the inflamed intestine, but not in the normal gut.

Clinical Metagenomics, Polymicrobial Sepsis and Dysbiosis

Although metagenomics is generally used to interrogate the functions of a whole bacterial community, in clinical microbiology it can also be used as a route to detection or discovery of a single pathogen in a sea of host or commensal genes. In a pioneering study, over 100,000 sequences derived from infected patients were screened by high-throughput sequencing; 14 were found to represent a novel arenavirus (Palacios et al., 2008). A similar approach was used to discover a novel Ebola virus associated with a haemorrhagic fever outbreak in Uganda (Towner et al., 2008). Direct metagenomics approaches have also been used for bacterial and viral pathogen detection in faecal and nasopharyngeal samples (Nakamura et al., 2008, 2009).

The polymicrobial nature of some infections was recognised long before the advent of genomics – for example, necrotising fasciitis often presents as a polymicrobial synergistic infection caused by a mixture of aerobic and anaerobic, Gram-positive and Gram-negative bacteria (Cherneski and Embil, 2001). However, the advent of sensitive culture-independent approaches has led to increased recognition of the complexity of bacterial populations in polymicrobial sepsis and to an acceptance of a population-based rather than single-pathogen view of pathogenesis.

For example, when applied to brain abscesses, culture-independent approaches identified over three times as many types of bacteria as did traditional approaches – a single patient could have up to 16 different bacterial species in a single abscess (Al Masalma et al., 2009). High-throughput sequencing of bacterial 16S rDNA sequences amplified from wound and ulcer samples has also revealed rich and diverse populations that include organisms not usually recognised as wound pathogens, such as *Abiotrophia paraadiacens* and *Rhodopseudomonas* spp. (Dowd et al., 2008a, b). As a result, Dowd and colleagues have proposed the concept of "functional equivalent pathogroups," which they define as "consortia of genotypically distinct bacteria that symbiotically produce a pathogenic community." Interestingly, this assignment of pathogenicity to a group of organisms neatly mirrors the previously mentioned idea of "microbial guilds" performing beneficial functions in the human microbiome.

A population-based view of pathogenesis can also be applied to sites where there is usually a rich microbiota, where disease-associated changes to the microbiome are sometimes called dysbiosis. Foremost here is bacterial vaginosis, a poorly understood condition, characterised by the loss of indigenous vaginal lactobacilli and massive polymicrobial anaerobic vaginal overgrowth. Under the auspices of the HMP, the vaginal microbiomes of 200 women will be characterised in depth using the state-of-the-art sequencing technologies, and vaginal microbiomes will be correlated with the occurrence of bacterial vaginosis (Peterson, 2009).

Conclusion

The next few years will bring an unprecedented leap forward in our attempts to catalogue and characterise the human microbiome, thanks largely to technical progress (high-throughput sequencing) and organisational advances (the HMP). This will be accompanied by changes in the conceptual landscape of host–pathogen interactions, including the recognition that pathogens, like Hamlet's sorrows, "come not single spies but in battalions" and that no pathogen is "an island, entire of itself" – each is instead tied to a continent of commensals. On this view, pathogenesis is more like guerrilla warfare or terrorism than a clash between standing armies – the success of the pathogen, like that of the partisan, is critically dependent on what is happening in the local community. With new catalogues and concepts, there is every hope that the next decade of "human microbiomics" will see basic science translated into new clinical interventions and outcomes.

References

Al Masalma M, Armougom F, Scheld WM, Dufour H, Roche PH, Drancourt M, Raoult D (2009) The expansion of the microbiological spectrum of brain abscesses with use of multiple 16S ribosomal DNA sequencing. Clin Infect Dis 48(9):1169–1178

Alverdy JC, Chang EB (2008) The re-emerging role of the intestinal microflora in critical illness and inflammation: why the gut hypothesis of sepsis syndrome will not go away. J Leukoc Biol 83(3):461–466

Are A, Aronsson L, Wang S, Greicius G, Lee YK, Gustafsson JA, Pettersson S, Arulampalam V (2008) *Enterococcus faecalis* from newborn babies regulate endogenous PPARgamma activity and IL-10 levels in colonic epithelial cells. Proc Natl Acad Sci USA 105(6): 1943–1948

Avila M, Ojcius DM, Yilmaz O (2009) The oral microbiota: living with a permanent guest. DNA Cell Biol 28(8):405–411

Bartlett JG (2009) *Clostridium difficile* infection: historic review. Anaerobe 15(6):227–229

Bik EM, Eckburg PB, Gill SR, Nelson KE, Purdom EA, Francois F, Perez-Perez G, Blaser MJ, Relman DA (2006) Molecular analysis of the bacterial microbiota in the human stomach. Proc Natl Acad Sci USA 103(3):732–737

Bjorksten B (2009) The hygiene hypothesis: do we still believe in it? Nestle Nutr Workshop Ser Pediatr Program 64(11–8); Discussion 18–22:251–257

Bonten MJ, Krueger WA (2006) Selective decontamination of the digestive tract: cumulating evidence, at last? Semin Respir Crit Care Med 27(1):18–22

Buddington KK, Donahoo JB, Buddington RK (2002) Dietary oligofructose and inulin protect mice from enteric and systemic pathogens and tumor inducers. J Nutr 132(3):472–477

Carroll IM, Threadgill DW, Threadgill DS (2009) The gastrointestinal microbiome: a malleable, third genome of mammals. Mamm Genome 20(7):395–403

Cherneski CL, Embil JM (2001) Necrotizing fasciitis. Saudi Med J 22(7):565–568

Cogen AL, Nizet V, Gallo RL (2008) Skin microbiota: a source of disease or defence? Br J Dermatol 158(3):442–455

Collins SM, Bercik P (2009) The relationship between intestinal microbiota and the central nervous system in normal gastrointestinal function and disease. Gastroenterology 136(6): 2003–2014

Corr SC, Li Y, Riedel CU, O'Toole PW, Hill C, Gahan CG (2007) Bacteriocin production as a mechanism for the anti-infective activity of *Lactobacillus salivarius* UCC118. Proc Natl Acad Sci USA 104(18):7617–7621

Coward C, Grant AJ, Swift C, Philp J, Towler R, Heydarian M, Frost JA, Maskell DJ (2006) Phase-variable surface structures are required for infection of *Campylobacter jejuni* by bacteriophages. Appl Environ Microbiol 72(7):4638–4647

Davies J (2009) Darwin and microbiomes. EMBO Rep 10(8):805

de Smet AM, Bonten MJ (2008) Selective decontamination of the digestive tract. Curr Opin Infect Dis 21(2):179–183

de Vrese M, Schrezenmeir J (2008) Probiotics, prebiotics, and synbiotics. Adv Biochem Eng Biotechnol 111:1–66

Dethlefsen L, Huse S, Sogin ML, Relman DA (2008) The pervasive effects of an antibiotic on the human gut microbiota, as revealed by deep 16S rRNA sequencing. PLoS Biol 6(11):e280

Dowd SE, Sun Y, Secor PR, Rhoads DD, Wolcott BM, James GA, Wolcott RD (2008a) Survey of bacterial diversity in chronic wounds using pyrosequencing, DGGE, and full ribosome shotgun sequencing. BMC Microbiol 8:43

Dowd SE, Wolcott RD, Sun Y, McKeehan T, Smith E, Rhoads D (2008b) Polymicrobial nature of chronic diabetic foot ulcer biofilm infections determined using bacterial tag encoded FLX amplicon pyrosequencing (bTEFAP). PLoS One 3(10):e3326

Dutta C, Pan A (2002) Horizontal gene transfer and bacterial diversity. J Biosci 27(1)(Suppl 1): 27–33

Eckburg PB, Bik EM, Bernstein CN, Purdom E, Dethlefsen L, Sargent M, Gill SR, Nelson KE, Relman DA (2005) Diversity of the human intestinal microbial flora. Science 308(5728):1635–1638

Falk PG, Hooper LV, Midtvedt T, Gordon JI (1998) Creating and maintaining the gastrointestinal ecosystem: what we know and need to know from gnotobiology. Microbiol Mol Biol Rev 62(4):1157–1170

Falkow S (1988) Molecular Koch's postulates applied to microbial pathogenicity. Rev Infect Dis 10(Suppl 2):S274–S276

Fitzgerald JR, Musser JM (2001) Evolutionary genomics of pathogenic bacteria. Trends Microbiol 9(11):547–553

Foxman B, Goldberg D, Murdock C, Xi C, Gilsdorf JR (2008) Conceptualizing human microbiota: from multicelled organ to ecological community. Interdiscip Perspect Infect Dis 2008:613979

Frank DN, Pace NR (2008) Gastrointestinal microbiology enters the metagenomics era. Curr Opin Gastroenterol 24(1):4–10

Frank DN, St Amand AL, Feldman RA, Boedeker EC, Harpaz N, Pace NR (2007) Molecular-phylogenetic characterization of microbial community imbalances in human inflammatory bowel diseases. Proc Natl Acad Sci USA 104(34):13780–13785

Gill SR, Pop M, Deboy RT, Eckburg PB, Turnbaugh PJ, Samuel BS, Gordon JI, Relman DA, Fraser-Liggett CM, Nelson KE (2006) Metagenomic analysis of the human distal gut microbiome. Science 312(5778):1355–1359

Grice EA, Kong HH, Conlan S, Deming CB, Davis J, Young AC, Bouffard GG, Blakesley RW, Murray PR, Green ED, Turner ML, Segre JA (2009) Topographical and temporal diversity of the human skin microbiome. Science 324(5931):1190–1192

Handelsman J, Rondon MR, Brady SF, Clardy J, Goodman RM (1998) Molecular biological access to the chemistry of unknown soil microbes: a new frontier for natural products. Chem Biol 5(10):R245–R249

Hoffmann C, Hill DA, Minkah N, Kirn T, Troy A, Artis D, Bushman F (2009) Community-wide response of the gut microbiota to enteropathogenic *Citrobacter rodentium* infection revealed by deep sequencing. Infect Immun 77(10):4668–4678

Hold GL, Pryde SE, Russell VJ, Furrie E, Flint HJ (2002) Assessment of microbial diversity in human colonic samples by 16S rDNA sequence analysis. FEMS Microbiol Ecol 39(1):33–39

Honda H, Dubberke ER (2009) *Clostridium difficile* infection: a re-emerging threat. Mo Med 106(4):287–291

Hooper LV, Gordon JI (2001) Commensal host–bacterial relationships in the gut. Science 292(5519):1115–1118

International Human Genome Sequencing Consortium (2004) Finishing the euchromatic sequence of the human genome. Nature 431(7011):931–945

Kelly D, Campbell JI, King TP, Grant G, Jansson EA, Coutts AG, Pettersson S, Conway S (2004) Commensal anaerobic gut bacteria attenuate inflammation by regulating nuclear–cytoplasmic shuttling of PPAR-gamma and RelA. Nat Immunol 5(1):104–112

Kim Y, Kim SH, Whang KY, Kim YJ, Oh S (2008) Inhibition of *Escherichia coli* O157:H7 attachment by interactions between lactic acid bacteria and intestinal epithelial cells. J Microbiol Biotechnol 18(7):1278–1285

Kinross JM, von Roon AC, Holmes E, Darzi A, Nicholson JK (2008) The human gut microbiome: implications for future health care. Curr Gastroenterol Rep 10(4):396–403

Kleessen B, Blaut M (2005) Modulation of gut mucosal biofilms. Br J Nutr 93(Suppl 1):S35–S40

Koch R (1884) Die Aetiologie der tuberculose. Mitt Kaiser Gesundh 2:1

Koch R (1892) Ueber bakteriologische Forschung. Verh. X. Int Med Congr, Berlin, p 35
Koval SF, Bayer ME (1997) Bacterial capsules: no barrier against Bdellovibrio. Microbiology 143(Pt 3):749–753
Kuhn R, Lohler J, Rennick D, Rajewsky K, Muller W (1993) Interleukin-10-deficient mice develop chronic enterocolitis. Cell 75(2):263–274
Kurokawa K, Itoh T, Kuwahara T, Oshima K, Toh H, Toyoda A, Takami H, Morita H, Sharma VK, Srivastava TP, Taylor TD, Noguchi H, Mori H, Ogura Y, Ehrlich DS, Itoh K, Takagi T, Sakaki Y, Hayashi T, Hattori M (2007) Comparative metagenomics revealed commonly enriched gene sets in human gut microbiomes. DNA Res 14(4):169–181
Lai Y, Di Nardo A, Nakatsuji T, Leichtle A, Yang Y, Cogen AL, Wu ZR, Hooper LV, Schmidt RR, von Aulock S, Radek KA, Huang CM, Ryan AF, Gallo RL (2009) Commensal bacteria regulate Toll-like receptor 3-dependent inflammation after skin injury. Nat Med 15(12):1377–1382
Lampe JW (2008) The human microbiome project: getting to the guts of the matter in cancer epidemiology. Cancer Epidemiol Biomarkers Prev 17(10):2523–2524
Leiman PG, Battisti AJ, Bowman VD, Stummeyer K, Muhlenhoff M, Gerardy-Schahn R, Scholl D, Molineux IJ (2007) The structures of bacteriophages K1E and K1-5 explain processive degradation of polysaccharide capsules and evolution of new host specificities. J Mol Biol 371(3):836–849
Lewis RE (2009) Overview of the changing epidemiology of candidemia. Curr Med Res Opin 25(7):1732–1740
Ley RE, Turnbaugh PJ, Klein S, Gordon JI (2006) Microbial ecology: human gut microbes associated with obesity. Nature 444(7122):1022–1023
Lupp C, Robertson ML, Wickham ME, Sekirov I, Champion OL, Gaynor EC, Finlay BB (2007) Host-mediated inflammation disrupts the intestinal microbiota and promotes the overgrowth of enterobacteriaceae. Cell Host Microbe 2(3):204
Mason KL, Huffnagle GB, Noverr MC, Kao JY (2008) Overview of gut immunology. Adv Exp Med Biol 635:1–14
Medellin-Pena MJ, Wang H, Johnson R, Anand S, Griffiths MW (2007) Probiotics affect virulence-related gene expression in *Escherichia coli* O157:H7. Appl Environ Microbiol 73(13): 4259–4267
Metchnikoff Elie, Chalmers Mitchell P (1908) The prolongation of life: optimistic studies. G. P. Putnam's sons, New York and London
Metzker ML (2010) Sequencing technologies – the next generation. Nat Rev Genet 11(1):31–46
Mihajlovski A, Alric M, Brugere JF (2008) A putative new order of methanogenic Archaea inhabiting the human gut, as revealed by molecular analyses of the mcrA gene. Res Microbiol 159(7–8):516–521
Mongodin EF, Emerson JB, Nelson KE (2005) Microbial metagenomics. Genome Biol 6(10):347
Nakamura S, Maeda N, Miron IM, Yoh M, Izutsu K, Kataoka C, Honda T, Yasunaga T, Nakaya T, Kawai J, Hayashizaki Y, Horii T, Iida T (2008) Metagenomic diagnosis of bacterial infections. Emerg Infect Dis 14(11):1784–1786
Nakamura S, Yang CS, Sakon N, Ueda M, Tougan T, Yamashita A, Goto N, Takahashi K, Yasunaga T, Ikuta K, Mizutani T, Okamoto Y, Tagami M, Morita R, Maeda N, Kawai J, Hayashizaki Y, Nagai Y, Horii T, Iida T, Nakaya T (2009) Direct metagenomic detection of viral pathogens in nasal and fecal specimens using an unbiased high-throughput sequencing approach. PLoS One 4(1):e4219
Nasidze I, Li J, Quinque D, Tang K, Stoneking M (2009a) Global diversity in the human salivary microbiome. Genome Res 19(4):636–643
Nasidze I, Quinque D, Li J, Li M, Tang K, Stoneking M (2009b) Comparative analysis of human saliva microbiome diversity by barcoded pyrosequencing and cloning approaches. Anal Biochem 391(1):64–68
Ott SJ, Kuhbacher T, Musfeldt M, Rosenstiel P, Hellmig S, Rehman A, Drews O, Weichert W, Timmis KN, Schreiber S (2008) Fungi and inflammatory bowel diseases: alterations of composition and diversity. Scand J Gastroenterol 43(7):831–841

Palacios G, Druce J, Du L, Tran T, Birch C, Briese T, Conlan S, Quan PL, Hui J, Marshall J, Simons JF, Egholm M, Paddock CD, Shieh WJ, Goldsmith CS, Zaki SR, Catton M, Lipkin WI (2008) A new arenavirus in a cluster of fatal transplant-associated diseases. N Engl J Med 358(10):991–998

Pallen MJ, Wren BW (2007) Bacterial pathogenomics. Nature 449(7164):835–842

Parahitiyawa NB, Scully C, Leung WK, Yam WC, Jin LJ, Samaranayake LP (2009) Exploring the oral bacterial flora: current status and future directions. Oral Dis 16(2):136–145

Pennisi M, Antonelli M (2009) Clinical aspects of invasive candidiasis in critically ill patients. Drugs 69(Suppl 1):21–28

Petersen A, Heegaard PM, Pedersen AL, Andersen JB, Sorensen RB, Frokiaer H, Lahtinen SJ, Ouwehand AC, Poulsen M, Licht TR (2009) Some putative prebiotics increase the severity of Salmonella enterica serovar Typhimurium infection in mice. BMC Microbiol 9:245

Peterson J (2009) The NIH human microbiome project. Genome Res 19(12):2317–2323

Preidis GA, Versalovic J (2009) Targeting the human microbiome with antibiotics, probiotics, and prebiotics: gastroenterology enters the metagenomics era. Gastroenterology 136(6): 2015–2031

Raqib R, Sarker P, Bergman P, Ara G, Lindh M, Sack DA, Nasirul Islam KM, Gudmundsson GH, Andersson J, Agerberth B (2006) Improved outcome in shigellosis associated with butyrate induction of an endogenous peptide antibiotic. Proc Natl Acad Sci USA 103(24): 9178–9183

Raza S, Baig MA, Russell H, Gourdet Y, Berger BJ (2009) *Clostridium difficile* infection following chemotherapy. Recent Pat Antiinfect Drug Discov 5(1):1–9

Reiff C, Kelly D (2009) Inflammatory bowel disease, gut bacteria and probiotic therapy. Int J Med Microbiol 300(1):25–33

Rendon MA, Saldana Z, Erdem AL, Monteiro-Neto V, Vazquez A, Kaper JB, Puente JL, Giron JA (2007) Commensal and pathogenic *Escherichia coli* use a common pilus adherence factor for epithelial cell colonization. Proc Natl Acad Sci USA 104(25):10637–10642

Rueping MJ, Vehreschild JJ, Cornely OA (2009) Invasive candidiasis and candidemia: from current opinions to future perspectives. Expert Opin Invest Drugs 18(6):735–748

Salzman NH, Bevins CL (2008) Negative interactions with the microbiota: IBD. Adv Exp Med Biol 635:67–78

Savage DC (1977) Microbial ecology of the gastrointestinal tract. Annu Rev Microbiol 31:107–133

Scanlan PD, Shanahan F, Marchesi JR (2008) Human methanogen diversity and incidence in healthy and diseased colonic groups using mcrA gene analysis. BMC Microbiol 8:79

Schellenberg J, Links MG, Hill JE, Dumonceaux TJ, Peters GA, Tyler S, Ball TB, Severini A, Plummer FA (2009) Pyrosequencing of the chaperonin-60 universal target as a tool for determining microbial community composition. Appl Environ Microbiol 75(9): 2889–2898

Schumann A, Nutten S, Donnicola D, Comelli EM, Mansourian R, Cherbut C, Corthesy-Theulaz I, Garcia-Rodenas C (2005) Neonatal antibiotic treatment alters gastrointestinal tract developmental gene expression and intestinal barrier transcriptome. Physiol Genom 23(2): 235–245

Searle LE, Best A, Nunez A, Salguero FJ, Johnson L, Weyer U, Dugdale AH, Cooley WA, Carter B, Jones G, Tzortzis G, Woodward MJ, La Ragione RM (2009) A mixture containing galactooligosaccharide, produced by the enzymic activity of *Bifidobacterium bifidum*, reduces *Salmonella enterica* serovar Typhimurium infection in mice. J Med Microbiol 58(Pt 1):37–48

Shanahan F (2002) The host-microbe interface within the gut. Best Pract Res Clin Gastroenterol 16(6):915–931

Sibartie S, O'Hara AM, Ryan J, Fanning A, O'Mahony J, O'Neill S, Sheil B, O'Mahony L, Shanahan F (2009) Modulation of pathogen-induced CCL20 secretion from HT-29 human intestinal epithelial cells by commensal bacteria. BMC Immunol 10:54

Sommer MO, Dantas G, Church GM (2009) Functional characterization of the antibiotic resistance reservoir in the human microflora. Science 325(5944):1128–1131

Srinivasan S, Fredricks DN (2008) The human vaginal bacterial biota and bacterial vaginosis. Interdiscip Perspect Infect Dis 2008:750479

Stecher B, Barthel M, Schlumberger MC, Haberli L, Rabsch W, Kremer M, Hardt WD (2008) Motility allows *S. typhimurium* to benefit from the mucosal defence. Cell Microbiol 10(5):1166–1180

Stecher B, Hardt WD (2008) The role of microbiota in infectious disease. Trends Microbiol 16(3):107–114

Stecher B, Robbiani R, Walker AW, Westendorf AM, Barthel M, Kremer M, Chaffron S, Macpherson AJ, Buer J, Parkhill J, Dougan G, von Mering C, Hardt WD (2007) *Salmonella enterica* serovar typhimurium exploits inflammation to compete with the intestinal microbiota. PLoS Biol 5(10):2177–2189

Steinberg KM, Levin BR (2007) Grazing protozoa and the evolution of the *Escherichia coli* O157:H7 Shiga toxin-encoding prophage. Proc Biol Sci 274(1621):1921–1929

Streit WR, Schmitz RA (2004) Metagenomics – the key to the uncultured microbes. Curr Opin Microbiol 7(5):492–498

Stummeyer K, Schwarzer D, Claus H, Vogel U, Gerardy-Schahn R, Muhlenhoff M (2006) Evolution of bacteriophages infecting encapsulated bacteria: lessons from *Escherichia coli* K1-specific phages. Mol Microbiol 60(5):1123–1135

Tanaka K, Sawamura S, Satoh T, Kobayashi K, Noda S (2007) Role of the indigenous microbiota in maintaining the virus-specific CD8 memory T cells in the lung of mice infected with murine cytomegalovirus. J Immunol 178(8):5209–5216

Theron J, Cloete TE (2000) Molecular techniques for determining microbial diversity and community structure in natural environments. Crit Rev Microbiol 26(1):37–57

Tobe T, Beatson SA, Taniguchi H, Abe H, Bailey CM, Fivian A, Younis R, Matthews S, Marches O, Frankel G, Hayashi T, Pallen MJ (2006) An extensive repertoire of type III secretion effectors in *Escherichia coli* O157 and the role of lambdoid phages in their dissemination. Proc Natl Acad Sci USA 103(40):14941–14946

Towner JS, Sealy TK, Khristova ML, Albarino CG, Conlan S, Reeder SA, Quan PL, Lipkin WI, Downing R, Tappero JW, Okware S, Lutwama J, Bakamutumaho B, Kayiwa J, Comer JA, Rollin PE, Ksiazek TG, Nichol ST (2008) Newly discovered Ebola virus associated with hemorrhagic fever outbreak in Uganda. PLoS Pathog 4(11):e1000212

Tringe SG, Rubin EM (2005) Metagenomics: DNA sequencing of environmental samples. Nat Rev Genet 6(11):805–814

Truusalu K, Mikelsaar RH, Naaber P, Karki T, Kullisaar T, Zilmer M, Mikelsaar M (2008) Eradication of *Salmonella typhimurium* infection in a murine model of typhoid fever with the combination of probiotic *Lactobacillus fermentum* ME-3 and ofloxacin. BMC Microbiol 8:132

Tschop MH, Hugenholtz P, Karp CL (2009) Getting to the core of the gut microbiome. Nat Biotechnol 27(4):344–346

Tuohy KM, Gougoulias C, Shen Q, Walton G, Fava F, Ramnani P (2009) Studying the human gut microbiota in the trans-omics era – focus on metagenomics and metabonomics. Curr Pharm Des 15(13):1415–1427

Turnbaugh PJ, Ley RE, Mahowald MA, Magrini V, Mardis ER, Gordon JI (2006) An obesity-associated gut microbiome with increased capacity for energy harvest. Nature 444(7122):1027–1031

Vaarala O, Atkinson MA, Neu J (2008) The "perfect storm" for type 1 diabetes: the complex interplay between intestinal microbiota, gut permeability, and mucosal immunity. Diabetes 57(10):2555–2562

Vaishnava S, Behrendt CL, Ismail AS, Eckmann L, Hooper LV (2008) Paneth cells directly sense gut commensals and maintain homeostasis at the intestinal host-microbial interface. Proc Natl Acad Sci USA 105(52):20858–20863

Vassallo MF, Walker WA (2008) Neonatal microbial flora and disease outcome. Nestle Nutr Workshop Ser Pediatr Program 61:211–224

Venter JC, Remington K, Heidelberg JF, Halpern AL, Rusch D, Eisen JA, Wu D, Paulsen I, Nelson KE, Nelson W, Fouts DE, Levy S, Knap AH, Lomas MW, Nealson K, White O, Peterson J, Hoffman J, Parsons R, Baden-Tillson H, Pfannkoch C, Rogers YH, Smith HO (2004) Environmental genome shotgun sequencing of the Sargasso Sea. Science 304(5667):66–74

Vieira LQ, dos Santos LM, Neumann E, da Silva AP, Moura LN, Nicoli JR (2008) Probiotics protect mice against experimental infections. J Clin Gastroenterol 42(Suppl 3 Pt 2):S168–S169

Wagner RD, Johnson SJ, Kurniasih Rubin D (2009) Probiotic bacteria are antagonistic to *Salmonella enterica* and *Campylobacter jejuni* and influence host lymphocyte responses in human microbiota-associated immunodeficient and immunocompetent mice. Mol Nutr Food Res 53(3):377–388

Waterfield NR, Wren BW, Ffrench-Constant RH (2004) Invertebrates as a source of emerging human pathogens. Nat Rev Microbiol 2(10):833–841

Woese CR (1987) Bacterial evolution. Microbiol Rev 51(2):221–271

Yap IK, Li JV, Saric J, Martin FP, Davies H, Wang Y, Wilson ID, Nicholson JK, Utzinger J, Marchesi JR, Holmes E (2008) Metabonomic and microbiological analysis of the dynamic effect of vancomycin-induced gut microbiota modification in the mouse. J Proteome Res 7(9):3718–3728

Chapter 4
The Human Virome

Matthew Haynes and Forest Rohwer

Background

Until fairly recently, it has been customary, in the absence of clinically significant infection, to view the human organism as an isolated entity. In fact, the healthy human body always contains a large number of foreign cells and viruses (Virgin et al., 2009; Dethlefsen et al., 2007; Relman, 2002). There are more viral particles in the human body than microbial cells, which are ten times more numerous than eukaryotic (human) cells. Similarly, only about 1.5% of the human genome encodes recognizable "human proteins," whereas approximately 45% our genome is retro-transposons, DNA transposons, and viral sequences. Most of the human-associated microbes and viruses, often found on "external" surfaces lining the lumens of organs such as the gut and oral/nasal cavities, participate in complex commensal or mutualistic relationships with their human host (Dethlefsen et al., 2007; Relman, 2002). Therefore, it is not advantageous to attempt to eradicate every virus and microbial cell from the body in response to infection. A new medical paradigm is emerging: an illness may be defined by a disruption of the normal "healthy" microbiome and/or virome, and that restoration of this state, not elimination of all nonhuman organisms, should be the goal of medical treatment (Harrison, 2007). Current interest in the human microbiome reflects the increasing acceptance of the view that the microbiota per se should not be seen merely as invasive disease vectors but are in fact an intrinsic part of the human supra-organism (Dethlefsen et al., 2007).

The classical method of viral isolation is by culturing. Koch's postulates (Rivers, 1937) dictate the conditions under which a virus cultured in vitro should be regarded as the cause of an infectious disease; human viruses are usually cultured only in this context. In addition, culturing will be successful only for the small fraction of viruses for which appropriate culture conditions can be determined. To break from the limited view that all viruses are intrinsically harmful requires new methodologies that enable us to characterize entire uncultured viral communities. A

M. Haynes (✉)
Department of Biology, San Diego State University, 5500 Campanile Dr., San Diego, CA 92182, USA
e-mail: mhaynes@projects.sdsu.edu

K.E. Nelson (ed.), *Metagenomics of the Human Body*,
DOI 10.1007/978-1-4419-7089-3_4,

culture-independent metagenomics approach to viral community analysis will yield a broader view of the human virome, just as metagenomic sequencing has revealed a wider range of bacteria in the human microbiome than culture-based methods (Harris et al., 2007; Rogers et al., 2004).

Metagenomics

A viral metagenome or virome is the total genetic (DNA and RNA) sequence derived from a viral community. Mathematically, the structure of a community may be represented by a graph whose functional form (lognormal, power function, etc.) reflects the relative abundance distribution of its members. The *evenness* of the distribution (fractional contribution of each genotype), along with the *richness* (the total number of genotypes), are often combined to denote the *diversity* of a community, as in the Shannon–Wiener index (H'),

$$H' = -\sum_{i=1}^{S} r_i \ln r_i \tag{4.1}$$

where S is the sample richness, and r_i is the relative abundance of genotype i. Viral communities tend to be unevenly distributed, with a small number of species or genotypes dominating in abundance (Fig. 4.1).

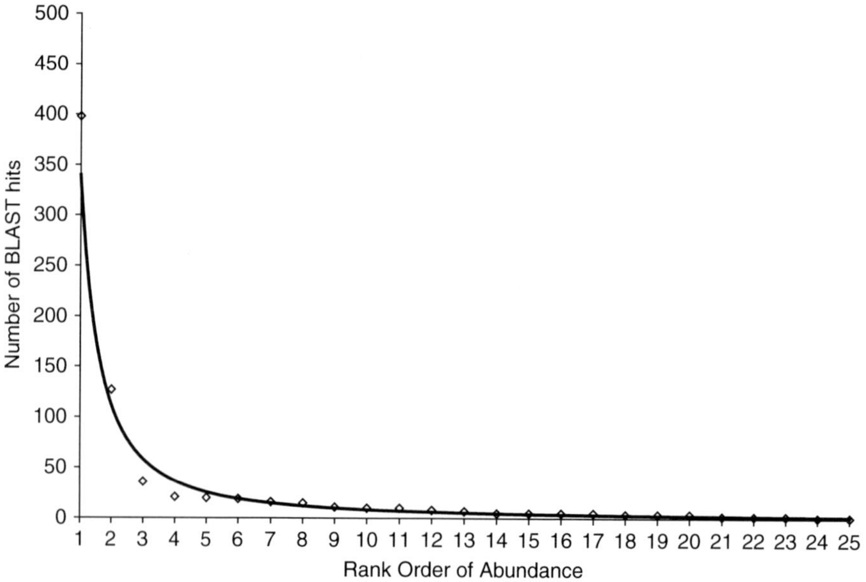

Fig. 4.1 Example of a human viral community rank-abundance curve using sequences from a human oropharyngeal metagenome. Here the relative frequencies of BLAST n hits to a viral sequence database follow a relationship that can be approximated by a power-law equation of the type $y = a\,x^{-b}$ (Willner et al., 2010)

Metagenomics has been greatly facilitated by recent advances in sequencing technology. Pyrosequencing (Roche/454 Life Sciences), as well as other technologies (e.g., Solexa, SOLiD), enable routine DNA sequencing on the scale of 10^8 bp. All of these new high-throughput methods replace traditional cloning in bacteria with mechanical separation of DNA molecules by some means (e.g., emPCR with DNA immobilized on beads for 454). This requires the creation of a minimally biased DNA library that better reflects the viral community in the sample. At present, the original DNA sample must often be amplified before sequencing, increasing the opportunity for artificial over- or underrepresentation of particular sequences. Despite this limitation, these methods appear to avoid most of the problems associated with conventional cloning, which is subject to strong sequence bias against some "unclonable" sequences. This phenomenon appears to be particularly pronounced in attempts to clone viral sequences. Microarrays, as well, remain semi-quantitative detection methods because it is impossible to simultaneously optimize the hybridization of thousands of individual sequences (Table 4.1).

Table 4.1 Examples of viruses detected in human samples by metagenomic methods

Method	Sample source	Examples of viruses found
DNA recovery from microarrays	Nasopharynx (Ksiazek et al., 2003)	SARS Coronavirus
	Nasopharynx (Kistler et al., 2007)	Coronavirus, Rhinovirus
Random RT-PCR	Plasma (Jones et al., 2005)	PARV4, SAV-1, SAV-2
	Nasopharynx (Allander et al., 2005)	HBoV
	Stool (Victoria et al., 2009)	Cosavirus HCoSV
	Nasopharynx (Nakamura et al., 2009)	Influenza A, polyomavirus
	Stool (Zhang et al., 2005)	PMMV
Metagenomic shotgun libraries	Stool (adult) (Breitbart et al., 2003)	phages
	Blood (Breitbart and Rohwer, 2005)	TTV, HHV3, SEN virus, phages
	Stool (infant) (Breitbart et al., 2008)	phages
	Respiratory tract (Willner et al., 2009a)	HHV-1, HHV-2, phages
	Oropharynx (Willner et al., 2010)	Epstein–Barr Virus
	Stool (Reyes et al., 2009)	Phages

For further details See "Viral Metagenomics Methods"

Approximate Number and Distribution of Viruses in the Human Body

How Many Viruses Are There in a Human?

We can approach this question from two directions: estimation of the number of phages expected based on the size of the human microbiome and the typical viral (phage):host ratio, or by direct counts of viruses in samples from healthy individuals. The human body is composed of about 10^{13} cells (Savage, 1977). There are about 10 times this number of microbial cells associated with the healthy human body (Savage, 1977). The observed ratio of 7–10 viral-like particles per microbial cell in environmental (Rohwer, 2003) and human samples (Furlan, 2009) means that we could expect to find about 10^{15} phages in the body. It is possible to compare this prediction with results from recent studies. The data in Table 4.2 are from direct counts of viruses using epifluorescence microscopy. These data indicate the presence of approximately 3×10^{12} viruses in the body.

Table 4.2 The diversity and estimated viral load of nonviremic humans

Organ system	VLPs per ml fluid	VLPs per cm^2	Total VLPs	Number of genotypes
Oral/nasal/pharynx	2.1×10^8		1.2×10^{10}	250
Tooth associated	8.5×10^7		3.0×10^9	?
Lower respiratory tract	2.8×10^8		1.4×10^{11}	250
Distal gut	4.8×10^8		3.0×10^{12}	1,000
Blood	1.0×10^5		1.0×10^8	20
Skin		1.2×10^6	2.1×10^{10}	?
Vagina	6.6×10^7		1.2×10^9	?
Urinary tract	3.7×10^6		1.9×10^9	?
Body total			3.0×10^{12}	1,500

Data sources: Willner et al. (2009a, 2010), Reyes et al. (2009), Furlan (2009), Furlan and Hanson (2009, personal communication)

Abundance of Viruses at Specific Body Sites

Wherever microbes (bacteria and archaea) are present, their viruses will be found. Thus in the human body, the regions of high microbial levels, in particular the gut, also have the highest abundance of viruses. Other organ systems with mucus membranes, such as the nasal and oral cavities and vagina, harbor a smaller but significant viral community.

What Types of Viruses Inhabit the Human Host?

Compared with environmental viral communities, the diversity of the human virome is low. We estimate that there are 1,500 viral genotypes in a typical healthy, human virome. By contrast, 1 kg of marine sediment will contain at least ten thousand, and perhaps a million, viral genotypes. The human-associated viruses are unevenly distributed, with the bulk of the virome composed of a handful of dominant species (Table 4.3). In the limited data available to date, it appears that a disease state is correlated with an increase in the diversity of the virome (Willner et al., 2009a).

Table 4.3 Comparison of diversity indices of typical environmental and human metagenomes. Values calculated using PHACCS (Angly et al., 2005) utilize all data, not merely identified sequences

Metagenome	Shannon index (nats)	Richness	Evenness	% most abundant genotype
Human oropharynx	5.0	248	0.92	6.5
Human lower resp. tract	4.8	243	0.92	6.2
Environmental (Angly et al., 2005, 2006)	6.0–10.8	>3000	0.85–1.00	2.3–13

Most of the viruses are phages. There are also certain eukaryotic viruses, such as herpesviruses, anelloviruses, and papillomaviruses, that are ubiquitous in the human virome and tend to cause few problems considering their abundance (Virgin et al., 2009). See also Fig. 4.3.

Phage Community

Commensal microbes are ubiquitous in the healthy human body (Dethlefsen et al., 2007; Wilson, 2005), occupying niches on skin (Grice et al., 2008, 2009), distal gut (Gill et al., 2006; Turnbaugh et al., 2009), vagina (Hyman et al., 2005). As a result, viruses that infect microbes (phages) are numerous (Letarov and Kulikov, 2009) and have been found in the gut (Reyes et al., 2009), nasopharynx (Allander et al., 2005), oropharynx (Willner et al., 2010), oral cavity (Hitch et al., 2004), blood (Breitbart and Rohwer, 2005), and lung secretions (Willner et al., 2009a).

Phages comprise by far the majority of the human virome (Willner et al., 2009a, 2010) and can be expected to exert an influence on the human microbial community (Gill et al., 2006; Hendrix, 2005) that parallels the interactions observed in a variety of environmental samples (Letarov and Kulikov, 2009; Weinbauer, 2006; Rodriguez-Mueller et al., 2010; Breitbart et al., 2005). By killing specific host organisms, phages regulate the absolute and relative abundance of microbial species (Breitbart et al., 2005). Genetic variation in the hosts is therefore favored as a means of escaping phage predation (Kunin et al., 2008). In addition, phages are major vehicles of DNA transfer to and from host cells (*horizontal gene transfer*) through both *lytic* and *lysogenic* pathways (Little, 2005), potentially conferring new phenotypes that can increase the pathogenicity (Breitbart et al., 2005) or the fitness (Sharon et al., 2009; Wagner and Waldor, 2002) of the host. Analysis of the phage metagenome can thus provide information not only about potential host taxonomy, but also reveal potential metabolic pathways available to the microbial community (Willner et al., 2009a; Sharon et al., 2009). Box 4.1 shows the "core" phage metagenome found in the human lower respiratory tract: 19 phage types that were all present in five normal control subjects and five cystic fibrosis patients (Willner et al., 2009a) (Fig. 4.2).

Box 4.1 Core phage community of the human lower respiratory tract. Phage types in the order of abundance that were present in all 10 samples regardless of disease state (tBLASTx hits with E < $10^{-5)}$ (Willner et al., 2009a)

Aeromonas hydrophila phi Aeh1
Aeromonas phi 31
Bacillus cereus phi phBC6A51
Bacillus subtilis phi 105
Bacillus subtilis phi SPBc2
Brucella melitensis 16 M phi Bruc 1 prophage
Escherichia coli phi CP073-4 prophage
Escherichia coli phi CP4-6 prophage
Escherichia coli phi QIN prophage
Escherichia coli phi Sp18 prophage
Haemophilus influenzae phi HP1
Lactobacillus plantarum phi LP65
Mycobacterium phi Bxz1
Mycobacterium phi CJW1
Pseudomonas phi KZ
Shigella flexneri phi Flex4 prophage
Staphylococcus phi Twort
Vibrio parahaemolyticus phi KVP40
Xylella fastidiosa phi Xpd5 prophage

Fig. 4.2 Virus-like particles (VLPs) from asymptomatic individuals. (**a**) respiratory tract; (**b**) gut. Viruses were purified and concentrated by CsCl density gradient centrifugation as described in Breitbart et al. (2003). The VLPs were visualized by capturing on a 0.02-μm Anodisc filter, SYBR Gold staining, and viewing by epifluorescence microscopy

Eukaryotic Viruses

Viruses capable of infecting the human host ("eukaryotic viruses"), while obviously present in diseased individuals, can also be found in healthy subjects (Virgin et al., 2009; Willner et al., 2009a, 2010). In asymptomatic subjects, the abundance of these viruses is far lower than that of phages in the healthy human body (Willner et al., 2009a, 2010). Depending on the area of the body under examination, the presence of eukaryotic viruses will be due to either transient environmental exposure of accessible regions (e.g., the lungs) or chronic infections that do not give rise to recognizable clinical symptoms. The lack of symptoms might reflect a low-level viral infection that is successfully suppressed by the immune system at an early stage, or perhaps a commensal virus that causes no apparent harm (Virgin et al., 2009; Stapleton et al., 2004; Okamoto, 2009; Antonsson et al., 2000). An example of the latter is Torque Teno Virus (TTV), which was originally thought to be associated with a form of hepatitis, but now seems likely to be a ubiquitous but benign commensal virus (Okamoto, 2009). Instances of true viral–human mutualism in this context are not yet well understood, but it has been suggested that co-infection with GB Virus Type C (originally termed Hepatitis G virus) reduces mortality in HIV-infected individuals (Stapleton et al., 2004). Box 4.2 shows the "core" eukaryotic viral metagenome found in the human lower respiratory tract: 20 viruses that were all present in five normal control subjects and five cystic fibrosis patients (Willner et al., 2009a).

Box 4.2 Core eukaryotic viral community of the human respiratory tract. Viruses that were present in all 10 samples regardless of disease state (tBLASTx hits with $E < 10^{-5}$) (Willner et al., 2009a)

Acanthamoeba polyphaga mimivirus
Aedes taeniophyncus iridescent virus
Amsacta moorei entomopoxvirus "L"
Bovine Adenovirus A
Bovine Adenovirus 5
Cercopithicine herpesvirus 1
Cercopithicine herpesvirus 16
Cercopithicine herpesvirus 2
Cercopithicine herpesvirus 9
Chlorella virus ATCV-1
Chlorella virus FR483
Ectocarpus siliculosis virus 1
Frog virus 3
Human herpesvirus 1
Human herpesvirus 2

Melanoplus sanguinipes entomopoxvirus
Paramecium bursaria Chlorella virus AR158
Suid herpesvirus 1
Taterapox virus
Trichoplusia ni ascovirus

Residence Time and Pathogenicity

We can characterize viruses by their persistence (residence time in the body) and the degree of mutualism they exhibit (Fig. 4.3). The viruses that comprise the core human virome are relatively persistent (never cleared from the body). This distinguished them from pathogenic viruses causing acute and short-lived infections. There are, however, a number of pathogenic viruses such as herpesviruses that may persist in the body in an intracellular form, only to cause sporadic shedding of viral particles. Still other viruses are transient but common members of the human virome. Plant viruses such as PMMV are taken in with food and pass directly through the digestive tract (Zhang et al., 2005).

Fig. 4.3 Residence time vs. symbiotic modality of selected viruses. *EBV* epstein–barr virus, *HIV* human immunodeficiency virus, *HSV* herpes simplex virus, *TTV* Torque Teno virus, *PMMV* pepper mild mottle virus

Viral Metagenomics Methods

Investigation of the human virome has recently been accelerated by technological and methodological developments. The methods fall into three categories: viral nucleic acid isolation, DNA sequencing, and data analysis. For a review of methods in viral metagenomics see Delwart (2007).

Recovery from Microarrays

The SARS coronavirus was discovered by hybridizing nucleic acids to an array (Virochip) that contained sequences representing all fully sequenced viruses, physically removing the annealed DNA from the array, and PCR amplifying this DNA using primers complementary to linkers that had been added (Kistler et al., 2007; Wang et al., 2002; Chiu et al., 2008). The prime example of this approach is the cloning and sequencing of the SARS coronavirus (Ksiazek et al., 2003). Limitations of the method are that it will only succeed with viruses that share significant homology with previously known viruses and that simultaneous optimization of multiple hybridizations on an array may be impossible.

Random RT-PCR

There are several variations of randomly primed reverse-transcription PCR (RT-PCR) for amplification of RNA viral sequences. Viral RNA is converted to cDNA using primers containing random octamers for both first- and second-strand synthesis, followed by PCR amplification. These methods have been successful in identifying many RNA viruses from human samples. Examples can be found in Victoria et al. (2009), Nakamura et al. (2009), and Jones et al. (2005). The method may be limited by PCR amplification bias, but it is highly sensitive.

Virus Purification and Phi29 Amplification

DNA viral metagenomes, including many phages, have been sequenced by purification of viral particles by CsCl density gradient centrifugation, DNase treatment, DNA isolation, and random amplification with Phi 29 DNA polymerase. Examples are respiratory tract metagenomes (mostly phages) from CF and non-CF subjects (Willner et al., 2009a) and an oropharyngeal metagenome from pooled samples from 19 healthy individuals (Willner et al., 2010). Limitations are potential amplification bias (Phi29 polymerase favors small circular and large linear genomes). This method has proved more successful for DNA than for RNA viruses.

Sequencing Methods

Due to the "untargeted" nature of metagenomics, and the often unavoidable contamination of viral nucleic acids with large amounts of human DNA, high-throughput sequencing has been essential. To date, the Roche/454 Life Sciences GS-FLX platform has been at the forefront of this technology, particularly because long sequence reads are necessary for shotgun sequencing. Sequencing technology is currently experiencing an unprecedented expansion, however, and it would not be surprising to see a series of further significant changes in sequencing methodology in the near future.

Bioinformatics

Data analysis is often the most challenging aspect of metagenomics research because the results are not pre-filtered by culturing or another selection process. The desired information must be extracted from a very large data set. Bioinformatics methods can be divided overall into two categories: similarity-based and similarity-independent approaches.

Similarity-Dependent Analyses

The original and more conventional means of sequence data analysis is to find segments of similarity to known sequences by searching databases. The most common tools are the various versions of BLAST (McGinnis and Madden, 2004), which will find local similarities based on the nucleic acid sequence or the deduced amino acid sequence. Microarray hybridization patterns have also been used to characterize novel viral nucleic acids (Urisman et al., 2005). These approaches are limited when the sample contains novel viruses that share little similarity with known viruses. Viruses in particular are subject to great variations in sequence composition. A large percentage of the sequences in a typical viral metagenome will not resemble any known sequences with any significance.

Metabolic Pathways

A metagenome can be characterized not only by taxonomy, but also by the cumulative metabolic potential encoded by the metagenome (Meyer et al., 2008). In the case of viral sequence data derived from lung sputum from CF patients and healthy subjects, the disease state of individuals correlated more strongly with the metabolic potential of viral metagenomes than with the taxonomic analysis (Willner et al., 2009a). In many cases the phage community appears to carry genes that complement the functions of the microbial community. In particular, phages often seem to use genes for proteins that will increase the short-term energy output of the host cells, either to increase viability (lysogeny) or to boost the production of viral particles (lytic). Some bacteria, such as cholera, are dependent on phage infection to achieve their virulence.

Similarity-Independent Analyses

More recently, similarity-independent methods have been developed that do not require database searches. For example, PHACCS (Angly et al., 2005) uses contig spectra derived from the sequence data to infer the diversity of genotypes present in the original sample. Other methods enable the comparison of one metagenome to another on the basis of relative abundance of shared sequences. These methods will not identify the unknown viruses, but they can help to characterize the sample by defining the overall complexity of the community. Other methods involve analysis based on the percent G/C content of genomes or the relative frequency of various dinucleotide combinations (Karlin et al., 1997; Burge et al., 1992; Karlin, 1998; Willner et al., 2009b), which in some cases is diagnostic of particular taxa.

Uncharacterized Viral Diversity

When viruses are purified from any human or environmental sample, the extracted DNA inevitably yields a large number of sequences (usually 70–99%) that show no significant similarity to any known sequences (Fig. 4.4) (Willner et al., 2009a, 2010; Jones et al., 2005).

Fig. 4.4 Unidentifiable sequences dominate the typical human viral metagenome. A similar phenomenon is observed in viral metagenomes from environmental samples

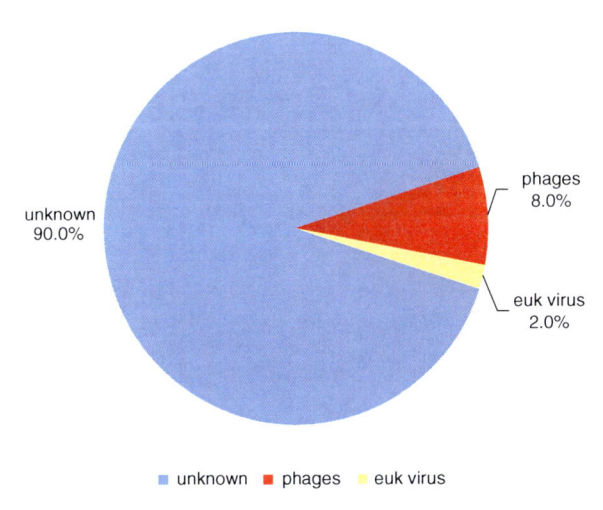

- unknown 90.0%
- phages 8.0%
- euk virus 2.0%

Provided that adequate precautions have been taken to avoid contamination with nonviral nucleic acids, this suggests that a very large fraction of the existing viral diversity remains uncharacterized. One of the strengths of the "untargeted" approach to viral metagenomics is that these sequences are obtained, but understanding the origin and significance of the "unknown" viral sequences is a substantial bioinformatic challenge that has yet to be solved. If a sequence has no similarity to the DNA of known organisms as defined by BLAST (McGinnis and Madden, 2004) or similar search algorithms, other methods must be developed for this purpose. For example, genome organization patterns such as large-scale arrangements of open reading frames or regulatory elements (promoters, enhancers, and origins of replication) may be signatures that would identify sequences as being of viral origin. This approach would likely require long sequences or even complete genomes to be successful.

Implications for Medical Care

An accurate assessment of the normal human virome provides a reference point from which to detect any novel viruses. This will serve as a background against which an emerging pathogen or bioterrorism agent would appear in the human population through suitable screening programs. The health of the human subject should be judged by variation from the true "community" that it is, not by the assumption that no nonhuman entities should be present. This is analogous to restoration of a disturbed ecosystem. Knowledge of the normal viral community and assessment of any perturbations found in patients may enable physicians to diagnose disturbances of the microbiome.

Glossary

amplification bias Inaccurate representation of the true relative abundances of genotypes in a DNA sample has been subjected to nonspecific amplification methods such as **MDA** or PCR.

BLAST (Basic Local Alignment Search Tool) An algorithm used to search nucleic acid and protein databases for sequences similar to a query sequence (McGinnis and Madden, 2004).

commensalism A form of symbiosis that benefits one partner while providing no apparent benefit to the other.

community A set of interacting populations in an ecosystem.

diversity A measure of the range of variation in a community, frequently represented as a combination of **richness** (number of variants) and **evenness** (skewness of the distribution).

emPCR PCR performed in a water-in-oil emulsion, so that each micelle functions as a microreactor containing a single amplicon.

evenness An index of the skewness of variation: an evenness value close to 0 implies that a community is dominated by one or very few members; a value of 1 implies equal abundance of every member.

genome The nucleic acid (DNA or RNA) that constitutes genetic information from a single organism.

genotype A genetic subtype that can be distinguished in a sample. In practical terms, two sequences will often be considered to legitimately represent the same genotype if they overlap at least 35 base pairs with 98% identity.

hybridization (molecular biology) The annealing of complementary single-stranded DNA or RNA.

MDA (multiple displacement amplification) DNA amplification using random primers in an isothermal reaction with a polymerase with helicase activity (Phi29 DNA polymerase), capable of nonspecific replication of double-stranded DNA.

metagenome The total genomic nucleic acid (DNA and/or RNA) derived from a community.

mutualism A form of symbiosis that benefits both partners.

population The total set of members of a genetically distinguishable species or genotypes in a defined biome.

sequence read A term frequently used to describe a sequence obtained by high-throughput methods

richness The total number of distinct species or genotypes that can be distinguished in a community.

Shannon–Wiener index One of the several measures of community diversity. A high value is associated with high richness and evenness values.

species A genomic subtype that constitutes a genetic lineage or **population** that exists in a sample or biome. Due to the genomic plasticity of viruses and microbes it can be challenging to define a species, hence the use of the term **genotype** in a DNA sample when species definition or identification is problematic.

symbiosis Any association between two organisms.

viremia The presence of viruses in the blood.

virome The cumulative viral community in an ecosystem.

References

Allander T, Tammi MT, Eriksson M, Bjerkner A, Tiveljung-Lindell A, Andersson B (2005) Cloning of a human parvovirus by molecular screening of respiratory tract samples. PNAS 102(36):12891–12896

Angly F, Rodriguez-Brito B, Bangor D, McNairnie P, Salamon P, Felts B, Nulton J, Mahaffy J, Rohwer F (2005) PHACCS, an online tool for estimating the structure and diversity of uncultured viral communities using metagenomic information. BMC Bioinformatics 2(6(1)):41

Angly FE, Felts B, Breitbart M, Salamon P, Edwards RA, Carlson C, Chan AM, Haynes M, Kelley S, Liu H, Mahaffy JM, Mueller JE, Nulton J, Olson R, Parsons R, Rayhawk S, Suttle CA, Rohwer F (2006) The marine viromes of four oceanic regions. PLoS Biol 4(11):2121–2131

Antonsson A, Forslund O, Ekberg H, Sterner G, Hansson BG (2000) The ubiquity and impressive genomic diversity of human skin papillomaviruses suggest a commensalic nature of these viruses. J Virol 74(24):11636–11641

Breitbart M, Rohwer F (2005) Method for discovering novel DNA viruses in blood using viral particle selection and shotgun sequencing. BioTechniques 39:729–736

Breitbart M, Rohwer F, Abedon ST (2005) Phage ecology and bacterial pathogenesis. In: Waldor MK, Friedman DI, Adhya SL (eds) Phages: their role in bacterial pathogenesis and biotechnolgy. ASM Press, Washington, DC, pp 66–92

Breitbart M, Haynes M, Kelley S, Angly F, Edwards RA, Felts B, Mahaffy JM, Mueller J, Nultonc J, Rayhawk S, Rodriguez-Brito B, Salamon P, Rohwer F (2008) Viral diversity and dynamics in an infant gut. Res Microbiol 159(5):367–373

Breitbart M, Hewson I, Felts B, Mahaffy JM, Nulton J, Salamon P, Rohwer F (2003) Metagenomic analyses of an uncultured viral community from human feces. J Bacteriol 185:6220–6223

Burge C, Campbell AM, Karlin S (1992) Over- and under-representation of short oligonucleotides in DNA sequences. PNAS 89(4):1358–1362

Chiu CY, Greninger AL, Kanada K, Kwok T, Fischer KF, Runckel C, Louie JK, Glaser CA, Yagi S, Schnurr DP, Haggerty TD, Parsonnet J, Ganem D, DeRisi JL (2008) Identification of

cardioviruses related to Theiler's murine encephalomyelitis virus in human infections. PNAS 105(37):14124–14129
Delwart EL (2007) Viral metagenomics. Rev Med Virol 17(2):115–131
Dethlefsen L, McFall-Ngai M, Relman DA (2007) An ecological and evolutionary perspective on human–microbe mutualism and disease. Nature 449:811–818
Furlan M (2009) Viral and microbial dynamics in the human respiratory tract. Biology. San Diego State University, San Diego, CA
Gill SR, Pop M, DeBoy RT, Eckburg PB, Turnbaugh PJ, Samuel BS, Gordon JI, Relman DA, Fraser-Liggett CM, Nelson KE (2006) Metagenomic analysis of the human distal gut microbiome. Science 312(5778):1355–1359
Grice E, Kong H, Renaud G, Young A, Bouffard G, Blakesley R, Wolfsberg T, Turner M, Segre J (2008) A diversity profile of the human skin microbiota. Genome Res 18(7):1043–1050
Grice E, Kong H, Conlan S, Deming C, Davis J, Young A, Bouffard G, Blakesley R, Murray P, Green E, Turner M, Segre J (2009) Topographical and temporal diversity of the human skin microbiome. Science 324(5931):1190–1192
Harris JK, Groote MAD, Sagel SD, Zemanick ET, Kapsner R, Penvari C, Kaess H, Deterding RR, Accurso FJ, Pace NR (2007) Molecular identification of bacteria in bronchoalveolar lavage fluid from children with cystic fibrosis. PNAS 104(51):20529–20533
Harrison F (2007) Microbial ecology of the cystic fibrosis lung. Microbiology 153(Part 4):917–923
Hendrix RW (2005) Bacteriophage evolution and the role of phages in host evolution. In: Waldor MK, Friedman DI, Adhya SL (eds) Phages: their role in bacterial pathogenesis and biotechnology. ASM Press, Washington, DC, pp 55–65
Hitch G, Pratten J, Taylor PW (2004) Isolation of bacteriophages from the oral cavity. Lett Appl Microbiol 39:215–219
Hyman RW, Fukushima M, Diamond L, Kumm J, Giudice LC, Davis RW (2005) Microbes on the human vaginal epithelium. PNAS 102(22):7952–7957
Jones MS, Kapoor A, Lukashov VV, Simmonds P, Hecht F, Delwart E (2005) New DNA viruses identified in patients with acute viral infection syndrome. J Virol 79:8230–8236
Karlin S (1998) Global dinucleotide signatures and analysis of genomic heterogeneity. Curr Opin Microbiol 1(5):598–610
Karlin S, Mrazek J, Campbell A (1997) Compositional biases of bacterial genomes and evolutionary implications. J Bacteriol 179(12):3899–3913
Kistler A, Avila PC, Rouskin S, Wang D, Ward T, Yagi S, Schnurr D, Ganem D, DeRisi JL, Boushey HA (2007) Pan-viral screening of respiratory tract infections in adults with and without asthma reveals unexpected human coronavirus and human rhinovirus diversity. J Infect Dis 196:817–825
Ksiazek TG, Erdman D, Goldsmith CS, Zaki SR, Peret T (2003) A novel coronavirus associated with severe acute respiratory syndrome. N Engl J Med 348:1953–1966
Kunin V, He S, Warnecke F, Peterson SB, Martin HG, Haynes M, Ivanova N, Blackall LL, Breitbart M, Rohwer F, McMahon KD, Hugenholtz P (2008) A bacterial metapopulation adapts locally to phage predation despite global dispersal. Genome Res 18:293–297
Letarov A, Kulikov E (2009) The bacteriophages in human- and animal body-associated microbial communities. J Appl Microbiol 107(1):1–13
Little JW (2005) Lysogeny, prophage induction, and lysogenic conversion. In: Waldor MK, Friedman DI, Adhya SL (eds) Phages: their role in bacterial pathogenesis and biotechnology. ASM Press, Washington, DC, pp 37–54
McGinnis S, Madden TL (2004) BLAST: at the core of a powerful and diverse set of sequence analysis tools. Nucleic Acids Res 32:W20–W25
Meyer F, Paarmann D, D'Souza M, Olson R, Glass EM, Kubal M, Paczian T, Rodriguez A, Stevens R, Wilke A, Wilkening J, Edwards RA (2008) The metagenomics RAST server – a public resource for the automatic phylogenetic and functional analysis of metagenomes. BMC Bioinformatics 9:386
Nakamura S, Yang C-S, Sakon N, Ueda M, Tougan T, Yamashita A, Goto N, Takahashi K, Yasunaga T, Ikuta K, Mizutani T, Okamoto Y, Tagami M, Morita R, Maeda N, Kawai J,

Hayashizaki Y, Nagai Y, Horii T, Iida T, Nakaya T (2009) Direct metagenomic detection of viral pathogens in nasal and fecal specimens using an unbiased high-throughput sequencing approach. PLoS One 4(1):e4219 [online only]

Okamoto H (2009) History of discoveries and pathogenicity of TT viruses. Curr Top Microbiol Immunol 331:1–201–220

Relman DA (2002) The human body as microbial observatory. Nat Genet 30:131–133

Reyes A, Haynes M, Hanson N, Angly FE, Heath AC, Rohwer F, Gordon J (2009) Phages in the distal human gut. Nature 466:334–338

Rivers TM (1937) Viruses and Koch's postulates. J Bacteriol 33(1):1–12

Rodriguez-Mueller B, Li LL, Wegley L, Furlan M, Angly F, Breitbart M, Buchanan J, Desnues C, Dinsdale E, Edwards R, Felts B, Haynes M, Liu H, Lipson D, Mahaffy J, Martin-Cuadrado AB, Mira A, Nulton J, Pasic L, Rayhawk S, Rodriguez-Mueller J, Rodriguez-Valera F, Salamon S, Thingstad TF, Tran T, Willner D, Youle M, Rohwer F (2010) Viral and microbial community dynamics in four aquatic environments. ISME J 4(6):739–751

Rogers GB, Carroll MP, Serisier DJ, Hockey PM, Jones G, Bruce KD (2004) Characterization of bacterial community diversity in cystic fibrosis lung infections by use of 16S ribosomal DNA terminal restriction fragment length polymorphism profiling. J Clin Microbiol 42(11): 5176–5183

Rohwer F (2003) Global phage diversity. Cell 113(2):141

Savage DC (1977) Microbial ecology of the gastrointestinal tract. Ann Rev Microbiol 31:107–133

Sharon I, Alperovitch A, Rohwer F, Haynes M, Glaser F, Atamna-Ismaeel N, Pinter RY, Partensky F, Koonin EV, Wolf YI, Nelson N, Béjà O (2009) Photosystem I gene cassettes are present in marine virus genomes. Nature 461:258–262

Stapleton JT, Williams CF, Xiang J (2004) GB virus type C: a beneficial infection? J Clin Microbiol 42(9):3915–3919

Turnbaugh PJ, Hamady M, Yatsunenko T, Cantarel BL, Duncan A, Ley RE, Sogin ML, Jones WJ, Roe BA, Affourtit JP, Egholm M, Henrissat B, Heath AC, Knight R, Gordon JI (2009) A core gut microbiome in obese and lean twins. Nature 457:480–484

Urisman A, Fischer KF, Chiu CY, Kistler AL, Beck S, Wang D, DeRisi JL (2005) E-Predict: a computational strategy for species identification based on observed DNA microarray hybridization patterns. Genome Biol 6: R78 [online only]

Victoria JG, Kapoor A, Li L, Blinkova O, Slikas B, Wang C, Naeem A, Zaidi S, Delwart E (2009) Metagenomic analyses of viruses in stool samples from children with acute flaccid paralysis. J Virol 83:4642–4651

Virgin HW, Wherry EJ, Ahmed R (2009) Redefining chronic viral infection. Cell 138:30–50

Wagner PL, Waldor MK (2002) Bacteriophage control of bacterial virulence. Infect Immun 70(8):3985–3993

Wang D, Coscoy L, Zylberberg M, Avila PC, Boushey HA, Ganem D, DeRisi JL (2002) Microarray-based detection and genotyping of viral pathogens. Proc Natl Acad Sci USA 99:15687–15692

Weinbauer MG (2006) Ecology of prokaryotic viruses. FEMS Microbiol Rev 28(2):127–181

Willner D, Furlan M, Haynes M, Schmieder R, Angly F, Silva J, Tammadoni S, Nosrat B, Conrad D, Rohwer F (2009a) Metagenomic analysis of respiratory tract DNA viral communities in cystic fibrosis and non cystic fibrosis individuals. PLoS One 4(10):e7370

Willner D, Furlan M, Schmieder R, Grasis J, Pride D, Relman D, Angly FE, McDole T, Mariella R, Rohwer F, Haynes M (2010) Metagenomic detection of phage-encoded platelet-binding factors in the human oral cavity. PNAS Early Edition.

Willner D, Thurber RV, Rohwer F (2009c) Metagenomic signatures of 86 microbial and viral metagenomes. Environ Microbiol 11(7):1752–1766

Wilson M (2005) Microbial inhabitants of humans: their ecology and role in health and disease. Cambridge University Press, New York, NY

Zhang T, Breitbart M, Lee WH, Run J-Q, Wei CL, Soh SWL, Hibberd ML, Liu ET, Rohwer F, Ruan Y (2005) RNA viral community in human feces: prevalence of plant pathogenic viruses. PLoS Biol 4(1):e3[online only]

Chapter 5
Selection and Sequencing of Strains as References for Human Microbiome Studies

Eline S. Klaassens, Mark Morrison, and Sarah K. Highlander

First- and Next-Generation Human Microbiome Projects

It has long been recognised that the human body is rapidly and extensively colonised by microbes soon after birth and that these host–microbe relationships are critical to health and well-being. Although these relationships have been extensively examined and characterised in relation to clinical and pathogenic microbiology, our understanding of the roles that "commensal" microorganisms might play in health and disease has been more difficult to reveal and understand. A key constraint was the limited abilities for research groups to isolate and culture fastidiously anaerobic microorganisms, which was not overcome until the late 1960s with the advent of the anaerobic culture techniques pioneered by Robert E. Hungate and Marvin P. Bryant. This breakthrough contributed shortly thereafter to what was really the first "human gut microbiome project," initiated in the USA at the Virginia Polytechnic Institute's anaerobe laboratory, with support from the US National Institutes of Health (NIH). Anyone with interest in the human microbiome project and especially gut microbiomes is encouraged to familiarise themselves with the detailed information provided in their "Anaerobe Laboratory Manual" (Anonymous, 1977); it still serves as an excellent reference list for many species of gut and oral bacteria, and also provides a wealth of information concerning media recipes and cultivation methods.

New opportunities are commensurate with stepwise advances in the technologies available to life scientists, and the continuing efficiency gains in cost and output being obtained in DNA sequencing have now presented the microbiological community with an unprecedented scope and detail of analysis of human microbiomes. Indeed the community is now faced with interesting and exciting challenges as a result of these technological advances. In order to facilitate the phylogenetic and functional analysis of the microbiomes associated with the human body, the Human Microbiome Project (HMP) was initiated with support from NIH

E.S. Klaassens (✉)
CSIRO Livestock Industries, Queensland Bioscience Precinct, 306 Carmody Road, St Lucia, QLD 4067, Australia
e-mail: eline.klaassens@csiro.au

K.E. Nelson (ed.), *Metagenomics of the Human Body*,
DOI 10.1007/978-1-4419-7089-3_5, © Springer Science+Business Media, LLC 2011

shortly after a workshop held in Washington DC in December 2007. The initial goal of the HMP was to sequence, or collect from publicly available sources, a total of at least 900 reference bacterial genomes, with relatively equal representation of the microbiomes associated with five body sites (the oral cavity, the gut, skin, nasopharynx, and the urogenital tract). The Washington DC workshop followed an organisational meeting in Europe 2 years earlier, which gave rise to the EU-funded project MetaHIT (Metagenomics of the Human Intestinal Tract) and also provided a forum for the presentation of other international initiatives in human microbiome research. The International Human Microbiome Consortium (IHMC; http://www.human-microbiome.org/) was formed in 2008. In addition to the leadership by US and European scientists, it includes scientific membership from Australia, Canada, China, Gambia, Japan, and Ireland.

Most IHMC member nations are contributing reference genomes to the project, with the majority of the DNA sequencing being undertaken by the J. Craig Venter Institute (JCVI), Baylor College of Medicine, Washington University and Broad Institute in North America, and the Sanger Institute in Europe. The data and annotations being produced for the microorganisms included in the projects are available through the NIH HMP Data Analysis and Coordination Centre (DACC) maintained by scientists at the University of Maryland, and at an equivalent centre maintained by the European Molecular Biology Laboratory (EMBL). The data will also be distributed to other public databases, including those supported by the National Centre for Biotechnology Information (NCBI) at the National Library of Medicine.

With this background, it is no longer a question of which microbes to prioritise for sequencing and analysis, but rather, to determine which microbes are missing or not represented, and how we can capture relevant genomic sequence data for these community members. This Chapter will provide an outline of the current approaches being employed within US and international initiatives to address these questions, will provide some suggestions in terms of what additional approaches might be employed to fully capture information pertaining to the human microbiome, and discuss opportunities that still remain open to the international community interested in strain selection, aimed at understanding and influencing the health implications associated with our microbiomes' interactions with the human host.

The Human Microbiome – Stability in Diversity

The characterisation of the human microbiome began in earnest in the late nineteenth century when Theodor Escherich isolated *Bacterium coli commune* from children's faeces, a bacterium that was later renamed *Escherichia coli* (Anonymous, 1981). Since then, and despite numerous cultivation studies, it is still widely accepted that the human microbiome is far from being fully described. While it is clear that there are site-specific communities on the skin and mucosal surfaces and in the intestinal tract, at lower levels of classification (phyla, families, and genera), there is also remarkable complexity and inter-subject variation when microbial communities are subjected to higher levels of classification (species – and with the advent of cultivation-independent approaches – operational taxonomic units [OTUs]). For instance, one study surveyed bacteria from up to 27 sites in seven to nine healthy adults on four occasions to elucidate the spatial and time variations in these bacterial communities (Costello et al., 2009). Each habitat harboured a characteristic microbiota and a relatively stable set of abundant taxa belonging to the Actinobacteria, Firmicutes, Proteobacteria, and Bacteroidetes phyla. However, higher taxonomic assignments demonstrated that there was a great degree of variability between individuals, at least in terms of the phylogenetic criteria used to measure and define bacterial diversity and complexity. Other studies that have examined the skin (Grice et al., 2009), the vagina (Hyman et al., 2005; Palmer et al., 2007), the oral cavity (Haffajee et al., 1998; Socransky et al., 1998; Haffajee et al., 2008) and the gastrointestinal tract (Eckburg et al., 2005; Gill et al., 2006; Dethlefsen et al., 2008) have given rise to similar findings. Such findings have also led to the consideration of whether there are "core microbiomes" associated with various sites of the human body. In studies of the human gut microbiome (principally via the analysis of stool samples), Tap et al. (2009) proposed that of the 3,180 OTUs identified in their studies, only 66 of these (~2%) were found in most of the subjects examined. Most of these "core" OTUs were also identified in other human gut microbiome datasets, produced in North America and Japan. The recent deep sequencing of metagenomic DNA recovered from the faecal samples of 124 individuals, as reported by Qin et al. (2010), expanded this core in relation to functional genes and bacterial species, with 75 being shared by more than 50% of the subjects examined, and 57 species common to more than 90% of the subjects studied. They further suggested that the gut microbiome of each individual is composed of at least 160 bacterial species but that the entire cohort of gut bacterial species probably does not exceed 1,150. Similar conclusions are likely to arise as next- and future-generation sequencing technologies are applied in a similar manner to the microbiomes within other body sites. These numbers fall well inside the current and projected capabilities worldwide dedicated to microbial genome sequencing projects related to the human microbiome.

Reference Genomes and Strain Selection

The current status of the major microbiome reference sequencing projects is illustrated in Fig. 5.1. More than 200 human gut bacterial genomes have now been sequenced, and the number from other body sites now approaches 500. Currently, the HMP reference genome catalogue (www.hmpdacc.org/reference_genomes.php/) includes over 1,000 strains, all of which are culturable, and sequencing is underway for more than 600 strains. Working groups comprised of international scientists with specific interests in specific body sites (oral cavity, skin and airways, gastrointestinal, and vaginal) coordinated the first rounds of strain selection for sequencing. These groups met by teleconference to identify strains and their sources for sequencing, based on a set of biological criteria determined by the

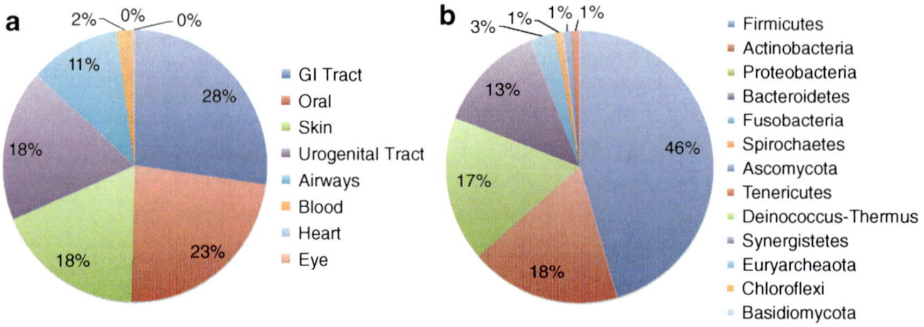

Fig. 5.1 (a) Distribution of HMP reference genomes in progress by body site. (**b**) Distribution of HMP reference genomes in progress by phylum

HMP Strains Working Group. These criteria are available online at the HMP DACC (hmp.dacc.org/doc/StrainSelection). Briefly, the criteria employed to date are:

(1) Phylogeny and uniqueness of the species to enable broad representation of as many lineages as possible, regardless of other criteria, and to provide improved scaffolding for the metagenomic data that are being produced.
(2) Established clinical significance. Any strain that has an established clinical significance to some health or disease condition should be included in the subset proposed to receive some level of improvement.
(3) Abundance (dominance) in a body site. Strains that are most abundant in a body site will contribute the largest number of metagenomic reads, so they should be sequenced. These are also likely to have a more significant impact on the metabolome of the community so they should be sequenced to finished status.
(4) Duplicate species but found in different body sites. Duplicate species found at different sites may have variable and interesting metabolic capabilities. An example is *Gardnerella vaginalis*, which has been collected from the vagina as well as the skin.
(5) Opportunity to explore pan-genomes. Isolates that have been closed as part of other sequencing projects may have been isolated from different environmental niches. The addition of closed isolates from the HMP can provide more information on the associated pan-genomes of such organisms.
(6) Strains having poor-quality draft assemblies that need improvement.

Despite these key considerations, it is important to consider that most of the OTUs identified via phylogenetic studies and many of the "species" identified from metagenomics studies are not yet cultured. As specific examples, only one-third of the "phylogenetic core" described by Tap et al. (2009) is currently represented by cultured bacteria, and only 12% of the sequence reads comprising the reference metagenome identified by Qin et al. (2010) can be assigned to the gut bacterial genomes so far sequenced. The following sections will provide an overview of approaches being employed to overcome these constraints.

Traditional Cultivation Methods

The roll tube techniques developed by Hungate in the 1950s and the development of anaerobic hoods and chambers (Aranki et al., 1969) vastly improved the cultivation of anaerobic bacteria from the gut and oral cavities in particular, and cultural counts relative to direct microscopic counts improved (Finegold et al., 1974; Moore and Holdeman, 1974). These methods remain somewhat cumbersome and time consuming, and compared to the PCR and shotgun sequencing approaches that have dominated microbial ecology research for the past two decades, there has been relatively little emphasis or interest in cultivation-based methods for bacterial isolation and the recovery of novel bacteria. However, microbiologists who have invested efforts in undertaking cultivation studies have been rewarded with the isolation of new organisms. For example, a variety of butyrate-producing bacteria belonging to the phylum Firmicutes (Barcenilla et al., 2000; Pryde et al., 2002) and bacteria belonging to the Bacteroidetes (Bakir et al., 2006; Hayashi et al., 2007a, b; Sakamoto et al., 2007) have now been isolated. The use of soft agar media as well as alternative carbon and nitrogen sources, such as mucins, led to the isolation of the first gastrointestinal tract representatives of the phylum Verrucomicrobia, *Akkermansia muciniphila*, and the Lentisphaerae, *Victivallis vadensis* (Zoetendal et al., 2003; Derrien et al., 2004). Interestingly, *A. mucinophila* is not only a specialist in mucin degradation, but it is also a relatively common and numerically abundant inhabitant of the human gut; its small size and specific carbon source requirements probably led to its recovery in earlier studies (Collado et al., 2007; Derrien et al., 2008). Similarly, *Bacteroides xylanisolvens* sp. nov. has now been proposed on the basis of the enrichment of six human gut bacterial isolates capable of using different xylans as their principal carbon source and their inability to use starch (Chassard et al., 2008). More recently, a high-throughput method to facilitate the detection of target bacteria by means of plate-wash PCR has been developed (Stevenson et al., 2004). Samples are plated in replicate on solid media and the colonies from one plate are then picked and propagated using the same media in a microtitre plate format, and then individual colonies are subjected to PCR. These methods resulted in the isolation of novel Acidobacteria and Verrucomicrobia and offer a time-effective way of screening literally thousands of colonies using a variety of different media.

Microscale Cultivation Methods

While novel bacterial isolates are still being obtained by the application of traditional cultivation techniques (Bakir et al., 2006; Mohan et al., 2006; Song et al., 2006), new technologies are also being developed that exploit microbeads or multiplexed solid surfaces and allow unconventional as well as very high-throughput culturing approaches. The microbead technique encapsulates bacterial cells within gel microdroplets suitable for massively parallel microbial cultivation under low nutrient flux conditions. As each cell is encapsulated, it can be cultured together with other organisms in a single vessel that allows the exchange of metabolites and other bioactive molecules. The microdroplets can be examined using flow cytometry or fluorescence in situ hybridization (FISH) methods and colony typing can be performed by PCR and sequencing of the 16S rRNA gene (Zengler et al., 2002; Ferrari et al., 2008; Alain and Querellou, 2009). Another promising application is micro-engineered mechanical systems (MEMS) that make complete systems-on-a-chip possible. The MEMS "micro-Petri" dish comprises a porous ceramic matrix subdivided into millions of compartments, in which cultures can be separately grown. This matrix reduces the need for gelling agents and assists with spent medium replacement and allows detection of growth by using microscale imaging detection (Ingham and van Hylckama Vlieg, 2008). Many otherwise intractable organisms undergo several rounds of division using these systems, even without forming visible colonies. Other methods to select the organism of interest involve use of substrate membrane systems as solid supports for growth and use of a sterile extract of environmental samples as a medium to mimic natural growth conditions (Ferrari et al., 2008). Such systems produce smaller colonies but increase recovered biodiversity, possibly due to the removal of inhibitors and prevention of overgrowth by bigger and faster growing organisms (Miteva and Brenchley, 2005). Addition of various electron acceptors and substrate gradients to the medium has allowed the isolation of novel bacteria from soil samples (Kopke et al., 2005).

Application of Sequencing from Single Cells and/or with Minute Quantities of DNA

The approaches described above offer useful ways to expand the collection of reference strains isolated from human microbiomes, though in many of these situations, the quality of cells and DNA recovered may not immediately facilitate more detailed biochemical or physiological studies. Nevertheless, the field of microbial genomics has benefited from new technologies that have made DNA sequencing per se cheaper and more productive and from methods that support the production of genome sequencing from minute quantities of DNA, including that from uncultured reference strains. Multiple displacement amplification (MDA) (Lasken, 2009) has been employed with single cells isolated from environmental microbiomes. MDA sequencing methods are rapidly improving and offer a high level of genome coverage (Ishoey et al., 2008; Rodrigue et al., 2009). These techniques have recently been used to sequence an uncultured commensal gut bacterium (Parkhill, pers. comm.). These methods have also been applied for the de novo genome assembly of the uncultured phytopathogen and insect symbiont, *Candidatus "Liberibacter asiaticus"* (Duan et al., 2009). Interestingly, a small amount of "metagenomic" DNA, which was dominated by this bacterium, but which included four other species of insect symbionts, was used as the template for this project, showing that MDA is not simply restricted in use to single-cell sequencing projects.

Reverse Isolation of Reference Strains Using (Meta)Genomic Data

Metagenomic data can also be used for the "reverse isolation" of novel reference species. High-throughput sequencing methods are now widely applied to DNA extracted from microbial communities and the metagenomic data can reveal key features of the community structure (Petrosino et al., 2009; Singh et al., 2009). In some instances, these data may also be used to support the enrichment and/or isolation of new microbes. For example, recent metagenomic studies of the foregut microbiome of the Tammar wallaby *(Macropus eugenii)* confirmed that novel not-yet-cultured clades of the Lachnospiraceae, Bacteroidales and Proteobacteriacea are present (Morrison, unpubl.). In collaboration with Alice McHardy's group (Max Planck Institute) we have used a combination of nucleotide composition-based sequence binning methods (Phylopythia) (McHardy et al., 2007) and fosmid libraries of metagenomic DNA to facilitate the assembly of more than 2 Mbp of genomic sequence data for the Proteobacterium (wallaby group 1 [WG-1]) and the Lachnospiraceae (WG-2) clades. The WG-1 assembly allowed the predictive reconstruction of key features of WG-1 metabolism, which were then used to devise cultural conditions that supported the purification of WG-1 and its whole genome sequencing (Pope et al., unpublished data).

Genome Finishing and Annotation

Currently, most reference genome sequencing is performed using the Roche 454 pyrosequencing platform. This requires high-quality, high-molecular-weight DNA to create 3 kbp paired-end libraries for subsequent emPCR and 454 sequencing. Four genomes can be sequenced using a single picotitre plate, resulting in 400–500 Mbp of sequence data or about 40-fold coverage for the "average" bacterial genome (ca. 3 Mb). The assembly of reads from a 3-kb paired-end library for a genome with moderate GC content (less than 50%) usually results in ca. 100 contigs forming tens of scaffolds. The increasing read lengths being achieved with the other next-generation sequencing platforms (e.g. Illumina and AB SOLiD) will certainly lead to changes in sequencing strategies in the near future, with the ultimate goal being a single scaffold from one sequencing run.

The criteria for finishing for the HMP and other microbial genome sequencing projects have been recently published (Chain et al., 2009). The 454 Newbler assembly is considered a "draft" sequence; each draft genome is subjected to one round of automated finishing, resulting in a "high-quality draft" sequence. This level of finishing is the default for HMP reference genomes. Of the 900 reference genomes that are targeted, about 85% will be sequenced to high-quality draft whereas the remaining 15% will be sequenced to further levels of finishing or improvement. Three additional levels of finishing were described. These are "improved high-quality draft," "non-contiguous finished" and "finished." Each of these levels requires manual finishing efforts that include sequence inspection and long- and short-range

PCR to span gaps. An improved high-quality draft sequence contains the complete sequence of at least one ribosomal (r)RNA operon, whereas a finished sequence contains full-length sequences of all rRNA operons in the strain. This latter level of finishing is very time consuming and is reserved for organisms of special interest.

The decision to elevate a strain from improved high-quality draft to one of the higher levels of finishing can be based on a number of criteria. For example, if a strain represents a new organism in a particular phylogeny, it may be of interest to have a fully finished sequence as a "gold standard." Strains with interesting biosynthetic or catabolic repertoires may also be good candidates for finishing or promotion from draft status. By contrast, species that have been sequenced multiple times (e.g. *Streptococcus pyogenes*) may be finished only to the high-quality-draft stage.

Genome annotation includes gene calling by any of a number of algorithms (e.g. Glimmer, GeneMark, RAST and Prodigal) (Borodovsky and McIninch, 1993; Delcher et al., 1999; Aziz et al., 2008) (http://compbio.ornl.gov/prodigal/). Gene calls can be further refined using other gene-finding pipelines such as GenePRIMP (http://geneprimp.jgi-psf.org/). Once open reading frames (ORFs) are established, the protein sequences can be annotated by a variety of methods, including BLAST (Altschul et al., 1990) versus the NCBI non-redundant database, nr, and other curated databases such as the Characterized Protein Database (CHAR at JCVI) and then by comparison using other annotation tools and databases such as the Conserved Domain Database (CDD) (Marchler-Bauer et al., 2005), the Kyoto Encyclopedia of Genes and Genomes (KEGG) (Kanehisa, 2002; Aoki-Kinoshita and Kanehisa, 2007), and Clusters of Orthologous Groups (COGs) (Tatusov et al., 2001, 2003). In the past, most annotation was performed manually or in a semi-automated method with significant manual curation. The JCVI has developed an automated annotation pipeline that is being used for HMP reference genomes. This pipeline includes BLAST searches, Hidden Markov Model searches against TIGRFAMs (Haft et al., 2003) and Pfam (Finn et al., 2008), and searches against PROSITE motifs (Bairoch, 1993; Mulder and Apweiler, 2007; Mulder et al., 2007), TMHMM (http://www.cbs.dtu.dk/services/TMHMM/), SignalP (Emanuelsson et al., 2007), LipoP (Juncker et al., 2003), PSORTb (Gardy et al., 2005) and other custom databases such as Transport DB (Ren et al., 2004). When possible, each protein is assigned a descriptive name, a gene symbol, and enzyme commission (EC) number (where applicable), a functional role category, gene ontology (GO) terms and GO-based evidence codes. Annotation also includes searches for RNA features such as tRNAs (Lowe and Eddy, 1997), rRNAs (Lagesen et al., 2007), ncRNAS and riboswitches (Rfam, http://www.sanger.ac.uk/Software/Rfam). CRISPR elements can be detected using the CRISPRfinder program (http://crispr.u-psud.fr/) (Grissa et al., 2007).

A critical component of development of a compendium of reference genome sequences is the availability of database resources for strain lists, sequencing status, DNA sequence, annotation, etc. Traditionally, single microbial genome sequences are deposited at NCBI. Reads are deposited to the Short Read Archive (SRA) and genomes with annotation are submitted to GenBank. In the EU, reads and genomes with annotation are submitted to EMBL-Bank. An important resource for the HMP has been the Genomes OnLine Database (GOLD) and the HMP-specific DACC. Both of these databases maintain lists of organisms that are finished or are in the process of being sequenced or are targeted for sequencing by a particular institution. The HMP DACC maintains significant detail about each project, including information on body site, collaborator, sequencing goals, sequencing statistics, sequencing status and links to the SRA and NCBI deposits. Both the DACC and GOLD maintain lists of ongoing international projects in addition to those in the US.

Prospects for the Future

The recent review by Kyrpides (2009) provides an excellent summary of the issues and advances likely to impact the fields of microbial genomics and metagenomics in the next decade. Some of the challenges raised include reference strain selection, maintenance of biological databases and their curation, and the establishment of standards supporting the exchange and integration of genome data. These have all been considered and addressed as part of the HMP and EU-MetaHIT projects. This leadership has also provided the guiding principles for the IHMC, and by doing so provides the scientific community interested in microbial biology and (meta)genomics with an inspirational target. It also provides scientists with interests in metagenomics of the human body with an excellent opportunity for pangenomic analyses. Creation and functional characterization of the pangenome of key reference species may be as important as trying to identify the "core microbiomes" associated with the human body. Pangenomics may lead to a more meaningful interpretation of epidemiological and clinical studies by revealing whether the functional diversity of reference strains varies for subjects of different cultural and/or ethnic backgrounds, as well as for persons genetically predisposed to disease (e.g. Crohn's disease or colorectal cancer). The IHMC provides an excellent opportunity for pangenomic assessment of human microbiomes and should be a core focus of its future efforts.

Concluding Remarks

Sequencing of reference genomes will continue to accelerate, driven by cultivation-dependent and -independent approaches that support a high-throughput screening for select strains, coupled with technologies that utilise minute quantities of DNA that can be recovered from small numbers of cells. Since the quantity of (meta)genomic DNA no longer appears to be rate limiting for producing draft genome sequence data, the need for strain purification and isolation might seem redundant. This might be true if the production of sequence data for human metagenomics is the final goal. However, it is not. Isolation of reference microbes remains relevant as long as one seeks to fully understand the structure–function relationships inherent to these microbiomes.

References

Alain K, Querellou J (2009) Cultivating the uncultured: limits, advances and future challenges. Extremophiles 13:583–594

Altschul SF, Gish W, Miller W, et al (1990) Basic local alignment search tool. J Mol Biol 215: 403–410

Anonymous (1977) Anaerobe laboratory manual. Anaerobe Laboratory, Blacksburg, VA.

Anonymous (1981) The pioneers of pediatric medicine: Teodor Escherich. Eur J Pediatr 137:131

Aoki-Kinoshita KF, Kanehisa M (2007) Gene annotation and pathway mapping in KEGG. Methods Mol Biol 396:71–91

Aranki A, Syed SA, Kenney EB et al (1969) Isolation of anaerobic bacteria from human gingiva and mouse cecum by means of a simplified glove box procedure. Appl Microbiol Biotechnol 17:568–576

Aziz RK, Bartels D, Best AA, et al (2008) The RAST server: rapid annotations using subsystems technology. BMC Genomics 9:75

Bairoch, A (1993) The PROSITE dictionary of sites and patterns in proteins, its current status. Nucleic Acids Res 21:3097–3103

Bakir MA, Kitahara M, Sakamoto M, et al (2006) *Bacteroides intestinalis* sp. nov., isolated from human faeces. Int J Syst Evol Microbiol 56:151–154

Barcenilla A, Pryde SE, Martin JC et al (2000) Phylogenetic relationships of butyrate-producing bacteria from the human gut. Appl Environ Microbiol 66:1654–1661

Borodovsky M, McIninch J (1993) GeneMark: parallel gene recognition for both DNA strands. Comput Chem 19:123–133

Chain PS, Grafham DV, Fulton RS et al (2009) Genomics. Genome project standards in a new era of sequencing. Science 326:236–237

Chassard C, Delmas E, Lawson PA et al (2008) *Bacteroides xylanisolvens* sp. nov., a xylan-degrading bacterium isolated from human faeces. Int J Syst Evol Microbiol 58:1008–1013

Collado MC, Derrien M, Isolauri E et al (2007) Intestinal integrity and *Akkermansia muciniphila*, a mucin-degrading member of the intestinal microbiota present in infants, adults, and the elderly. Appl Environ Microbiol 73:7767–7770

Costello EK, Lauber CL, Hamady M et al (2009) Bacterial community variation in human body habitats across space and time. Science 326:1694–1697

Delcher AL, Harmon D, Kasif S et al (1999) Improved microbial gene identification with GLIMMER. Nucleic Acids Res 27:4636–4641

Derrien M, Collado MC, Ben-Amor K et al (2008) The mucin degrader *Akkermansia muciniphila* is an abundant resident of the human intestinal tract. Appl Environ Microbiol 74: 1646–1648

Derrien M, Vaughan EE, Plugge CM et al (2004) *Akkermansia muciniphila* gen. nov., sp. nov., a human intestinal mucin-degrading bacterium. Int J Syst Evol Microbiol 54:1469–1476

Dethlefsen L, Huse S, Sogin ML et al (2008) The pervasive effects of an antibiotic on the human gut microbiota, as revealed by deep 16S rRNA sequencing. PLoS Biol 6:2383–2400

Duan Y, Zhou L, Hall DG et al (2009) Complete genome sequence of citrus huanglongbing bacterium, '*Candidatus* Liberibacter asiaticus' obtained through metagenomics. Mol Plant Microbe Interact 22:1011–1020

Eckburg PB, Bik EM, Bernstein CN, et al (2005) Diversity of the human intestinal microbial flora. Science 308:1635–1638

Emanuelsson O, Brunak S, von Heijne G et al (2007) Locating proteins in the cell using TargetP, SignalP and related tools. Nat Protoc 2:953–971

Ferrari BC, Winsley T, Gillings M et al (2008) Cultivating previously uncultured soil bacteria using a soil substrate membrane system. Nat Protoc 3:1261–1269

Finegold SM, Attebery HR, Sutter VL (1974) Effect of diet on human fecal flora: comparison of Japanese and American diets. Am J Clin Nutr 27:1456–1469

Finn RD, Tate J, Mistry J et al (2008) The Pfam protein families database. Nucleic Acids Res 36:D281–288

Gardy JL, Laird MR, Chen F et al (2005) PSORTb v.2.0: expanded prediction of bacterial protein subcellular localization and insights gained from comparative proteome analysis. Bioinformatics 21:617–623

Gill SR, Pop M, DeBoy RT et al (2006) Metagenomic analysis of the human distal gut microbiome. Science 312:1355–1359

Grice EA, Kong HH, Conlan S et al (2009) Topographical and temporal diversity of the human skin microbiome. Science 324:1190–1192

Grissa I, Vergnaud G, Pourcel C (2007) CRISPRFinder: a web tool to identify clustered regularly interspaced short palindromic repeats. Nucleic Acids Res 35:W52–57

Haffajee AD, Cugini MA, Tanner A et al (1998) Subgingival microbiota in healthy, well-maintained elder and periodontitis subjects. J Clin Periodontol 25:346–353

Haffajee AD, Socransky SS, Patel MR et al (2008) Microbial complexes in supragingival plaque. Oral Microbiol Immunol 23:196–205

Haft DH, Selengut JD, White O (2003) The TIGRFAMs database of protein families. Nucleic Acids Res 31:371–373

Hayashi H, Shibata K, Bakir MA et al (2007a) *Bacteroides coprophilus* sp. nov., isolated from human faeces. Int J Syst Evol Microbiol 57:1323–1326

Hayashi H, Shibata K, Sakamoto M et al (2007b) *Prevotella copri* sp. nov. and *Prevotella stercorea* sp. nov., isolated from human faeces. Int J Syst Evol Microbiol 57:941–946

Hyman RW, Fukushima M, Diamond L et al (2005) Microbes on the human vaginal epithelium. Proc Natl Acad Sci USA 102:7952–7957

Ingham CJ, van Hylckama Vlieg JE (2008) MEMS and the microbe. Lab Chip 8:1604–1616

Ishoey T, Woyke T, Stepanauskas R et al (2008) Genomic sequencing of single microbial cells from environmental samples. Curr Opin Microbiol 11:198–204

Juncker AS, Willenbrock H, Von Heijne G et al (2003) Prediction of lipoprotein signal peptides in Gram-negative bacteria. Protein Sci 12:1652–1662

Kanehisa M (2002) The KEGG database. Novartis Found Symp 247:91–101

Kopke B, Wilms R, Engelen B et al (2005) Microbial diversity in coastal subsurface sediments: a cultivation approach using various electron acceptors and substrate gradients. Appl Environ Microbiol 71:7819–7830

Kyrpides NC (2009) Fifteen years of microbial genomics: meeting the challenges and fulfilling the dream. Nat Biotechnol 27:627–632

Lagesen K, Hallin P, Rodland EA et al (2007) RNAmmer: consistent and rapid annotation of ribosomal RNA genes. Nucleic Acids Res 35:3100–3108

Lasken RS (2009) Genomic DNA amplification by the multiple displacement amplification (MDA) method. Biochem Soc Trans 37:450–453

Lowe TM, Eddy SR (1997) tRNAscan-SE: a program for improved detection of transfer RNA genes in genomic sequence. Nucleic Acids Res 25:955–964

Marchler-Bauer A, Anderson JB, Cherukuri PF et al (2005) CDD: a Conserved Domain Database for protein classification. Nucleic Acids Res 33:D192–196

McHardy AC, Martin HG, Tsirigos A et al (2007) Accurate phylogenetic classification of variable-length DNA fragments. Nat Methods 4:63–72

Miteva VI, Brenchley JE (2005) Detection and isolation of ultrasmall microorganisms from a 120,000-year-old Greenland glacier ice core. Appl Environ Microbiol 71: 7806–7818

Mohan R, Namsolleck P, Lawson PA et al (2006) *Clostridium asparagiforme* sp. nov., isolated from a human faecal sample. Syst Appl Microbiol 29:292–299

Moore WE, Holdeman LV (1974) Human fecal flora: the normal flora of 20 Japanese-Hawaiians. Appl Microbiol 27:961–979
Mulder N, Apweiler R (2007) InterPro and InterProScan: tools for protein sequence classification and comparison. Methods Mol Biol 396:59–70
Mulder NJ, Apweiler R, Attwood TK, et al (2007) New developments in the InterPro database. Nucleic Acids Res 35:D224–228
Palmer C, Bik EM, Digiulio DB et al (2007) Development of the human infant intestinal microbiota. PLoS Biol 5:e177
Petrosino JF, Highlander S, Luna RA et al (2009) Metagenomic pyrosequencing and microbial identification. Clin Chem 55:856–866
Pryde SE, Duncan SH, Hold GL, et al (2002) The microbiology of butyrate formation in the human colon. FEMS Microbiol Lett 217:133–139
Qin J, Li R, Raes J et al (2010) A human gut microbial gene catalogue established by metagenomic sequencing. Nature 464:59–65
Ren Q, Kang KH, Paulsen IT (2004) TransportDB: a relational database of cellular membrane transport systems. Nucleic Acids Res 32:D284–288
Rodrigue S, Malmstrom RR, Berlin AM, et al (2009) Whole genome amplification and de novo assembly of single bacterial cells. PLoS ONE 4:e6864
Sakamoto M, Kitahara M, Benno Y (2007) *Parabacteroides johnsonii* sp. nov., isolated from human faeces. Int J Syst Evol Microbiol 57:293–296
Singh J, Behal A, Singla N et al (2009) Metagenomics: concept, methodology, ecological inference and recent advances. Biotechnol J 4:480–494
Socransky SS, Haffajee AD, Cugini MA et al (1998) Microbial complexes in subgingival plaque. J Clin Periodontol 25:134–144
Song Y, Kononen E, Rautio M et al (2006) *Alistipes onderdonkii* sp. nov. and *Alistipes shahii* sp. nov., of human origin. Int J Syst Evol Microbiol 56:1985–1990
Stevenson BS, Eichorst SA, Wertz JT et al (2004) New strategies for cultivation and detection of previously uncultured microbes. Appl Environ Microbiol 70:4748–4755
Tap J, Mondot S, Levenez F et al (2009) Towards the human intestinal microbiota phylogenetic core. Environ Microbiol 11:2574–2584
Tatusov RL, Fedorova ND, Jackson JD et al (2003) The COG database: an updated version includes eukaryotes. BMC Bioinformatics 4:41
Tatusov RL, Natale DA, Garkavtsev IV et al (2001) The COG database: new developments in phylogenetic classification of proteins from complete genomes. Nucleic Acids Res 29:22–28
Zengler K, Toledo G, Rappe M et al (2002) Cultivating the uncultured. Proc Natl Acad Sci USA 99:15681–15686
Zoetendal EG, Plugge CM, Akkermans AD et al (2003) *Victivallis vadensis* gen. nov., sp. nov., a sugar-fermenting anaerobe from human faeces. Int J Syst Evol Microbiol 53:211–215

Chapter 6
The Human Vaginal Microbiome

Brenda A. Wilson, Susan M. Thomas, and Mengfei Ho

Humans live in association with abundant, complex, and dynamic microbial populations (the microbiome) that colonize many body sites, including the vaginal tract. Interactions between the host and the vaginal microbiota greatly affect women's health, where they often serve a protective role in maintaining vaginal health. Disruption of the microbial composition can lead to increased susceptibility to various urogenital diseases, including bacterial vaginosis (BV), vulvovaginal candidiasis (VVC), pelvic inflammatory disease (PID), and sexually transmitted diseases (STDs) such as infection with *Chlamydia*, *Trichomonas*, and human immunodeficiency virus (HIV). The composition of the vaginal microbiota also has a notable impact on pregnancy and neonatal outcome, including complications such as preterm labor and delivery. Understanding the composition and dynamics of the vaginal ecosystem as well as the involvement of metabolic and immunologic components is an area of increasing interest and research. However, unlike other body sites such as the gut and oral cavities, only limited information is available on the vaginal metagenome. In this chapter, we provide an overview of the data available to date on the microbial composition of the vaginal tract and discuss the importance of metagenomic data in understanding the relationship between the vaginal microbiome and women's health.

The Impact of Vaginal Infections on Women's Health

Vaginal infections such as BV and VVC afflict over 1 billion women annually (Allsworth and Peipert, 2007; Darroch and Frost, 1999; Fathalla, 2003; Honest et al., 2009, Muller et al., 1999, Oleen-Burkey and Hillier, 1995; Reid and Bruce, 2003; Reid et al., 2009; Van der Veen and Fransen, 1998). BV is the most common vaginal infection in women worldwide (Allsworth and Peipert, 2007; Darroch and Frost, 1999; Eschenbach, 1993; Fathalla, 2003; Gibbs, 2001; Hay, 2009; Oleen-Burkey

---

B.A. Wilson (✉)
Host-Microbe Systems Theme, Department of Microbiology, Institute for Genomic Biology, University of Illinois at Urbana-Champaign, 601 South Goodwin Ave., Urbana, IL 61801, USA
e-mail: bawilson@life.illinois.edu

K.E. Nelson (ed.), *Metagenomics of the Human Body*,
DOI 10.1007/978-1-4419-7089-3_6,

and Hillier, 1995; Steer, 2005; Verstraelen and Verhelst, 2009). The prevalence of BV is estimated to range from 10 to 40%, depending on the population studied (Allsworth and Peipert, 2007; Koumans et al., 2007; O'Brien, 2005; Sobel, 2005; Weir, 2004), with 50–80% of women being asymptomatically infected (McGregor and French, 2000). Although the episode cure rate is relatively high at 60–70% (Bunge et al., 2009; Eschenbach, 2007; Hay, 2009; Wilson, 2004), the rate of recurrence within 1 year is also high at 30–70% (Bradshaw et al., 2006a, b; Hay, 2009; Wilson, 2004).

Similar statistics can be found for fungal vaginitis (Fidel, 2004). Approximately three out of four women of childbearing age experience at least one episode of VVC, while 20% of healthy women are asymptomatically colonized with *Candida* (Sobel et al., 1998). Recurrent VVC occurs in about 5–10% of women (Fidel, 2005). *Candida albicans* is responsible for the majority of VVC cases, but other *Candida* species can also cause infection (Reid and Bruce, 2003; Reid et al., 2009). In many cases, BV is misdiagnosed as yeast vaginitis by the patient or even the physician and is treated empirically with over-the-counter antifungal drugs (Ferris et al., 1996, 2002).

BV as a Risk Factor for Urogenital Infections and STDs

BV is associated with urogenital infections, pelvic inflammatory disease (PID), ectopic pregnancy, tubal factor infertility, and increased susceptibility to STDs. Urinary tract infections (UTIs) afflict millions of women each year and cost healthcare providers billions of dollars (Foxman, 2003; Reid and Bruce, 2003; Reid et al., 2009). Women experiencing UTIs often exhibit prior colonization of the vagina by pathogenic bacteria (Stamey, 1973), highlighting the health consequences of disrupting the normally protective vaginal bacteria. Pregnant women with BV are also at increased risk for UTIs (Sharami et al., 2007). The cervical mucus and secretions, which normally protect against the spread of microbes to the upper genital tract, may be altered during ovulation or menses (Eschenbach et al., 2000; Morison et al., 2005; Schwebke et al., 1999), which in turn may predispose an individual to infection. It is still not clear how the vaginal microbiota changes during menses or how it is influenced by contraception or sexual practices. Damage to the cervix can lead to entry of organisms into the uterus and fallopian tubes. The resulting inflammatory response, termed PID, is the leading cause of infertility and ectopic pregnancy in American women (Mania-Pramanik et al., 2009; Paavonen, 1998; Soper et al., 1994; Sweet, 2000). Although sexually acquired organisms often cause the original damage, the vaginal microbiota clearly contributes to the overall infection (Paavonen, 1998; Schwebke and Weiss, 2002; Soper et al., 1994). Moreover, there is now strong evidence that the presence of BV or VVC, or the absence of vaginal lactobacilli, increases the risk of acquiring STDs such as HIV (Al-Harthi et al., 1999; Allsworth et al., 2008; Martin et al., 1999; Peipert et al., 2008; Schwebke, 2001, 2005; Sewankambo et al., 1997), cytomegalovirus (Ross et al., 2005), human papilloma virus (Allsworth et al., 2008; Mania-Pramanik et al., 2009; Murta et al., 2000), herpes simplex virus (Allsworth et al., 2008; Cherpes et al., 2008, 2003, 2006; Kaul et al., 2007), and other sexually acquired diseases (Cherpes et al., 2006; Yoshimura et al., 2009). The composition of the vaginal microbiota therefore has clear clinical impact on vaginal and reproductive health.

BV as a Risk Factor for Pregnancy Complications and Preterm Birth

In the United States alone, preterm birth (PTB) defined as birth before 37 weeks of gestation, is associated with an estimated annual societal cost of $26.2 billion (Melnyk and Feinstein, 2009), with a cost per case of $15,000–31,000 (Kirkby et al., 2007; Russell et al., 2007). The cost of known BV-related pregnancy complications in the United States alone is nearly $1 billion annually (Honest et al., 2009; Kekki et al., 2004; Muller et al., 1999; Myers, 2004; Oleen-Burkey and Hillier, 1995). Affecting nearly one in eight births in the United States, PTB is the single most significant factor contributing to neonatal morbidity and mortality (Ashton et al., 2009; Green et al., 2005; Williamson et al., 2008). One of the four major pathophysiologic pathways attributed to the outcome of PTB is maternal infection or abnormal vaginal microbial community composition (i.e., BV) and the associated inflammatory immune response (Ashton et al., 2009; Green et al., 2005). Women with BV have a two- to five-fold increased risk of PTB (Donders et al., 2009; Flynn et al., 1999; Srinivasan et al., 2009). This is supported by microbiologic evidence that infection contributes to 25% of preterm births, with bacterial colonization rates as high as 80% for birth at 23 weeks of gestation and declining to 11% for births at 31–34 weeks (Digiulio et al., 2010; Gibbs et al., 1982; Watts et al., 1992).

Mounting evidence points to BV as a leading cause of premature rupture of membranes, intrauterine growth restriction, intrauterine fetal demise, chorioamnionitis, endometritis, chronic pelvic pain, preterm labor and delivery, and postpartum infection (Darwish et al., 2007; Denney and Culhane, 2009; Donati et al., 2010; Donders et al., 2000; Flynn et al., 1999; Hay, 2009; Kekki et al., 2004; McGregor and French, 2000; Myers, 2004; Oleen-Burkey and Hillier, 1995; Srinivasan et al., 2009; Verstraelen and Verhelst, 2009). The role of BV in PTB is apparent from studies such as the recent meta-analysis that demonstrated that BV more than doubles the risk of preterm delivery in asymptomatic patients and in patients with symptoms of preterm labor, as well as increases the risk of late miscarriages and maternal infection in asymptomatic patients (Leitich and Kiss, 2007). Moreover, studies have shown specific associations between certain vaginal microbes linked to BV (e.g., *Gardnerella vaginalis*, *Atopobium vaginae*, or *Mycoplasma hominis*) and *Trichomonas vaginalis*, alone or together, and PTB (Bradshaw et al., 2006b; Denney and Culhane, 2009; Donati et al., 2010; Donders et al., 2000, 2009; Menard et al., 2008, 2010). BV has also been associated with shortening of the cervix in pregnancy, which is a known risk factor for preterm delivery (Donders et al., 2010).

A conundrum still exists, however, about the relationship between the vaginal microbiota and PTB, as treatment for BV with antimicrobials during pregnancy has had mixed results in reducing the risk of preterm delivery (Hendler et al., 2007; Kekki et al., 2004; Mitchell et al., 2009b; Okun et al., 2005; Simcox et al., 2007). In several studies, women with a history of preterm delivery were found to have a decreased risk of PTB with treatment of BV during early pregnancy, but in low risk women or women with asymptomatic BV treatment with antibiotics had no impact on the rate of PTB (Hendler et al., 2007; Kekki et al., 2004; Leitich et al., 2003). In other studies, treatment of BV with antibiotics had no impact on the incidence of PTB in either low- or high-risk women (Leitich et al., 2003; Mitchell et al., 2009b; Okun et al., 2005; Simcox et al., 2007). A Cochrane Review of 15 trials for the use of antibiotics to treat BV during pregnancy (McDonald et al., 2007) revealed that although antibiotic treatment was effective at eradicating BV during pregnancy, there was little evidence that treating women with asymptomatic BV had any impact on PTB or premature rupture of membranes. In women with previous history of PTB, there was some evidence that treatment might reduce the risk of premature membrane rupture and low birth weight. In women with BV, there was some evidence that treatment before 20 weeks of gestation may reduce the risk of PTB.

Thus, while BV occurs in 10–50% of pregnant women (Srinivasan et al., 2009), and depending on clinical setting, diagnostic criteria, and demographic factors, is clearly associated with PTB and adverse pregnancy outcome (Donati et al., 2010; Donders et al., 2009; O'Brien, 2005), the reason why BV causes preterm labor or birth and pregnancy complications in some women, but not in others, is still unknown. The mechanism by which BV may result in PTB is also still poorly understood. A hypothesis has been advanced that development of symptomatic BV and adverse sequelae in some individuals but not in others is due to alterations in host immune responses to the presence of certain vaginal microbial communities (Romero et al., 2004; Witkin et al., 2007b), and testing this model as discussed below is a promising area for future research.

Microbial Diversity in the Vaginal Tract

It is commonly believed that a predominance of lactobacilli in the vaginal microbiota is an indicator of health, whereas women with an altered composition of vaginal microbes, specifically a lack of lactobacilli, are more likely to exhibit symptomatic BV (Nugent et al., 1991) and are therefore at increased risk of sexually acquired diseases or pregnancy complications (Allsworth et al., 2008; Denney and Culhane, 2009; Donati et al., 2010; Hillier, 1998; Schwebke, 2005; Srinivasan et al., 2009). Results from previous culture-based methods suggest that *Lactobacillus* species are the most abundant bacterial species present in most healthy women. In one study (Antonio et al., 1999), lactobacilli were recovered from 215 (71%) of 302 women screened, with the predominant species being hydrogen-peroxide-producing

L. crispatus (32%) and *L. jensenii* (23%). A wide variety of other microbes were also present, albeit in much lower numbers. Interestingly, 29% of the women lacked culturable lactobacilli. Indeed, several studies have now shown that a significant number (10–40%) of women whose vaginal ecosystems are apparently healthy, usually defined as asymptomatic, lack appreciable numbers of lactobacilli (Antonio et al., 1999; Hyman et al., 2005; Kim et al., 2009; Larsen and Monif, 2001; Marrazzo et al., 2002; Monif and Carson, 1998; Zhou et al., 2004, 2007).

Recent molecular-based studies have allowed for more detailed examination of the vaginal microbiota and have clearly shown that identification and enumeration of vaginal microbes strictly by cultivation methods has provided only an incomplete and biased understanding of the vaginal microbiome. Indeed, studies performed using culture-independent molecular methods have revealed a vastly more complicated story for the vaginal ecosystem, with significant microbial diversity even within a healthy vaginal environment (Ferris et al., 2007; Fredricks et al., 2005; Hyman et al., 2005; Kim et al., 2009; Verhelst et al., 2004; Verstraelen et al., 2004; Zhou et al., 2004, 2007).

In one of the first of such culture-independent studies using 16S rRNA gene-based sequencing analysis to identify microbes present in the vaginal ecosystem, Burton et al. (2003), studied 19 premenopausal white women from Canada who had no symptoms or signs of vaginal or UTI and found that 79% of the women possessed *Lactobacillus* species, with 42% having a previously unreported strain of *L. iners*, which does not grow on common media selective for other *Lactobacillus* species (Falsen et al., 1999).

Zhou et al. (2004) examined five healthy white women between the ages of 28 and 44 from Idaho. Swabs were taken from the mid-vaginal region, 16S rRNA gene libraries from each sample were generated, and ~200 clones from each library were sequenced. They found that the diversity and types of bacteria that comprised the vaginal microbiota varied among the women. *Lactobacillus* species were the dominant vaginal microbes in four of the five women, with two of the women colonized mostly with *L. crispatus* and the other two with *L. iners*. *Atopobium vaginae* dominated the fifth woman's vaginal microbiome. *Atopobium*, *Megasphaera*, and *Leptotrichia* species, none of which had previously been reported as being present in a vaginal ecosystem, were also found in one of the women in addition to *Lactobacillus*. It was suggested that since these three species are lactic acid-producing bacteria, perhaps they were able to substitute for lactobacilli in maintaining the appropriate acidity of the vagina. Recently, Zhou et al. (2007) expanded their study to include 144 healthy Caucasian and black women. Using T-RFLP as well as 16S rRNA gene-based sequence analysis, they determined that the vaginal communities found in healthy women could be divided into eight distinct "supergroups" that differed in the type and abundance of bacterial species. *Lactobacillus* species were found to dominate in some subgroups, whereas other subgroups had few lactobacilli, but instead had *Atopobium,* streptococcal species, and other Firmicutes. Moreover, the rank abundance of the bacterial communities differed significantly between Caucasian and black women, leading the authors to propose that the differences in their vaginal microbial communities could in part account for disparities between the two racial groups in susceptibility to vaginal infections.

In a more in-depth sequencing study by Hyman et al. (2005), ~2,000 sequence-reads were collected per sample from 20 healthy premenopausal women (ages 27–44) from California. The subjects were at various self-reported times in their menstrual cycle. Sample collection was performed using a sterile cryoloop, which was passed across the vaginal epithelium in the posterior vaginal fornix. Four subjects had only lactobacilli. However, the lactobacilli were not clonal but instead a mixture of species and strains. Eight subjects had complex mixtures of *Lactobacillus* and other microbes, including *Bifidobacterium*, *Gardnerella*, *Atopobium*, *Corynebacterium*, and *Janthinobacterium*. Eight subjects had no *Lactobacillus*, but instead had *Bifidobacterium*, *Gardnerella*, *Gemella*, *Prevotella*, *Pseudomonas*, or *Streptococcus* as predominant bacteria. In several individuals, the predominant bacterial species were previously unidentified.

A recent study of eight healthy, asymptomatic individuals (ages 20–45) by Kim et al. (2009) aimed at evaluating the influence of sampling technique and anatomical site of sample collection on the composition of the vaginal microbiota revealed that there is considerable heterogeneity within an individual. In this study, six of the eight subjects had low Nugent scores indicative of non-BV status, whereas two individuals had high Nugent scores indicative of BV. Of the six non-BV individuals, two had microbiomes dominated by *Lactobacillus* (>90%), whereas the other four had microbiomes with mostly *Lactobacillus* (60–70%) and *Pseudomonas* (20–30%). Although the two individuals with high Nugent scores had some *Lactobacillus* present (20–50%), their microbiomes were much more diverse with higher proportions of *G. vaginalis* (30–40%), *Leptotrichia amnionii* (20%), and *Pseudomonas* species (20%). Of particular interest is the finding that the microbial community compositions of swab and scrape samples were considerably different from each other, with much greater proportion of *Pseudomonas* species present in the scrape samples. Inclusion of scrape samples significantly increased the diversity of the overall vaginal microbiome within an individual, suggesting that combining results for both overall swabs and scrapes might provide a better representation of the vaginal microbiome of an individual.

Taken together, these studies demonstrate that not only are the vaginal microbiomes among different healthy women highly varied, but also there is a striking diversity of microbiota present within individual women, suggesting that there may be no single definition of a healthy vaginal microbiota. *Lactobacillus* species are often, but not always, the major bacterial component present in the microbiomes of apparently healthy women. Indeed, *Lactobacillus* is not always found in a given sample. Other bacterial species often found include *Gardnerella*, *Atopobium*, *Prevotella*, *Streptococcus*, and *Pseudomonas*.

To date, Koch's postulates for establishing disease causation have not been satisfactorily fulfilled for any bacterium or group of bacteria associated with BV. Evidence so far suggests that BV is a complex condition that occurs as a result of a shift in the vaginal microbial community from the normally dominant *Lactobacillus* species to a polymicrobial community, with elevated concentrations of other aerobes and anaerobes (Donati et al., 2010; Larsson and Forsum, 2005; Nugent et al., 1991). One group of researchers (Verhelst et al., 2004; Verstraelen et al., 2004) sequenced 16S rRNA gene clones from eight healthy nonpregnant Belgian women. In four women, *Lactobacillus* species, *L. crispatus*, *L. gasseri*, or *L. iners*, accounted for 85–99% of the clones. Of the remaining four women with abnormal microbiota based on Gram-stain, one had predominantly *A. vaginae* with *L. iners*, two had equal portions of *A. vaginae* and *Prevotella* species, and one had mostly *Peptostreptococcus* with *Peptoniphilus*.

Using bacteria-specific PCR primers to study 73 subjects, 27 with BV and 46 without BV, Fredricks et al. (2005) identified a large number of new bacterial species in the women with BV. Oakley et al. (2008) compared the composition of the vaginal bacterial communities of 13 healthy subjects without BV and 28 subjects with BV. Sequencing of 16S rRNA gene libraries revealed that *Lactobacillus* species accounted for 86% of the sequences derived from healthy women. The profile of the vaginal microbiota from women with BV was more diverse and comprised a greater number of taxa. Although there was high variability in vaginal bacterial composition among subjects, the authors observed a strong association between BV and the presence of *Actinobacteria* and *Bacteriodetes*.

Two recent longitudinal studies of the dynamics of vaginal microbiota, one with seven nonpregnant African-American women over the course of one menstrual cycle (Wertz et al., 2008) and one with 100 pregnant women during the course of their pregnancy (Verstraelen et al., 2009a), showed that women without symptoms of BV had microbiomes dominated by lactobacilli that were stable over time, whereas those with BV had greater diversity and decreased stability in the composition of their microbiomes over time. Results from the Verstraelen et al. study (2009a) suggested that the presence of certain *Lactobacillus* species contributes to the microbial stability during pregnancy, with *L. crispatus* promoting a more stable, normal microbial composition, whereas *L. gasseri* and *L. iners* promoted an abnormal BV-like vaginal microbiota. Five of the seven women in the Wertz et al. study (2008) had low Nugent scores indicative of non-BV status, whereas two had high Nugent scores indicative of BV. Although *L. crispatus*, *L. gasseri*, and *L. iners* were found to dominate the microbiomes of the women without BV, only *L. iners* was detected in the microbiomes of the two women with BV. Significant portions of *L. amnionii*, *Prevotella buccalis*, and *Megasphaera* species were also found in the BV subjects, and the dominant species in each case varied over time.

The significant microbial diversity seen in these studies emphasizes that a single microbial taxonomic profile may not be useful in defining either a healthy or diseased vaginal state. However, while culture-independent methods have allowed for a less biased assessment of the vaginal ecosystem, our knowledge of the pathogenesis of vaginal infections like BV continues to be limited. Apart from two recent studies by Zhou et al. (2007) and Ryckman et al. (2009b) that examined racial differences in vaginal microbiomes and two studies that examined possible temporal effects (Verstraelen et al., 2009a; Wertz et al., 2008), few of these studies have rigorously addressed variables that may contribute to microbiome composition and diversity, including factors such as age, race/ethnicity, hormone fluctuations (stage of menstrual cycle), sexual practices, geographic location, or other demographic or medical conditions, including genetic or immunologic status. This suggests that the vaginal microbiome has not yet been adequately characterized to fully evaluate the role of the microbiome and its dynamics in health and disease.

Advances in DNA sequencing technologies have enabled a comprehensive examination of the genomes from entire microbial communities isolated directly from the host environment and have established a new field called metagenomics. Metagenomics, the study of metagenomes, employs high-throughput sequencing technologies, such as "shotgun" Sanger sequencing and/or massively parallel pyrosequencing platforms such as 454 or Illumina, to directly obtain genomic information from all members of a microbial community, without the biases often introduced through PCR and cloning of gene libraries. Consequently, metagenomics allows direct comparison of genetic content from samples, such that gene expression profiles and metabolic physiologic profiles can be analyzed in unprecedented detail. This, in turn, enables analysis and characterization of the myriad of microbial inter-species interactions and microbe–host interactions involved in healthy and diseased states. Elucidating the composition of the vaginal microbiome will thus be critical for understanding the role that the normal vaginal microbial communities play in health and disease and the impact of abnormal or perturbed microbiomes in disease progression, severity, and recovery.

The functional contribution of the entire microbial community needs to be considered as an important parameter in the etiology of disease. Metagenomic analysis of the vaginal environment would contribute greatly to our understanding of vaginal health and the changes that result in the shift to a disease status. However, such data are noticeably lacking for the vaginal system. Targeted pyrosequencing methods have been utilized to analyze microbial diversity in vaginal samples (Spear et al., 2008; Sundquist et al., 2007), but these studies have been limited to assessment of microbial composition based on the 16S rRNA gene sequence. Examination of the metabolic contribution of the vaginal microbiota as well as the immunologic response of the human host, as discussed further below, will greatly increase our understanding of the transition from a healthy to a diseased vaginal state. Backed by the wealth of taxonomic data already available about the vaginal microbiomes of healthy and diseased states, we are now poised to exploit the power of metagenomics to tackle the metabolic and immunologic aspects of these issues.

Effect of Antibiotic and Probiotic Treatment on Vaginal Microbiota

Although oral or vaginal metronidazole and clindamycin, sometimes in conjunction with erythromycin, are the treatments of choice for BV, cure rates vary widely, ranging from 60 to 70% for 1 month after treatment to 30–50% for 1 year after treatment (Bradshaw et al., 2006a; Bunge et al., 2009; Larsson and Forsum, 2005; Thulkar et al., 2010). Recurrence rates of BV are also high with as much as 70% recurrence documented 1 year after treatment with metronidazole or clindamycin

(Bradshaw et al., 2006a; Larsson and Forsum, 2005). A recent review of antibiotic therapy for BV reported on various in vitro and in vivo studies of the efficacy of antimicrobials in eradicating BV-associated bacteria (Sobel, 2009). *G. vaginalis*, *Mobiluncus* species, and *A. vaginae* were all found to have high metronidazole minimum inhibitory concentrations. Moreover, both *G. vaginalis* and *A. vaginae* are found in vaginal biofilms, which allow for some protection against antimicrobials. This is evident in the high (83%) recurrence rate of BV in women harboring both *G. vaginalis* and *A. vaginae* (Sobel, 2009).

Two recent studies have evaluated the efficacy of metronidazole in pregnancy and analyzed the effect of the treatment on microbiota present in the vaginal tract (Mitchell et al., 2009a, b). It was hypothesized that the failure of metronidazole to decrease PTB rates might result from the route of delivery of the treatment and that oral metronidazole treatment may not result in high enough concentrations of the antibiotic to treat fastidious bacteria found in the vaginal tract; however, the route of administration (oral versus vaginal) was not found to have a significant impact on the effect of metronidazole treatment on the bacterial content (Mitchell et al., 2009b).

Although oral or vaginal metronidazole treatment in early pregnancy decreased colonization with bacteria associated with BV, it was not effective in restoring the healthy levels of vaginal lactobacilli and it was not effective in reducing the rates of preterm delivery in women at low risk for PTB (Mitchell et al., 2009a). At the start of the study, 98% of the subjects were colonized with *G. vaginalis*, 73% with *Atopobium* species, 70% with *Megasphaera*, and 55% with *Leptotrichia/Sneathia*. The concentration of *Leptotrichia/Sneathia* decreased significantly following oral metronidazole treatment, but not with vaginal treatment. *Atopobium* species, which have been reported to be resistant to metronidazole in vitro, also decreased significantly with both oral and vaginal treatment. The authors suggest that these results show the inter-dependence of the vaginal bacterial community, with the decrease in *Atopobium* concentrations in these subjects possibly resulting from the dependence of this bacterium on the presence of other microbes for its survival. The concentration of *L. crispatus*, which is associated with a healthy vaginal state, did not change in either treatment group, suggesting a slow return of the normal vaginal flora.

Probiotics have been a more recent avenue for treatment and prevention of vaginal infections. The rationale behind probiotic use is that recolonization of the vagina with microbes normally associated with a healthy state will out-compete the pathogenic bacteria responsible for the diseased state and this, in turn, will allow for the return to a *Lactobacillus*-dominated healthy state. Probiotics have been used for the prevention and treatment of BV, UTI, and VVC (Reid et al., 2009). Several studies have shown cure of BV and VVC using a combination of metronidazole treatment along with administration of probiotics (Anukam et al., 2006; Martinez et al., 2009a, b). The success of this approach has been attributed to the observation that metronidazole treatment decreases the numbers of pathogenic bacteria while leaving the population of lactobacilli unaffected, thereby allowing the lactobacilli to recolonize the vagina. The species of *Lactobacillus* used for probiotic treatment have specific characteristics that enable them to out-compete pathogenic bacteria. These species are characterized by high adhesion to the vaginal epithelium; ability to produce lactic acid, bacteriocins, and hydrogen peroxide for antibiotic activity; production of biosurfactants; and ability to induce pathogen coaggregation (Hay, 2009).

Overall, antibacterial therapy of BV remains inadequate, largely owing to the uncertainty surrounding the bacterial species responsible for triggering BV and for causing the symptoms associated with BV. Understanding the community structure of the vaginal microbiota and the interactions between the various species, as well as deciphering the metabolic contributions of the different microbes to the vaginal environment, will aid in better targeting and treating vaginal infections.

The Role of Host Response in BV

Although the final outcome of a microbial infection likely depends on many factors, the host's genetic makeup is also thought to play a major role in predisposition or susceptibility to BV (Bailey et al., 2009; Cauci et al., 2007; Genc et al., 2007; Srinivasan et al., 2009; Verstraelen et al., 2009b). For example, studies have indicated that black women are more likely to have BV than are non-Hispanic white women and that known risk factors do not explain the observed racial disparities (Allsworth and Peipert, 2007; Cherpes et al., 2008; Ness et al., 2003; Paul et al., 2008; Peipert et al., 2008; Royce et al., 1999; Ryckman et al., 2009b; Trabert and Misra, 2007). Yeast vaginitis also affects black women disproportionately, with about 1 in 5 black women experiencing an episode of VVC in a given period compared to 1 in 10 white women (Foxman et al., 2000; Reid and Bruce, 2003).

Recent evidence suggests that there is a strong host immune response component to development of symptomatic BV (Beigi et al., 2007; Cauci, 2004; Cauci and Culhane, 2007; Hedges et al., 2006; Ryckman et al., 2009a; Witkin et al., 2007a, b; Xu et al., 2008), and several groups have begun to examine the inflammatory cytokine and chemokine responses to vaginal microbes (Beigi et al., 2007; Cauci and Culhane, 2007; Libby et al., 2008; Ryckman et al., 2009a; Srinivasan et al., 2009; Xu et al., 2008). Certain genetic markers indicative of differences in immune responses appear to correlate with predisposition to BV (Bailey et al., 2009; Beigi et al., 2007; Cauci, 2004; Cauci and Culhane, 2007; Genc et al., 2007; Ryckman et al., 2009a, c; Srinivasan et al., 2009; Verstraelen et al., 2009b; Witkin et al., 2007b). For instance, gene polymorphisms in TLRs and other molecular recognition receptors, which are involved in innate immune responses, have been identified in relation to the presence of BV in pregnant women (Genc et al., 2004a, 2007; Ryckman et al., 2009c; Verstraelen et al., 2009b). These findings are in agreement with observations that vaginal epithelial cells secrete IL-6 and IL-8 upon exposure to *G. vaginalis* or *A. vaginae*, but not to *L. crispatus*, and that this response was mediated by TLR-2 signaling (Libby et al., 2008). Host genetic variants at the IL-1b locus were found to predispose Caucasian nonpregnant women to BV, whereas a rare allele of the IL-1 receptor antagonist (IL-1ra) gene showed a trend toward protection against BV (Cauci et al., 2007). Again, these findings are consistent with other studies that showed significantly higher concentrations of IL-1β in vaginal washes of women with BV (Hedges et al., 2006). Proinflammatory cytokines IL-1b, IL-6, and IL-8 are significantly elevated in pregnant women with BV than in nonpregnant women (Beigi et al., 2007), particularly in pregnant women with *T. vaginalis* coinfection (Cauci and Culhane, 2007). In one study, pregnant women with one of two alleles of a polymorphism in the promoter region of the tumor necrosis factor α (TNFα) gene were shown to be at more than double the risk for premature membrane rupture and subsequent PTB, but those individuals with the "susceptible" TNFα genotype and with symptomatic BV were at even greater risk for PTB (Macones et al., 2004).

Immunologic Impact on the Vaginal Microbiome

Similar to the gut, the vaginal tract consists of a layer of epithelial cells covered with mucus, which is the first point of contact between the vaginal tract and microbes. This site is protected by a number of specialized molecules of the innate immune system, including mannose-binding lectin, defensins, lysozyme, and nitric oxide (Johansson and Lycke, 2003; Witkin et al., 2007a). The stratified squamous epithelial cells of the vaginal tract express a number of pattern recognition receptors including TLRs 1–9, which recognize bacterial components and mediate antibacterial responses (Wira et al., 2005). Vaginal epithelial cells are also known to produce powerful nonspecific antimicrobials including defensins, which bind the surface of bacteria and disrupt the bacterial membrane resulting in cell lysis (Witkin et al., 2007a). Mannose-binding lectin (MBL) is another antimicrobial protein found in vaginal secretions (Witkin et al., 2007a). MBL binds mannose of the bacterial surface, which leads to the activation of the complement system and subsequent bacterial lysis. Heat shock protein 70 (Hsp70) is an additional antimicrobial protein found in the vaginal tract (Giraldo et al., 1999a, b, c). Extracellular Hsp70 defends the vaginal tract from pathogens by binding to TLRs and inducing an immune response against the pathogens (Genc et al., 2005). Further studies have also suggested that nitric oxide, a potent antimicrobial agent, is released in response to extracellular Hsp70 (Genc et al., 2006).

Unlike other mucosal sites, the genital tract lacks organized lymphoid follicles and does not appear to possess inductive sites for adaptive immunity (Russell and Mestecky, 2002). Within the genital mucosa, classical antigen-presenting cells such as macrophages and dendritic cells can be found. Additionally, epithelial cells of the genital tract have been reported to possess antigen-presenting ability (Wira et al., 2005). The lack of organized immune architecture and the presence of soluble immunoglobulin G (IgG) and immunoglobulin A (IgA) antibodies, as opposed to dimeric sIgA, suggests a lack of local adaptive responses and significant involvement of serum, and therefore, systemic immunity (Johansson and Lycke, 2003; Russell and Mestecky, 2002). The rectum, small intestine, regional lymph nodes, and perhaps also the nasal cavity have all been implicated as sources of antibody-producing B cells. Cell-mediated responses in the form of effector CD4 and CD8 T cells are also observed following infection with sexually transmitted pathogens. The true origins of these effector B and T cells in the genital tract remain somewhat enigmatic.

Little is known about the nature of early innate immune responses that take place during inflammation in the genital tract. A wide range of cytokine concentrations is found in vaginal fluids, not only in women who exhibit BV, but also in healthy subjects (Cauci et al., 2002a, b; Johansson and Lycke, 2003; Russell and Mestecky, 2002; Yudin et al., 2003). A handful of previous studies have correlated cervical and vaginal cytokine expression with a specific disease state. For example, vaginal fluid from women with BV exhibit increased concentrations of pro-inflammatory cytokines, such as IL-1 in the cervical mucus (Basso et al., 2005; Mattsby-Baltzer et al., 1998; Valore et al., 2002, 1998). Additionally, among women colonized with *G. vaginalis* or anaerobic Gram-negative bacteria such as *Prevotella,* elevated pro inflammatory IL-1β and lower anti-inflammatory IL-1 receptor antagonist (IL-1ra) are observed in vaginal samples (Genc et al., 2004b; Hedges et al., 2006). The increased vaginal IL-1β observed in BV is not paralleled by an increase in IL-8 levels, suggesting that BV-associated factors specifically dampen IL-8. The impairment of an IL-8 increase may explain the absence of increased neutrophils in most women exposed to vaginal colonization with abnormal anaerobic bacteria. Neither altered levels of TNFα nor IL-6 appear to be associated with BV. Moreover, concentrations of IL-1β, IL-6, and IL-8 in the cervix were significantly reduced in women after successful treatment of BV (Yudin et al., 2003). Increases in vaginal pH, considered a hallmark of chronic infections including conditions of BV, are positively associated with higher IL-10 and IL-12 concentrations in cervical secretions (Gravitt et al., 2003). Thus, as might be expected, a number of innate immune and inflammatory markers exhibit correlates with the condition of BV.

These observations have led some researchers to advance a hypothesis that alterations in innate immunity with changes in proinflammatory cytokine levels trigger the transition from healthy vaginal microbiota to the development of symptomatic BV (Witkin et al., 2007b). However, the nature and type of the inflammatory response observed in this condition is not clearly understood. Moreover, although clinically suggestive, no studies to date have attempted to systematically define correlates between the phylogenetic composition of the vaginal microbiota and any specific immune response markers.

Immunoglobulins in the cervical mucus provide a potentially important barrier function (Johansson and Lycke, 2003). Antibody-producing B cells are present principally not only in the endocervix but also in the vagina. Both IgG and IgA antibodies are locally produced in the vagina, and it is possible to identify antibodies in cervico-vaginal secretions that are not detectable in peripheral blood (Witkin et al., 2007a). As observed with cytokines, total IgG and total IgA in lavage samples vary greatly among subjects even after correction for total protein (Snowhite et al., 2002). This observation is important, as one of the major functions of secreted antibody is the control of commensal microbes through inhibition of adherence and uptake and neutralization of adhesins and toxins.

Microbial Metabolic Function and Vaginal Health

As discussed above, several mechanisms are in place to prevent the colonization of the vaginal tract by pathogenic bacteria. These include maintenance of a low vaginal pH, presence of the cervicovaginal mucus and the vaginal mucosal barrier, and secretion of antimicrobial substances by the host (such as nitric oxide, defensins, or other antimicrobial peptides) as well as by the normal resident bacteria (such as hydrogen peroxide, hemolysins, mucolytic enzymes, or lytic phage) (Famularo et al., 2001; Roberton et al., 2005; Wiggins et al., 2001). Defense mechanisms also include coaggregation, and bacteria to bacteria signaling (Reid et al., 2009). Each of these protective strategies requires specific functional capabilities by the colonizing microbes, and changes in the metabolic potential of the vaginal microbial populations will clearly impact the dynamics of the vaginal environment.

Lactic acid, produced from glycogen by the metabolism of *Lactobacillus*, is a significant source of acidity in the vaginal tract and is thought to be a major deterrent to colonization by pathogens (Atassi et al., 2006a, b, c; Boris and Barbes, 2000; Boskey et al., 2001, 1999; Juarez Tomas et al., 2003; Pybus and Onderdonk, 1999). The inhibitory effect of lactic acid was demonstrated in a study by McLean et al. (McLean and McGroarty, 1996), in which the in vitro growth of *G. vaginalis* was inhibited when co-cultured with lactobacilli. A low pH and presence of lactic acid were found to account for 60–95% of the inhibitory activity. Juarez Tomas et al. (2003), also reported on the growth inhibition of various pathogenic bacteria by lactic acid-producing *L. acidophilus*. Lactic acid production is not limited to *Lactobacillus*, but is also found in *Atopobium, Leptotrichia*, and *Megasphaera* (Rodriguez Jovita et al., 1999; Witkin et al., 2007b; Zhou et al., 2004). Therefore, it is possible that lactic acid-producing bacteria other than *Lactobacillus* might be able to serve this function in a microbial community.

Hydrogen peroxide, another antimicrobial substance produced by *Lactobacillus,* inhibits the growth of microbes such as *G. vaginalis* and *Bacteroides,* which lack the hydrogen peroxide-scavenging enzyme catalase (Al-Mushrif and Jones, 1998; Boris and Barbes, 2000; Eschenbach et al., 1989; Klebanoff et al., 1991; Vallor et al., 2001). Several studies have reported on the antimicrobial effects of hydrogen peroxide-producing lactobacilli isolated from healthy women. Eschenbach et al. (1989) found hydrogen peroxide-producing lactobacilli present in 27 out of 28 (96%) healthy women, but in only 4 of 67 (6%) women with BV. Al-Mushrif and Jones (1998) reported similar results, but found that 75% of the lactobacilli isolated from the healthy individuals produced hydrogen peroxide, compared with 13% isolated from BV patients. Atassi et al. (2006b) attributed the *in vitro* killing activity of *G. vaginalis* and *P. bivia* by several *Lactobacillus* strains, including *L. acidophilus*, *L. gasseri*, *L. jensenii*, and *L. crispatus*, primarily to the production of hydrogen peroxide, and not low pH or the presence of lactic acid alone. More recently, Atassi and Servin (2010) demonstrated that for hydrogen peroxide-producing *Lactobacillus* strains, the main metabolites lactic acid and hydrogen peroxide act co-operatively to kill enteric, BV-associated, and uropathogenic pathogens.

The vaginal mucosal barrier is another major impediment to colonization by harmful microbes (Cauci, 2004; Cole, 2006; Culhane et al., 2006; Forsum et al., 2005; Johansson and Lycke, 2003; Russell and Mestecky, 2002; Witkin et al., 2000, 2007b). Mucins are the major components of mucus, and many bacterial species possess mucinases, hydrolyzing enzymes with the ability to degrade mucins (Howe et al., 1999; McGregor et al., 1994; Wiggins et al., 2001). Mucolytic enzymes have been established as virulence factors in many pathogenic bacteria, e.g., *Shigella* sp., *Pseudomonas aeruginosa*, and group B *Streptococcus* (Brown and Straus, 1987; Chiarini et al., 1989; Haider et al., 1993). The presence of mucolytic enzymes has also been described in several BV-associated microbes, including *Prevotella* species, *Mobiluncus* species, and *G. vaginalis* (Briselden et al., 1992; Cauci et al., 1998, 2002c, 1996; McGregor et al., 1994; Roberton et al., 2005). However, mucolytic activity is not necessarily restricted to pathogenic bacteria. Even some *Lactobacillus* species produce mucinases, and these enzymes may contribute to the normal mucin regulation in the vagina (Wiggins et al., 2001, 2002). Indeed, there is some evidence that mucinase activity is not necessarily correlated with BV (Wiggins et al., 2002).

Interference with the local host immune response is a proposed mechanism employed by vaginal microbiota in the pathogenesis of BV (Cauci, 2004; Cauci et al., 1998; Forsum et al., 2005; Genc et al., 2006, 2004a, b; Witkin et al., 2000, 2007a, b). Along with mucinases, other enzymes such as sialidases, glycosidases, sulphatases, and proteases contribute to the impairment of the host defense component of the cervicovaginal mucus (Cauci et al., 1998, 2002c, 2005; Howe et al., 1999; McGregor et al., 1994; Roberton et al., 2005; Wiggins et al., 2001). McGregor et al. (1994) showed that sialidase activity was present mainly in BV-associated microbes such as *G. vaginalis* and *Mobiluncus* species. Sialidases cleave sialic acid residues of IgA molecules, leaving them susceptible to degradation by proteases. Cauci et al. (1998) found high levels of sialidase activity in cultures of *G. vaginalis*, which in turn was correlated with a lack of an IgA immune response to the cytolysin toxin produced by the microbe. Cauci et al. (2002c) also found much higher levels of both sialidase and prolidase in women with BV than in women with healthy vaginal microbiota. This same group later found that a combination of high vaginal pH (≥ 5) with sialidase and prolidase activities were predictors of PTB (Cauci et al., 2005). More recently, the group found that a strong induction of IL-1β in women with BV was correlated with production of sialidase and prolidase by BV-associated bacteria (Cauci et al., 2008). The presence of another new mucolytic enzyme, glycosulfatase, was also demonstrated in women with BV (Roberton et al., 2005). Glycosulfatase activity was correlated with sialidase activity in these women, and it was suggested that both enzymes might contribute to the pathogenesis of BV.

Other metabolic processes correlated with the presence of BV are also important to consider. The malodor (fishy odor) characteristic of BV has been attributed to the presence of amine-containing metabolites with several diagnostic tests based on the detection of these metabolites (Chaim et al., 2003; Cook et al., 1992; Famularo et al., 2001; Jones et al., 1994; O'Dowd et al., 1996; Rodrigues et al., 1999; Wolrath et al., 2002, 2001, 2005). However, amines and amine-producing microbes have been found in asymptomatic women as well (Fredricks et al., 2005; Hyman et al.,

2005; Kubota et al., 1995; Verhelst et al., 2004; Verstraelen et al., 2004; Witkin et al., 2007a; Zhou et al., 2004), bringing into question the association between BV and amine production. It will be important for proper diagnosis of BV to determine whether microbes present in a healthy vagina also produce amine-containing metabolites that may cause malodor.

Future Prospectus

Clearly, accessing tangible connections between the vaginal microbiota and their metabolic potential and the changes that occur during progression from a healthy state to a diseased state such as in BV has been challenging. Yet, understanding the metabolic contributions of vaginal microbiota will be critical to deciphering the pathophysiology of BV and its role in susceptibility to STDs and PTB and other pregnancy complications. A limitation of the approach of taxonomic profiling of vaginal microbiomes to assess healthy versus diseased states is its failure to generate functional genomic information for determining the metabolic contributions of microbes to the vaginal ecosystem. Metagenomic studies using recently available sequencing technologies enable us to take advantage of the plethora of complete genome sequences acquired through the NIH-sponsored Human Microbiome Project and the associated tools to discover novel genes involved in metabolic functions and to survey the structure and function of complex microbial communities for their metabolic potential.

The microbes present in the vaginal tract undoubtedly vary in their metabolic properties and therefore contribute differently to the status of the vaginal environment of a healthy individual versus that of an individual predisposed to disease or already suffering from disease. Metagenomic analyses of vaginal samples from non-BV and BV patients will provide unprecedented levels of information on the microbial diversity and metabolic capabilities of microbial communities present under different environmental conditions. Importantly, the capacity to assign metabolic pathways based on the presence of their corresponding genes in the composite microbiomes will enable evaluation of changes accompanying onset of disease. This will in turn provide a tremendous contribution to improving the management and treatment of complex vaginal diseases such as BV and WC and their roles in susceptibility to STDs and PTB.

References

Al-Harthi L, Roebuck KA, Olinger GG, Landay A, Sha BE, Hashemi FB, Spear GT (1999) Bacterial vaginosis-associated microflora isolated from the female genital tract activates HIV-1 expression. J Acquir Immune Defic Syndr 21:194–202

Al-Mushrif S, Jones BM (1998) A study of the prevalence of hydrogen peroxide-generating lactobacilli in bacterial vaginosis: the determination of H_2O_2 concentrations generated, in vitro, by isolated strains and the levels found in vaginal secretions of women with and without infection. J Obstet Gynaecol 18:63–67

Allsworth JE, Lewis VA, Peipert JF (2008) Viral sexually transmitted infections and bacterial vaginosis: 2001–2004 National health and nutrition examination survey data. Sex Transm Dis 35:791–796

Allsworth JE, Peipert JF (2007) Prevalence of bacterial vaginosis:2001–2004 National health and nutrition examination survey data. Obstet Gynecol 109:114–120

Antonio MA, Hawes SE, Hillier SL (1999) The identification of vaginal lactobacillus species and the demographic and microbiologic characteristics of women colonized by these species. J Infect Dis 180:1950–1956

Anukam K, Osazuwa E, Ahonkhai I, Ngwu M, Osemene G, Bruce AW, Reid G (2006) Augmentation of antimicrobial metronidazole therapy of bacterial vaginosis with oral probiotic *Lactobacillus rhamnosus* GR-1 and *Lactobacillus reuteri* RC-14: randomized, double-blind, placebo controlled trial. Microbes Infect 8:1450–1454

Ashton DM, Lawrence HC 3rd, Adams NL 3rd, Fleischman AR (2009) Surgeon general's conference on the prevention of preterm birth. Obstet Gynecol 113:925–930

Atassi F, Brassart D, Grob P, Graf F, Servin AL (2006a) In vitro antibacterial activity of *Lactobacillus helveticus* strain KS300 against diarrhoeagenic, uropathogenic and vaginosis-associated bacteria. J Appl Microbiol 101:647–654

Atassi F, Brassart D, Grob P, Graf F, Servin AL (2006b) Lactobacillus strains isolated from the vaginal microbiota of healthy women inhibit *Prevotella bivia* and *Gardnerella vaginalis* in coculture and cell culture. FEMS Immunol Med Microbiol 48:424–432

Atassi F, Brassart D, Grob P, Graf F, Servin AL (2006c) Vaginal lactobacillus isolates inhibit uropathogenic *Escherichia coli*. FEMS Microbiol Lett 257:132–138

Atassi F, Servin AL (2010) Individual and co-operative roles of lactic acid and hydrogen peroxide in the killing activity of enteric strain *Lactobacillus johnsonii* NCC933 and vaginal strain *Lactobacillus gasseri* KS120.1 against enteric, uropathogenic and vaginosis-associated pathogens. FEMS Microbiol Lett 304:29–38

Bailey RL, Natividad-Sancho A, Fowler A, Peeling RW, Mabey DC, Whittle HC, Jepson AP (2009) Host genetic contribution to the cellular immune response to *Chlamydia trachomatis*: heritability estimate from a gambian twin study. Drugs Today (Barc) 45 (Suppl B): 45–50

Basso B, Gimenez F, Lopez C (2005) IL-1beta, IL-6 and IL-8 levels in gyneco-obstetric infections. Infect Dis Obstet Gynecol 13:207–211

Beigi RH, Yudin MH, Cosentino L, Meyn LA, Hillier SL (2007) Cytokines, pregnancy, and bacterial vaginosis: comparison of levels of cervical cytokines in pregnant and non-pregnant women with bacterial vaginosis. J Infect Dis 196:1355–1360

Boris S, Barbes C (2000) Role played by lactobacilli in controlling the population of vaginal pathogens. Microbes Infect 2:543–546

Boskey ER, Cone RA, Whaley KJ, Moench TR (2001) Origins of vaginal acidity: high D/L lactate ratio is consistent with bacteria being the primary source. Hum Reprod 16:1809–1813

Boskey ER, Telsch KM, Whaley KJ, Moench TR, Cone RA (1999) Acid production by vaginal flora in vitro is consistent with the rate and extent of vaginal acidification. Infect Immun 67:5170–5175

Bradshaw CS, Morton AN, Hocking J, Garland SM, Morris MB, Moss LM, Horvath LB, Kuzevska I, Fairley CK (2006a) High recurrence rates of bacterial vaginosis over the course of 12 months after oral metronidazole therapy and factors associated with recurrence. J Infect Dis 193: 1478–1486

Bradshaw CS, Tabrizi SN, Fairley CK, Morton AN, Rudland E, Garland SM (2006b) The association of *Atopobium vaginae* and *Gardnerella vaginalis* with bacterial vaginosis and recurrence after oral metronidazole therapy. J Infect Dis 194:828–836

Briselden AM, Moncla BJ, Stevens CE, Hillier SL (1992) Sialidases (neuraminidases) in bacterial vaginosis and bacterial vaginosis-associated microflora. J Clin Microbiol 30:663–666

Brown JG, Straus DC (1987) Characterization of neuraminidases produced by various serotypes of group B streptococci. Infect Immun 55:1–6

Bunge KE, Beigi RH, Meyn LA, Hillier SL (2009) The efficacy of retreatment with the same medication for early treatment failure of bacterial vaginosis. Sex Transm Dis 36:711–713

Burton JP, Cadieux PA, Reid G (2003) Improved understanding of the bacterial vaginal microbiota of women before and after probiotic instillation. Appl Environ Microbiol 69:97–101

Cauci S (2004) Vaginal immunity in bacterial vaginosis. Curr Infect Dis Rep 6:450–456

Cauci S, Culhane JF (2007) Modulation of vaginal immune response among pregnant women with bacterial vaginosis by *Trichomonas vaginalis*, *Chlamydia trachomatis*, *Neisseria gonorrhoeae*, and yeast. Am J Obstet Gynecol 196:133 e1–137

Cauci S, Culhane JF, Di Santolo M, McCollum K (2008) Among pregnant women with bacterial vaginosis, the hydrolytic enzymes sialidase and prolidase are positively associated with interleukin-1beta. Am J Obstet Gynecol 198:132 e1–137

Cauci S, Di Santolo M, Casabellata G, Ryckman K, Williams SM, Guaschino S (2007) Association of interleukin-1beta and interleukin-1 receptor antagonist polymorphisms with bacterial vaginosis in non-pregnant italian women. Mol Hum Reprod 13:243–250

Cauci S, Driussi S, Guaschino S, Isola M, Quadrifoglio F (2002a) Correlation of local interleukin-1beta levels with specific IgA response against *Gardnerella vaginalis* cytolysin in women with bacterial vaginosis. Am J Reprod Immunol 47:257–264

Cauci S, Driussi S, Monte R, Lanzafame P, Pitzus E, Quadrifoglio F (1998) Immunoglobulin a response against *Gardnerella vaginalis* hemolysin and sialidase activity in bacterial vaginosis. Am J Obstet Gynecol 178:511–515

Cauci S, Guaschino S, Driussi S, De Santo D, Lanzafame P, Quadrifoglio F (2002b) Correlation of local interleukin-8 with immunoglobulin a against *Gardnerella vaginalis* hemolysin and with prolidase and sialidase levels in women with bacterial vaginosis. J Infect Dis 185: 1614–1620

Cauci S, Hitti J, Noonan C, Agnew K, Quadrifoglio F, Hillier SL, Eschenbach DA (2002c) Vaginal hydrolytic enzymes, immunoglobulin a against *Gardnerella vaginalis* toxin, and risk of early preterm birth among women in preterm labor with bacterial vaginosis or intermediate flora. Am J Obstet Gynecol 187:877–881

Cauci S, McGregor J, Thorsen P, Grove J, Guaschino S (2005) Combination of vaginal pH with vaginal sialidase and prolidase activities for prediction of low birth weight and preterm birth. Am J Obstet Gynecol 192:489–496

Cauci S, Scrimin F, Driussi S, Ceccone S, Monte R, Fant L, Quadrifoglio F (1996) Specific immune response against *Gardnerella vaginalis* hemolysin in patients with bacterial vaginosis. Am J Obstet Gynecol 175:1601–1605

Chaim W, Karpas Z, Lorber A (2003) New technology for diagnosis of bacterial vaginosis. Eur J Obstet Gynecol Reprod Biol 111:83–87

Cherpes TL, Hillier SL, Meyn LA, Busch JL, Krohn MA (2008) A delicate balance: risk factors for acquisition of bacterial vaginosis include sexual activity, absence of hydrogen peroxide-producing lactobacilli, black race, and positive herpes simplex virus type 2 serology. Sex Transm Dis 35:78–83

Cherpes TL, Meyn LA, Krohn MA, Lurie JG, Hillier SL (2003) Association between acquisition of herpes simplex virus type 2 in women and bacterial vaginosis. Clin Infect Dis 37:319–325

Cherpes TL, Wiesenfeld HC, Melan MA, Kant JA, Cosentino LA, Meyn LA, Hillier SL (2006) The associations between pelvic inflammatory disease, *Trichomonas vaginalis* infection, and positive herpes simplex virus type 2 serology. Sex Transm Dis 33:747–752

Chiarini F, Mastromarino P, Orsi GB, Riscaldati T (1989) Adhesiveness of *Pseudomonas aeruginosa* to rabbit vesical mucosa effect of glycosidases on cellular binding. Ann Ig 1:399–408

Cole AM (2006) Innate host defense of human vaginal and cervical mucosae. Curr Top Microbiol Immunol 306:199–230

Cook RL, Redondo-Lopez V, Schmitt C, Meriwether C, Sobel JD (1992) Clinical, microbiological, and biochemical factors in recurrent bacterial vaginosis. J Clin Microbiol 30: 870–877

Culhane JF, Nyirjesy P, McCollum K, Goldenberg RL, Gelber SE, Cauci S (2006) Variation in vaginal immune parameters and microbial hydrolytic enzymes in bacterial vaginosis positive pregnant women with and without *Mobiluncus* species. Am J Obstet Gynecol 195: 516–521

Darroch JE, Frost JJ (1999) Women's interest in vaginal microbicides. Fam Plann Perspect 31: 16–23

Darwish A, Elnshar EM, Hamadeh SM, Makarem MH (2007) Treatment options for bacterial vaginosis in patients at high risk of preterm labor and premature rupture of membranes. J Obstet Gynaecol Res 33:781–787

Denney JM, Culhane JF (2009) Bacterial vaginosis: a problematic infection from both a perinatal and neonatal perspective. Semin Fetal Neonatal Med 14:200–203

Digiulio DB, Romero R, Kusanovic JP, Gomez R, Kim CJ, Seok KS, Gotsch F, Mazaki-Tovi S, Vaisbuch E, Sanders K, Bik EM, Chaiworapongsa T, Oyarzun E, Relman DA (2010) Prevalence and diversity of microbes in the amniotic fluid, the fetal inflammatory response, and pregnancy outcome in women with preterm pre-labor rupture of membranes. Am J Reprod Immunol 64:38–57

Donati L, Di Vico A, Nucci M, Quagliozzi L, Spagnuolo T, Labianca A, Bracaglia M, Ianniello F, Caruso A, Paradisi G (2010) Vaginal microbial flora and outcome of pregnancy. Arch Gynecol Obstet 281:589–600

Donders GG, Van Bulck B, Caudron J, Londers L, Vereecken A, Spitz B (2000) Relationship of bacterial vaginosis and mycoplasmas to the risk of spontaneous abortion. Am J Obstet Gynecol 183:431–437

Donders GG, Van Calsteren C, Bellen G, Reybrouck R, Van den Bosch T, Riphagen I, Van Lierde S (2010) Association between abnormal vaginal flora and cervical length as risk factors for preterm birth. Ultrasound Obstet Gynecol (in press)

Donders GG, Van Calsteren K, Bellen G, Reybrouck R, Van den Bosch T, Riphagen I, Van Lierde S (2009) Predictive value for preterm birth of abnormal vaginal flora, bacterial vaginosis and aerobic vaginitis during the first trimester of pregnancy. BJOG 116:1315–1324

Eschenbach DA (1993) History and review of bacterial vaginosis. Am J Obstet Gynecol 169: 441–445

Eschenbach DA (2007) Bacterial vaginosis: resistance, recurrence, and/or reinfection? Clin Infect Dis 44:220–221

Eschenbach DA, Davick PR, Williams BL, Klebanoff SJ, Young-Smith K, Critchlow CM, Holmes KK (1989) Prevalence of hydrogen peroxide-producing lactobacillus species in normal women and women with bacterial vaginosis. J Clin Microbiol 27:251–256

Eschenbach DA, Thwin SS, Patton DL, Hooton TM, Stapleton AE, Agnew K, Winter C, Meier A, Stamm WE (2000) Influence of the normal menstrual cycle on vaginal tissue, discharge, and microflora. Clin Infect Dis 30:901–907

Falsen E, Pascual C, Sjoden B, Ohlen M, Collins MD (1999) Phenotypic and phylogenetic characterization of a novel lactobacillus species from human sources: description of *Lactobacillus iners* sp. nov. Int J Syst Bacteriol 49 Pt 1:217–221

Famularo G, Perluigi M, Coccia R, Mastroiacovo P, De Simone C (2001) Microecology, bacterial vaginosis and probiotics: perspectives for bacteriotherapy. Med Hypotheses 56: 421–430

Fathalla MF (2003) Vaginal microbicides: a priority need for women's health. Curr Womens Health Rep 3:263–264

Ferris DG, Dekle C, Litaker MS (1996) Women's use of over-the-counter antifungal medications for gynecologic symptoms. J Fam Pract 42:595–600

Ferris DG, Nyirjesy P, Sobel JD, Soper D, Pavletic A, Litaker MS (2002) Over-the-counter antifungal drug misuse associated with patient-diagnosed vulvovaginal candidiasis. Obstet Gynecol 99:419–425

Ferris MJ, Norori J, Zozaya-Hinchliffe M, Martin DH (2007) Cultivation-independent analysis of changes in bacterial vaginosis flora following metronidazole treatment. J Clin Microbiol 45:1016–1018

Fidel PL Jr (2004) History and new insights into host defense against vaginal candidiasis. Trends Microbiol 12:220–227

Fidel PL Jr (2005) Immunity in vaginal candidiasis. Curr Opin Infect Dis 18:107–111

Flynn CA, Helwig AL, Meurer LN (1999) Bacterial vaginosis in pregnancy and the risk of prematurity: a meta-analysis. J Fam Pract 48:885–892

Forsum U, Holst E, Larsson PG, Vasquez A, Jakobsson T, Mattsby-Baltzer I (2005) Bacterial vaginosis–a microbiological and immunological enigma. Apmis 113:81–90

Foxman B (2003) Epidemiology of urinary tract infections: incidence, morbidity, and economic costs. Dis Mon 49:53–70

Foxman B, Barlow R, D'Arcy H, Gillespie B, Sobel JD (2000) Candida vaginitis: self-reported incidence and associated costs. Sex Transm Dis 27:230–235

Fredricks DN, Fiedler TL, Marrazzo JM (2005) Molecular identification of bacteria associated with bacterial vaginosis. N Engl J Med 353:1899–1911

Genc MR, Delaney ML, Onderdonk AB, Witkin SS (2006) Vaginal nitric oxide in pregnant women with bacterial vaginosis. Am J Reprod Immunol 56:86–90

Genc MR, Karasahin E, Onderdonk AB, Bongiovanni AM, Delaney ML, Witkin SS (2005) Association between vaginal 70-kd heat shock protein, interleukin-1 receptor antagonist, and microbial flora in mid trimester pregnant women. Am J Obstet Gynecol 192: 916–921

Genc MR, Vardhana S, Delaney ML, Onderdonk A, Tuomala R, Norwitz E, Witkin SS (2004a) Relationship between a toll-like receptor-4 gene polymorphism, bacterial vaginosis-related flora and vaginal cytokine responses in pregnant women. Eur J Obstet Gynecol Reprod Biol 116:152–156

Genc MR, Vardhana S, Delaney ML, Witkin SS, Onderdonk AB (2007) TNFA-308G>A polymorphism influences the TNF-alpha response to altered vaginal flora. Eur J Obstet Gynecol Reprod Biol 134:188–191

Genc MR, Witkin SS, Delaney ML, Paraskevas LR, Tuomala RE, Norwitz ER, Onderdonk AB (2004b) A disproportionate increase in IL-1beta over IL-1ra in the cervicovaginal secretions of pregnant women with altered vaginal microflora correlates with preterm birth. Am J Obstet Gynecol 190:1191–1197

Gibbs RS (2001) Impact of infectious diseases on women's health: 1776–2026. Obstet Gynecol 97:1019–1023

Gibbs RS, Blanco JD, St Clair PJ, Castaneda YS (1982) Quantitative bacteriology of amniotic fluid from women with clinical intraamniotic infection at term. J Infect Dis 145:1–8

Giraldo P, Neuer A, Korneeva IL, Ribeiro-Filho A, Simoes JA, Witkin SS (1999a) Vaginal heat shock protein expression in symptom-free women with a history of recurrent vulvovaginitis. Am J Obstet Gynecol 180:524–529

Giraldo P, Neuer A, Ribeiro-Filho A, Linhares I, Witkin SS (1999b) Detection of the human 70-kD and 60-kD heat shock proteins in the vagina: relation to microbial flora, vaginal pH, and method of contraception. Infect Dis Obstet Gynecol 7:23–25

Giraldo PC, Ribeiro-Filho AD, Simoes JA, Neuer A, Feitosa SB, Witkin SS (1999c) Circulating heat shock proteins in women with a history of recurrent vulvovaginitis. Infect Dis Obstet Gynecol 7:128–132

Gravitt PE, Hildesheim A, Herrero R, Schiffman M, Sherman ME, Bratti MC, Rodriguez AC, Morera LA, Cardenas F, Bowman FP, Shah KV, Crowley-Nowick PA (2003) Correlates of IL-10 and IL-12 concentrations in cervical secretions. J Clin Immunol 23:175–183

Green NS, Damus K, Simpson JL, Iams J, Reece EA, Hobel CJ, Merkatz IR, Greene MF, Schwarz RH (2005) Research agenda for preterm birth: recommendations from the march of dimes. Am J Obstet Gynecol 193:626–635

Haider K, Hossain A, Wanke C, Qadri F, Ali S, Nahar S (1993) Production of mucinase and neuraminidase and binding of shigella to intestinal mucin. J Diarrhoeal Dis Res 11:88–92

Hay P (2009) Recurrent bacterial vaginosis. Curr Opin Infect Dis 22:82–86

Hedges SR, Barrientes F, Desmond RA, Schwebke JR (2006) Local and systemic cytokine levels in relation to changes in vaginal flora. J Infect Dis 193:556–562

Hendler I, Andrews WW, Carey CJ, Klebanoff MA, Noble WD, Sibai BM, Hillier SL, Dudley D, Ernest JM, Leveno KJ, Wapner R, Iams JD, Varner M, Moawad A, Miodovnik M, O'Sullivan MJ, Van Dorsten PJ (2007) The relationship between resolution of asymptomatic bacterial vaginosis and spontaneous preterm birth in fetal fibronectin-positive women. Am J Obstet Gynecol 197:488 e1–5

Hillier SL (1998) The vaginal microbial ecosystem and resistance to HIV. AIDS Res Hum Retroviruses 14 (Suppl 1):S17–21

Honest H, Forbes CA, Duree KH, Norman G, Duffy SB, Tsourapas A, Roberts TE, Barton PM, Jowett SM, Hyde CJ, Khan KS (2009) Screening to prevent spontaneous preterm birth: systematic reviews of accuracy and effectiveness literature with economic modelling. Health Technol Assess 13:1–627

Howe L, Wiggins R, Soothill PW, Millar MR, Horner PJ, Corfield AP (1999) Mucinase and sialidase activity of the vaginal microflora: implications for the pathogenesis of preterm labour. Int J STD AIDS 10:442–447

Hyman RW, Fukushima M, Diamond L, Kumm J, Giudice LC, Davis RW (2005) Microbes on the human vaginal epithelium. Proc Natl Acad Sci USA 102:7952–7957

Johansson M, Lycke NY (2003) Immunology of the human genital tract. Curr Opin Infect Dis 16:43–49

Jones BM, al-Fattani M, Gooch H (1994) The determination of amines in the vaginal secretions of women in health and disease. Int J STD AIDS 5:52–55

Juarez Tomas MS, Ocana VS, Wiese B, Nader-Macias ME (2003) Growth and lactic acid production by vaginal *Lactobacillus acidophilus* CRL 1259, and inhibition of uropathogenic *Escherichia coli*. J Med Microbiol 52:1117–1124

Kaul R, Nagelkerke NJ, Kimani J, Ngugi E, Bwayo JJ, Macdonald KS, Rebbaprgada A, Fonck K, Temmerman M, Ronald AR, Moses S (2007) Prevalent herpes simplex virus type 2 infection is associated with altered vaginal flora and an increased susceptibility to multiple sexually transmitted infections. J Infect Dis 196:1692–1697

Kekki M, Kurki T, Kotomaki T, Sintonen H, Paavonen J (2004) Cost-effectiveness of screening and treatment for bacterial vaginosis in early pregnancy among women at low risk for preterm birth. Acta Obstet Gynecol Scand 83:27–36

Kim TK, Thomas SM, Ho M, Sharma S, Reich CI, Frank JA, Yeater KM, Biggs DR, Nakamura N, Stumpf R, Leigh SR, Tapping RI, Blanke SR, Slauch JM, Gaskins HR, Weisbaum JS, Olsen GJ, Hoyer LL, Wilson BA (2009) Heterogeneity of vaginal microbial communities within individuals. J Clin Microbiol 47:1181–1189

Kirkby S, Greenspan JS, Kornhauser M, Schneiderman R (2007) Clinical outcomes and cost of the moderately preterm infant. Adv Neonatal Care 7:80–87

Klebanoff SJ, Hillier SL, Eschenbach DA, Waltersdorph AM (1991) Control of the microbial flora of the vagina by H_2O_2-generating lactobacilli. J Infect Dis 164:94–100

Koumans EH, Sternberg M, Bruce C, McQuillan G, Kendrick J, Sutton M, Markowitz LE (2007) The prevalence of bacterial vaginosis in the United States, 2001–2004; associations with symptoms, sexual behaviors, and reproductive health. Sex Transm Dis 34:864–869

Kubota T, Sakae U, Takeuchi H, Usui M (1995) Detection and identification of amines in bacterial vaginosis. J Obstet Gynaecol (Tokyo 1995) 21:51–55

Larsen B, Monif GR (2001) Understanding the bacterial flora of the female genital tract. Clin Infect Dis 32:e69–77

Larsson PG, Forsum U (2005) Bacterial vaginosis–a disturbed bacterial flora and treatment enigma. Apmis 113:305–316

Leitich H, Brunbauer M, Bodner-Adler B, Kaider A, Egarter C, Husslein P (2003) Antibiotic treatment of bacterial vaginosis in pregnancy: a meta-analysis. Am J Obstet Gynecol 188:752–758

Leitich H, Kiss H (2007) Asymptomatic bacterial vaginosis and intermediate flora as risk factors for adverse pregnancy outcome. Best Pract Res Clin Obstet Gynaecol 21:375–390

Libby EK, Pascal KE, Mordechai E, Adelson ME, Trama JP (2008) *Atopobium vaginae* triggers an innate immune response in an in vitro model of bacterial vaginosis. Microbes Infect 10:439–446

Macones GA, Parry S, Elkousy M, Clothier B, Ural SH, Strauss JF 3rd (2004) A polymorphism in the promoter region of TNF and bacterial vaginosis: preliminary evidence of gene–environment interaction in the etiology of spontaneous preterm birth. Am J Obstet Gynecol 190:1504–1508: Discussion 3A

Mania-Pramanik J, Kerkar SC, Salvi VS (2009) Bacterial vaginosis: a cause of infertility? Int J STD AIDS 20:778–781

Marrazzo JM, Koutsky LA, Eschenbach DA, Agnew K, Stine K, Hillier SL (2002) Characterization of vaginal flora and bacterial vaginosis in women who have sex with women. J Infect Dis 185:1307–1313

Martin HL, Richardson BA, Nyange PM, Lavreys L, Hillier SL, Chohan B, Mandaliya K, Ndinya-Achola JO, Bwayo J, Kreiss J (1999) Vaginal lactobacilli, microbial flora, and risk of human immunodeficiency virus type 1 and sexually transmitted disease acquisition. J Infect Dis 180:1863–1868

Martinez RC, Franceschini SA, Patta MC, Quintana SM, Candido RC, Ferreira JC, De Martinis EC, Reid G (2009a) Improved treatment of vulvovaginal candidiasis with fluconazole plus probiotic *Lactobacillus rhamnosus* GR-1 and *Lactobacillus reuteri* RC-14. Lett Appl Microbiol 48:269–274

Martinez RC, Franceschini SA, Patta MC, Quintana SM, Gomes BC, De Martinis EC, and Reid G (2009b) Improved cure of bacterial vaginosis with single dose of tinidazole (2 g), *Lactobacillus rhamnosus* GR-1, and *Lactobacillus reuteri* RC-14: a randomized, double-blind, placebo-controlled trial. Can J Microbiol 55:133–138

Mattsby-Baltzer I, Platz-Christensen JJ, Hosseini N, Rosen P (1998) IL-1beta, IL-6, TNFalpha, fetal fibronectin, and endotoxin in the lower genital tract of pregnant women with bacterial vaginosis. Acta Obstet Gynecol Scand 77:701–706

McDonald HM, Brocklehurst P, Gordon A (2007) Antibiotics for treating bacterial vaginosis in pregnancy. Cochrane Database Syst Rev (1):CD000262

McGregor JA, French JI (2000) Bacterial vaginosis in pregnancy. Obstet Gynecol Surv 55: S1–19

McGregor JA, French JI, Jones W, Milligan K, McKinney PJ, Patterson E, Parker R (1994) Bacterial vaginosis is associated with prematurity and vaginal fluid mucinase and sialidase: results of a controlled trial of topical clindamycin cream. Am J Obstet Gynecol 170:1048–1059: Discussion 1059–1060

McLean NW, McGroarty JA (1996) Growth inhibition of metronidazole-susceptible and metronidazole-resistant strains of *Gardnerella vaginalis* by lactobacilli in vitro. Appl Environ Microbiol 62:1089–1092

Melnyk BM, Feinstein NF (2009) Reducing hospital expenditures with the COPE (Creating Opportunities for Parent Empowerment) program for parents and premature infants: an analysis of direct healthcare neonatal intensive care unit costs and savings. Nurs Adm Q 33: 32–37

Menard JP, Fenollar F, Henry M, Bretelle F, Raoult D (2008) Molecular quantification of *Gardnerella vaginalis* and *Atopobium vaginae* loads to predict bacterial vaginosis. Clin Infect Dis 47:33–43

Menard JP, Mazouni C, Salem-Cherif I, Fenollar F, Raoult D, Boubli L, Gamerre M, Bretelle F (2010) High vaginal concentrations of *Atopobium vaginae* and *Gardnerella vaginalis* in women undergoing preterm labor. Obstet Gynecol 115:134–140

Mitchell C, Balkus J, Agnew K, Lawler R, Hitti J (2009a) Changes in the vaginal microenvironment with metronidazole treatment for bacterial vaginosis in early pregnancy. J Womens Health (Larchmt) 18:1817–1824

Mitchell CM, Hitti JE, Agnew KJ, Fredricks DN (2009b) Comparison of oral and vaginal metronidazole for treatment of bacterial vaginosis in pregnancy: impact on fastidious bacteria. BMC Infect Dis 9:89

Monif GR, Carson HJ (1998) Female genital tract bacterial coisolates with *Candida albicans* in patients without clinical vaginitis. Infect Dis Obstet Gynecol 6:52–56

Morison L, Ekpo G, West B, Demba E, Mayaud P, Coleman R, Bailey R, Walraven G (2005) Bacterial vaginosis in relation to menstrual cycle, menstrual protection method, and sexual intercourse in rural gambian women. Sex Transm Infect 81:242–247
Muller E, Berger K, Dennemark N, Oleen-Burkey M (1999) Cost of bacterial vaginosis in pregnancy. Decision analysis and cost evaluation of a clinical study in Germany. J Reprod Med 44:807–814
Murta EF, Souza MA, Araujo Junior E, Adad SJ (2000) Incidence of *Gardnerella vaginalis*, Candida sp and human papilloma virus in cytological smears. Sao Paulo Med J 118:105–108
Myers ER (2004) Screening for bacterial vaginosis to prevent preterm birth: assessing effectiveness and cost-effectiveness. Acta Obstet Gynecol Scand 83:2–3
Ness RB, Hillier S, Richter HE, Soper DE, Stamm C, Bass DC, Sweet RL, Rice P (2003) Can known risk factors explain racial differences in the occurrence of bacterial vaginosis? J Natl Med Assoc 95:201–212
Nugent RP, Krohn MA, Hillier SL (1991) Reliability of diagnosing bacterial vaginosis is improved by a standardized method of gram stain interpretation. J Clin Microbiol 29:297–301
O'Brien RF (2005) Bacterial vaginosis: many questions–any answers? Curr Opin Pediatr 17: 473–479
O'Dowd TC, West RR, Winterburn PJ, Hewlins MJ (1996) Evaluation of a rapid diagnostic test for bacterial vaginosis. Br J Obstet Gynaecol 103:366–370
Oakley BB, Fiedler TL, Marrazzo JM, Fredricks DN (2008) Diversity of human vaginal bacterial communities and associations with clinically defined bacterial vaginosis. Appl Environ Microbiol 74:4898–4909
Okun N, Gronau KA, Hannah ME (2005) Antibiotics for bacterial vaginosis or *Trichomonas vaginalis* in pregnancy: a systematic review. Obstet Gynecol 105:857–868
Oleen-Burkey MA, Hillier SL (1995) Pregnancy complications associated with bacterial vaginosis and their estimated costs. Infect Dis Obstet Gynecol 3:149–157
Paavonen J (1998) Pelvic inflammatory disease. from diagnosis to prevention. Dermatol Clin 16:747–756, xii
Paul K, Boutain D, Manhart L, Hitti J (2008) Racial disparity in bacterial vaginosis: the role of socioeconomic status, psychosocial stress, and neighborhood characteristics, and possible implications for preterm birth. Soc Sci Med 67:824–833
Peipert JF, Lapane KL, Allsworth JE, Redding CA, Blume JD, Stein MD (2008) Bacterial vaginosis, race, and sexually transmitted infections: does race modify the association? Sex Transm Dis 35:363–367
Pybus V, Onderdonk AB (1999) Microbial interactions in the vaginal ecosystem, with emphasis on the pathogenesis of bacterial vaginosis. Microbes Infect 1:285–292
Reid G, Bruce AW (2003) Urogenital infections in women: can probiotics help? Postgrad Med J 79:428–432
Reid G, Dols J, Miller W (2009) Targeting the vaginal microbiota with probiotics as a means to counteract infections. Curr Opin Clin Nutr Metab Care 12:583–587
Roberton AM, Wiggins R, Horner PJ, Greenwood R, Crowley T, Fernandes A, Berry M, Corfield AP (2005) A novel bacterial mucinase, glycosulfatase, is associated with bacterial vaginosis. J Clin Microbiol 43:5504–5508
Rodrigues AG, Mardh PA, Pina-Vaz C, Martinez-de-Oliveira J, da Fonseca AF (1999) Is the lack of concurrence of bacterial vaginosis and vaginal candidosis explained by the presence of bacterial amines? Am J Obstet Gynecol 181:367–370
Rodriguez Jovita M, Collins MD, Sjoden B, Falsen E (1999) Characterization of a novel Atopobium isolate from the human vagina: description of *Atopobium vaginae* sp. nov. Int J Syst Bacteriol 49 Pt 4:1573–1576
Romero R, Chaiworapongsa T, Kuivaniemi H, Tromp G (2004) Bacterial vaginosis, the inflammatory response and the risk of preterm birth: a role for genetic epidemiology in the prevention of preterm birth. Am J Obstet Gynecol 190:1509–1519
Ross SA, Novak Z, Ashrith G, Rivera LB, Britt WJ, Hedges S, Schwebke JR, Boppana AS (2005) Association between genital tract cytomegalovirus infection and bacterial vaginosis. J Infect Dis 192:1727–1730

Royce RA, Jackson TP, Thorp JM Jr, Hillier SL, Rabe LK, Pastore LM, Savitz DA (1999) Race/ethnicity, vaginal flora patterns, and pH during pregnancy. Sex Transm Dis 26:96–102
Russell MW, Mestecky J (2002) Humoral immune responses to microbial infections in the genital tract. Microbes Infect 4:667–677
Russell RB, Green NS, Steiner CA, Meikle S, Howse JL, Poschman K, Dias T, Potetz L, Davidoff MJ, Damus K, Petrini JR (2007) Cost of hospitalization for preterm and low birth weight infants in the United States. Pediatrics 120:e1–9
Ryckman KK, Simhan HN, Krohn MA, Williams SM (2009a) Cervical cytokine network patterns during pregnancy: the role of bacterial vaginosis and geographic ancestry. J Reprod Immunol 79:174–182
Ryckman KK, Simhan HN, Krohn MA, Williams SM (2009b) Predicting risk of bacterial vaginosis: the role of race, smoking and corticotropin-releasing hormone-related genes. Mol Hum Reprod 15:131–137
Ryckman KK, Williams SM, Krohn MA, Simhan HN (2009c) Genetic association of Toll-like receptor 4 with cervical cytokine concentrations during pregnancy. Genes Immun 10:636–640
Schwebke JR (2001) Role of vaginal flora as a barrier to HIV acquisition. Curr Infect Dis Rep 3:152–155
Schwebke JR (2005) Abnormal vaginal flora as a biological risk factor for acquisition of HIV infection and sexually transmitted diseases. J Infect Dis 192:1315–1317
Schwebke JR, Richey CM, Weiss HL (1999) Correlation of behaviors with microbiological changes in vaginal flora. J Infect Dis 180:1632–1636
Schwebke JR, Weiss HL (2002) Interrelationships of bacterial vaginosis and cervical inflammation. Sex Transm Dis 29:59–64
Sewankambo N, Gray RH, Wawer MJ, Paxton L, McNaim D, Wabwire-Mangen F, Serwadda D, Li C, Kiwanuka N, Hillier SL, Rabe L, Gaydos CA, Quinn TC, Konde-Lule J (1997) HIV-1 infection associated with abnormal vaginal flora morphology and bacterial vaginosis. Lancet 350:546–550
Sharami SH, Afrakhteh M, Shakiba M (2007) Urinary tract infections in pregnant women with bacterial vaginosis. J Obstet Gynaecol 27:252–254
Simcox R, Sin WT, Seed PT, Briley A, Shennan AH (2007) Prophylactic antibiotics for the prevention of preterm birth in women at risk: a meta-analysis. Aust N Z J Obstet Gynaecol 47:368–377
Snowhite IV, Jones WE, Dumestre J, Dunlap K, Braly PS, Hagensee ME (2002) Comparative analysis of methods for collection and measurement of cytokines and immunoglobulins in cervical and vaginal secretions of HIV and HPV infected women. J Immunol Methods 263:85–95
Sobel JD (2005) What's new in bacterial vaginosis and trichomoniasis? Infect Dis Clin North Am 19:387–406
Sobel JD (2009) Antibiotic consideration in bacterial vaginosis. Curr Infect Dis Rep 11:471–475
Sobel JD, Faro S, Force RW, Foxman B, Ledger WJ, Nyirjesy PR, Reed BD, Summers PR (1998) Vulvovaginal candidiasis: epidemiologic, diagnostic, and therapeutic considerations. Am J Obstet Gynecol 178:203–211
Soper DE, Brockwell NJ, Dalton HP, Johnson D (1994) Observations concerning the microbial etiology of acute salpingitis. Am J Obstet Gynecol 170:1008–1014; Discussion 1014–1017
Spear GT, Sikaroodi M, Zariffard MR, Landay AL, French AL, Gillevet PM (2008) Comparison of the diversity of the vaginal microbiota in HIV-infected and HIV-uninfected women with or without bacterial vaginosis. J Infect Dis 198:1131–1140
Srinivasan U, Misra D, Marazita ML, Foxman B (2009) Vaginal and oral microbes, host genotype and preterm birth. Med Hypotheses 73:963–975
Stamey TA (1973) The role of introital enterobacteria in recurrent urinary infections. J Urol 109:467–472
Steer P (2005) The epidemiology of preterm labor–a global perspective. J Perinat Med 33:273–276
Sundquist A, Bigdeli S, Jalili R, Druzin ML, Waller S, Pullen KM, El-Sayed YY, Taslimi MM, Batzoglou S, Ronaghi M (2007) Bacterial flora-typing with targeted, chip-based Pyrosequencing. BMC Microbiol 7:108

Sweet RL (2000) Gynecologic conditions and bacterial vaginosis: implications for the non-pregnant patient. Infect Dis Obstet Gynecol 8:184–190
Thulkar J, Kriplani A, Agarwal N (2010) Probiotic and metronidazole treatment for recurrent bacterial vaginosis. Int J Gynaecol Obstet 108:251–252
Trabert B, Misra DP (2007) Risk factors for bacterial vaginosis during pregnancy among African American women. Am J Obstet Gynecol 197:477 e1–8
Vallor AC, Antonio MA, Hawes SE, Hillier SL (2001) Factors associated with acquisition of, or persistent colonization by, vaginal lactobacilli: role of hydrogen peroxide production. J Infect Dis 184:1431–1436
Valore EV, Park CH, Igreti SL, Ganz T (2002) Antimicrobial components of vaginal fluid. Am J Obstet Gynecol 187:561–568
Valore EV, Park CH, Quayle AJ, Wiles KR, McCray PB, Jr., Ganz T (1998) Human beta-defensin-1: an antimicrobial peptide of urogenital tissues. J Clin Invest 101:1633–1642
Van der Veen F, Fransen L (1998) Drugs for STD management in developing countries: choice, procurement, cost, and financing. Sex Transm Infect 74(Suppl 1):S166–174
Verhelst R, Verstraelen H, Claeys G, Verschraegen G, Delanghe J, Van Simaey L, De Ganck C, Temmerman M, Vaneechoutte M (2004) Cloning of 16S rRNA genes amplified from normal and disturbed vaginal microflora suggests a strong association between *Atopobium vaginae*, *Gardnerella vaginalis* and bacterial vaginosis. BMC Microbiol 4:16
Verstraelen H, Verhelst R (2009) Bacterial vaginosis: an update on diagnosis and treatment. Exp Rev Anti Infect Ther 7:1109–1124
Verstraelen H, Verhelst R, Claeys G, De Backer E, Temmerman M, Vaneechoutte M (2009a) Longitudinal analysis of the vaginal microflora in pregnancy suggests that *L. crispatus* promotes the stability of the normal vaginal microflora and that *L. gasseri* and/or *L. iners* are more conducive to the occurrence of abnormal vaginal microflora. BMC Microbiol 9:116
Verstraelen H, Verhelst R, Claeys G, Temmerman M, Vaneechoutte M (2004) Culture-independent analysis of vaginal microflora: the unrecognized association of *Atopobium vaginae* with bacterial vaginosis. Am J Obstet Gynecol 191:1130–1132
Verstraelen H, Verhelst R, Nuytinck L, Roelens K, De Meester E, De Vos D, Van Thielen M, Rossau R, Delva W, De Backer E, Vaneechoutte M, Temmerman M (2009b) Gene polymorphisms of toll-like and related recognition receptors in relation to the vaginal carriage of *Gardnerella vaginalis* and *Atopobium vaginae*. J Reprod Immunol 79:163–173
Watts DH, Krohn MA, Hillier SL, Eschenbach DA (1992) The association of occult amniotic fluid infection with gestational age and neonatal outcome among women in preterm labor. Obstet Gynecol 79:351–357
Weir E (2004) Bacterial vaginosis: more questions than answers. Cmaj 171:448
Wertz J, Isaacs-Cosgrove N, Holzman C, Marsh TL (2008) Temporal shifts in microbial communities in nonpregnant african-american women with and without bacterial vaginosis. Interdiscip Perspect Infect Dis 2008:181253
Wiggins R, Hicks SJ, Soothill PW, Millar MR, Corfield AP (2001) Mucinases and sialidases: their role in the pathogenesis of sexually transmitted infections in the female genital tract. Sex Transm Infect 77:402–408
Wiggins R, Millar MR, Soothill PW, Hicks SJ, Corfield AP (2002) Application of a novel human cervical mucin-based assay demonstrates the absence of increased mucinase activity in bacterial vaginosis. Int J STD AIDS 13:755–760
Williamson DM, Abe K, Bean C, Ferre C, Henderson Z, Lackritz E (2008) Current research in preterm birth. J Womens Health (Larchmt) 17:1545–1549
Wilson J (2004) Managing recurrent bacterial vaginosis. Sex Transm Infect 80:8–11
Wira CR, Grant-Tschudy KS, Crane-Godreau MA (2005) Epithelial cells in the female reproductive tract: a central role as sentinels of immune protection. Am J Reprod Immunol 53:65–76
Witkin SS, Linhares I, Giraldo P, Jeremias J, Ledger WJ (2000) Individual immunity and susceptibility to female genital tract infection. Am J Obstet Gynecol 183:252–256

Witkin SS, Linhares IM, Giraldo P (2007a) Bacterial flora of the female genital tract: function and immune regulation. Best Pract Res Clin Obstet Gynaecol 21:347–354

Witkin SS, Linhares IM, Giraldo P, Ledger WJ (2007b) An altered immunity hypothesis for the development of symptomatic bacterial vaginosis. Clin Infect Dis 44:554–557

Wolrath H, Boren H, Hallen A, Forsum U (2002) Trimethylamine content in vaginal secretion and its relation to bacterial vaginosis. Apmis 110:819–824

Wolrath H, Forsum U, Larsson PG, Boren H (2001) Analysis of bacterial vaginosis-related amines in vaginal fluid by gas chromatography and mass spectrometry. J Clin Microbiol 39:4026–4031

Wolrath H, Stahlbom B, Hallen A, Forsum U (2005) Trimethylamine and trimethylamine oxide levels in normal women and women with bacterial vaginosis reflect a local metabolism in vaginal secretion as compared to urine. Apmis 113:513–516

Xu J, Holzman CB, Arvidson CG, Chung H, Goepfert AR (2008) Midpregnancy vaginal fluid defensins, bacterial vaginosis, and risk of preterm delivery. Obstet Gynecol 112:524–531

Yoshimura K, Yoshimura M, Kobayashi T, Kubo T, Hachisuga T, Kashimura M (2009) Can bacterial vaginosis help to find sexually transmitted diseases, especially chlamydial cervicitis? Int J STD AIDS 20:108–111

Yudin MH, Landers DV, Meyn L, Hillier SL (2003) Clinical and cervical cytokine response to treatment with oral or vaginal metronidazole for bacterial vaginosis during pregnancy: a randomized trial. Obstet Gynecol 102:527–534

Zhou X, Bent SJ, Schneider MG, Davis CC, Islam MR, Forney LJ (2004) Characterization of vaginal microbial communities in adult healthy women using cultivation-independent methods. Microbiology 150:2565–2573

Zhou X, Brown CJ, Abdo Z, Davis CC, Hansmann MA, Joyce P, Foster JA, Forney LJ (2007) Differences in the composition of vaginal microbial communities found in healthy caucasian and black women. Isme J 1:121–133

Chapter 7
The Human Lung Microbiome

Liliana Losada, Elodie Ghedin, Alison Morris, Hong Wei Chu, and William C. Nierman

The Human Lung

Lung Microbiome Overview

The human LOWER respiratory tract is considered sterile in normal healthy individuals (Flanagan et al., 2007; Speert, 2006) despite the fact that every day we breathe in multiple microorganisms present in the air and aspirate thousands of organisms from the mouth and nasopharynx. This apparent sterility is maintained by numerous interrelated components of the lung physical structures such as the mucociliary elevator and components of the innate and adaptive immune systems (discussed below) (reviewed in (Diamond et al., 2000; Gerritsen, 2000)). However, it is possible that the observed sterility might be a result of the laboratory practices applied to study the flora of the lungs. Historically, researchers faced with a set of diseases characterized by a changing and largely cryptic lung microbiome have lacked tools to study lung ecology as a whole and have concentrated on familiar, cultivatable candidate pathogens. With the availability of new technologies for cultivation-independent analysis of microbial populations, it is now possible to follow individuals by sampling their lung microbiome sequentially during episodes of disease and recovery in order to identify associations between the lung microbiome and health and disease.

Any respired contaminating particles or pathogens that evade the lung's physical and immune barriers are usually eliminated by dendritic cells and alveolar macrophages that deliver them into local draining lymph nodes. Macrophages kill invading microorganisms while *en route* to the draining lymph nodes, and in some cases at the nodes themselves (Bozza et al., 2002; Kirby et al., 2009). Thus, it is not unusual to isolate viable bacteria or fungi from "normal" lung tissue (Lass-Florl et al., 1999), and the term "sterile" should be applied with caution. It is perhaps

L. Losada (✉)
J. Craig Venter Institute, 9704 Medical Center Drive, Rockville, MD 20850, USA
e-mail: llosada@jcvi.org

K.E. Nelson (ed.), *Metagenomics of the Human Body*,
DOI 10.1007/978-1-4419-7089-3_7,

more accurate to say that there is no resident flora that permanently colonizes normal lungs.

Even normal healthy lungs are not microbe free all the time. Lower airway infections by bacteria, viruses, or fungi are among the most prevalent causes of transmissible disease in humans, with two to three million community-acquired (non-hospital-acquired) cases per year in the United States (Segreti et al., 2005). In 2006, the number of deaths attributed to pneumonia (bacterial and viral) and influenza in the United States was 60,000 (Gao et al., 2008) www.cdc.gov/nchs/fastats/deaths.htm). In 2009, nearly 9.3 million new cases of tuberculosis were reported around the world (http://www.who.int/mediacentre/factsheets/fs104/en/index.html). With proper treatment, the lungs of individuals with these infectious diseases will revert to their normal "sterile" state.

Little is known about the composition of the microbial population of the upper and lower airways in health or disease. It is likely, given the multiple microorganisms already implicated in chronic lung diseases such as chronic obstructive pulmonary disease (COPD), that there are other undetected organisms and that there are complex relationships between multiple pathogens involved that are not currently understood. A few studies have examined microbial species in limited numbers of normal subjects and patients with various respiratory disorders. One study using 16S rDNA clone libraries and microarrays did not detect any bacteria in the lungs of patients without respiratory disease who were briefly intubated for surgery (Flanagan et al., 2007). The same study reported that all patients intubated for longer periods had detectable 16S rDNA and that the bacterial diversity present decreased during antibiotic usage. Another study used 16S rDNA amplification to identify bacterial species in 16 patients with ventilator-associated pneumonia (Bahrani-Mougeot et al., 2007). This study identified bacterial pathogens not seen using conventional culture techniques, especially anaerobes, and found that oral bacteria could be detected in the lung. In one study, sputum samples from 25 cystic fibrosis (CF) patients were analyzed using 16S gene profiling and the authors identified an average of 7.2 species present per subject (Bittar et al., 2008). Viruses have also been examined in nasal lavages in both asthmatic and normal subjects with cold symptoms using the Virochip microarray (Kistler et al., 2007). The microarray technology identified more viruses than conventional culture methods and had excellent sensitivity and specificity compared with pathogen-specific polymerase chain reaction (PCR). An unexpected diversity of human coronavirus and rhinovirus strains was discovered in the subjects.

So if the lung is generally sterile, why do some individuals become chronically colonized? What organisms colonize the lungs? Those with physically compromised airways or immune system deficiencies are subject to chronic microbial colonization of their airways and to high-frequency episodes of viral, bacterial, or fungal lower respiratory infections. Perhaps in no other body site is the direct relationship between disease and microbiome more explicit than in the lungs where there is a distinct and obvious microbial difference between normal and diseased individuals.

Anatomical and Immunological Setting

The lower respiratory tract, composed of the trachea and lungs, is quite different in structure and function from the upper respiratory tract, which is highly colonized by microorganisms. The lungs themselves are divided into different sections according to their function and structures: the bronchi, bronchioles, and alveoli. Bronchi and bronchioles are primarily conductive airways surrounded by thick cartilage that allow easy airflow into the parenchyma (or alveolar tissue) of the lungs, where gas exchange occurs. Conductive airways are covered in ciliated epithelium interspersed with different types of secretory cells that release mucins, immunomodulatory proteins, surfactants, and proteases. Together, the physical and chemical barriers protect against physical and biological damage by establishing a mucociliary elevator, which brings about an upward transport of a mucus stream for the lungs (Fraser, 2005). The secretory cells decrease in proportion from 20 to 30% in the trachea to less than 1% in the distal and alveolar parts of the lungs. In addition to the physicochemical protection provided by cilia and mucus, the epithelium is also protected by several immune cells, including dendritic, Langerhans, T lymphocytes, and mast cells, that respond to inhaled antigens establishing a robust immunity (Fraser, 2005). It is thought that in conjunction with the physical barriers provided by the nose and upper respiratory mucosa, these defenses are enough to maintain sterility in the lower respiratory tract.

The vast majority of the lung surface epithelium, however, is alveolar. It is estimated that 87% of the total volume of the lungs is alveolar, with only 6% of this being tissue and the remainder gas (Stone et al., 1992). The primary role of this tissue is to carry out gas exchange. The epithelium is mostly a continuous single layer of cells overlying a thin interstitium, which contains numerous capillaries that supply ample blood for gas exchange (Fraser, 2005; Stone et al., 1992). Unlike the epithelium in the conducting airways, the respiratory epithelium is not ciliated or protected by mucus. Instead, it is covered by surfactant proteins that maintain the surface tension for efficient gas exchange. The lack of mucus or secretory cells is compensated by the presence of alveolar macrophages, mast cells, lymphocytes, dendritic cells, and other monocyte-like cells that protect the epithelium from potential pathogens and help maintain sterility.

Diseases of the Lower Respiratory Tract and Their Impact on the Lung Microbiome

Recent World Health Organisation (WHO) figures rank lower respiratory diseases second in an assessment of the burden of disease worldwide (http://www.who.int/respiratory/en/). In 2006, 124, 500 people died in the US due to chronic lower respiratory disease (www.cdc.gov/nchs/fastats/deaths.htm). Chronic respiratory diseases include: asthma, COPD, CF, and bronchiectasis. These diseases generally lead to impaired clearance and function of the mucociliary elevator and/or the immune protection of the lung. In addition, immune deficiency such as that caused by the human immunodeficiency virus (HIV) also disrupts the typical immune homeostasis in the lungs. Without the normal protective barriers, the lungs fall victim to persistent and severe colonization that can ultimately lead to death, particularly for CF patients. As discussed below, the lung microbiome in each of these diseases is very different from normal individuals. The data discussed in the following sections demonstrate a clear link between microbial colonization and severity of disease symptoms. It is unclear, however, what exact role these different microbial populations play in initiating and enhancing the progress of such chronic respiratory diseases. Lastly, in some cases, the data hint that some population structures might actually be protective against further decline, but much more research needs to be conducted in this area to make a definitive claim.

This chapter discusses the methods for sampling and characterizing the microbiome of the lungs. In addition, it reviews the current status of our understanding of the lung microbiome in asthma, idiopathic bronchiectasis, CF, COPD, and during immune deficiency due to HIV infection.

Microbiome Characterization

Human Lung Sampling Methods

Several procedures have been developed for sampling the microbial population of the human lung econiche. In order of increasing invasiveness they are sputum induction, bronchoalveolar lavage (BAL), bronchial brushing, endobronchial biopsy, and transbronchial biopsy. Sputum induction by inhalation of hypertonic saline is a non-invasive method to obtain samples from the lower respiratory tract for cell and microbial analysis (Bickerman et al., 1958). The quality of samples varies and can be scored on the volume of the obtained sputum plugs and the level of salivary contamination as measured by squamous cells observed by microscopy.

BAL is a procedure in which a bronchoscope is passed through the mouth or nose into the lungs and saline is instilled into a segment of the lung and then recollected for examination (Henderson, 1994; Reynolds and Chretien, 1984). BAL is most commonly used to diagnose infections in both immunocompetent and immunosuppressed patients. BAL is the most common procedure for sampling the lower respiratory system microbial colonization/infection status, to sample the components of the epithelial lining fluid, and to determine the protein composition of the airways. It is often used in evaluating the patient's lung immunological status by sampling cells and pathogen levels. BAL is an invasive procedure and thus is less ideal for research purposes.

Bronchial brushing provides access to cells and microbes that are adherent to the luminal surfaces of the lower airways. In this procedure, a flexible fiber optic bronchoscope is used for brushing a targeted lesion or site (Fennessy, 1967; Zavala et al., 1973), where induced sputum and BAL procedures will allow sampling of cells and microbes that can be washed from the lumen surface, brushing will recover adherent cells (e.g., bronchial epithelial cells) and microbes. Recently, brushing techniques have been developed to sample distal lung (i.e., small airway) epithelial cells and associated microbes (Ammous et al., 2008). This technique will enable investigators to further study the microbiome in lung diseases such as COPD.

Endobronchial biopsy involves using the fiber optic bronchoscope to identify appropriate target sites in the lung and obtain large airway tissue samples using inserted alligator forceps, cup forceps, or curette passed through the endoscope's central channel. This procedure poses a higher risk than BAL but allows sampling the invasive microbes within the airway tissue (Scott et al., 1991; Trulock et al., 1992).

Transbronchial biopsy, the most invasive of these sampling procedures, is routinely performed for clinical care and allows clinicians and researchers to obtain distal (small) airway tissues as well as alveolar tissues. This procedure has been safely done by several research groups (Balzar et al., 2005) and will likely further our understanding of the microbiome in human distal lung tissues, but carries a significant risk of complications.

High-Throughput Sequencing of Bacterial Ribosomal RNA Subunits

Standard microbiological and virologic methods detect only a small proportion of the bacteria and viruses present in various body sites because the great majority of these organisms are uncharacterized or uncultivable. To understand the real diversity, culture-independent methods, such as sequencing, are thus a necessity. Sequenced-based identification of microbial species is facilitated by decreased costs of sequencing, and the availability of next-generation sequencing technologies, further enhances the capacity to generate large amounts of data.

For the identification of bacterial species within an environment, the amplification of 16S rRNA genes (or 16S rDNA) using universal primers are useful for diversity characterization because this genetic locus is present in all bacterial species (Relman et al., 1991). The nine hypervariable regions of the 16S rDNA can be used for bacterial species identification (Chakravorty et al., 2007; Rokas et al., 2007) with some regions having better discriminatory value than others. The sequencing and phylogenetic analysis of bacterial 16S rRNA derived from microbiome samples has been the primary method used to investigate bacterial diversity in the human body (Bik et al., 2006; Dekio et al., 2005; Eckburg et al., 2005; Gao et al., 2007; Hugenholtz et al., 1998; Hyman et al., 2005; Zhang et al., 2006). These studies have revealed a far higher level of diversity than conventional culture techniques (Aly et al., 1976; Bik et al., 2006; Dekio et al., 2005; Kazor et al., 2003; Korting et al., 1988; Kroes et al., 1999; Paster et al., 2001). These studies revealed that the majority of bacterial sequences correspond to uncultivated species and novel organisms. There was significant intersubject variability and variability between stool and mucosal microbial populations. For example, recent studies by Blaser and colleagues at New York University have demonstrated substantial changes in the ratio of the genus *Streptococcus* to *Propionibacterium* in skin samples from healthy persons and in normal skin of patients with psoriasis (ratio $= 0.4$; $n = 2,649$ clones), and from psoriatic lesion samples (ratio $=$ 5.0; $n = 1,314$ clones; $P = 0.01$) (Gao et al., 2008).

In a lung study, bacterial diversity was analyzed in the endotrachael aspirates from seven intubated patients colonized with *Pseudomonas aeruginosa* using both sequencing from 16S rRNA clone libraries and an oligonucleotide microarray termed as the PhyloChip (Flanagan et al., 2007). Controls were subjects briefly intubated for elective surgery. Bacteria were not detected by either method in samples from the controls. Sequencing from the clone libraries detected the presence of many orally, nasally, and gastrointestinal associated bacteria including known pathogens. The Phylochip detected the same organisms and many additional bacterial groups present at low abundance. Following antibiotic therapy, the bacterial populations' diversity decreased and was dominated by a single respiratory pathogen. In six of the seven patients, the dominant species was *P. aeruginosa* in spite of targeting this organism with antibiotics to which it was reportedly sensitive. The authors hypothesize that the loss of population diversity may directly contribute to pathogenicity, persistence, and development of pneumonia.

Ribosomal RNA ITS Typing of Fungal Populations

Similarly, amplification of regions from the 18S and internal transcribed spacer (ITS) regions of the rRNA, a conserved fungal gene, allows discrimination of fungal species (Fujita et al., 2001; Makimura, 2001). A preliminary study was undertaken to examine the efficacy of a community sequencing method to identify the fungal species in BAL lung samples from 23 human subjects. The fungal ITS1-5.8S-ITS2 region was amplified and the results showed that 4 of 23 patients (17%) had fungal DNA levels that could be reproducibly detected by PCR. The detected fungi included *Aspergillus fumigatus*, *Candida tropicalis,* and *Penicillium digitatum*, among others (Denning, unpublished). These data agree with a culture-based study that showed that 63% of their sample population had evidence of pulmonary fungal colonization (Lass-Florl et al., 1999), most commonly with *A. fumigatus* and other *Candida* species, and also Zygomycetes. The results also demonstrate that rRNA sequencing is a viable platform for characterization of fungal communities in the human body.

Viral Identification via Genome Sequencing for Population Analysis

Although there are no conserved genes that can be targeted for determination of viral diversity, whole genome shotgun of a sample enriched for viruses (such as by filtering) can lead to an effective characterization of viral communities (Angly et al., 2006). Hundreds of viral genome sequences can be completed in a single sequencing reaction run using the GS-FLX (454/Roche) sequencing platform. Using this technology and a random priming-based method, referred to as sequence independent single primer amplification (SISPA), near-full-length genomes of RNA or DNA viruses can be sequenced. SISPA can be used to sequence known and unknown viral genomes (Djikeng et al., 2008). This viral sequencing methodology can potentially be adapted for the determination of viruses within BAL by enrichment using nuclease treatment and filtration followed by taking the extracted total RNA and DNA through the SISPA process followed by sequence comparisons to known viruses.

Metagenomics

The initial studies of small 16S rRNA described above hinted at great diversity within the human microbiome, yet it left important questions unanswered such as the identity of the nondominant community members and their biological roles. The applications of shotgun techniques to the study of the human microbiome (Kurokawa et al., 2007; Manichanh et al., 2006; Zhang et al., 2006) again highlight the extent of microbial diversity associated with the human body while revealing much more of the identity and biology of nonculturable microorganisms. As a result of reduced costs and improved sequencing technologies, it is possible to perform in-depth metagenomic surveys of the human body's microbial diversity beyond the 16S rRNA surveys. Metagenomics, a term introduced in 1998, describes the functional and sequence-based analysis of total microbial genomes from environmental samples (Handelsman et al., 1998). Metagenomics uses techniques that resemble the "Whole Genome Shotgun" approach of single genome sequencing, but it is not limited to a single species.

Human metagenomics has provided insight into the complex composition of the microbiome of these several body sites, and this information has allowed us to draw tentative conclusions about the relationship between specific microbiomes and health. The human microbiome is composed of multiple "ecological niches", including the mouth (Kroes et al., 1999; Paster et al., 2001), esophagus (Zhang et al., 2006), stomach (Bik et al., 2006), intestine (Gill et al., 2006), skin (Gao et al., 2007), and vagina (Zhou et al., 2004). Our understanding of the overlap and the degree of communication between them is rudimentary at best. Perhaps the most extensively studied has been the human gut microbiome where the interaction of the gut microflora, independently or through interaction with the genetic makeup of the host plays a role in obesity, Crohn's disease, and ulcerative colitis (Frank et al., 2007; Gophna et al., 2006; Turnbaugh et al., 2006).

The Lung Microbiome in Asthma

Asthma is a complex disease characterized by chronic inflammation in the lungs and reversible narrowing of the airways. Symptoms include dyspnea, coughing, wheezing, airway hyper–reactivity, chronic eosinophilic atopy, and mucus hyper secretion (Busse and Lemanske, 2001). About 20 million people in the US have been diagnosed with asthma; 9 million of them are children. Asthma causes 4,000 deaths per year in the US and 11 million exacerbations. Asthma is caused by environmental and genetic factors (Martinez, 2007), with asthma attacks resulting from immune responses to inhaled allergens. The majority of asthma exacerbations are caused by viral infections (Krishnan et al., 2006). Atypical bacterial infections have also been associated with asthma exacerbations and with chronic asthma (Johnston and Martin, 2005; Martin et al., 2001). In susceptible individuals, the development of asthma has been associated with bacterial colonization in neonates (Bisgaard et al., 2007) and viral and bacterial infections (Wu and Chu, 2009). There is abundant evidence testifying to the importance of microbes to the development and maintenance of asthma. A recent publication using a bacterial gene sequencing method suggests a disordered microbiome in asthmatic airways (Hilty et al., 2010).

In the developed world there has been an increased focus on predisposing factors for asthma due to its rapidly increasing prevalence, now affecting up to a quarter of urban children (Lilly, 2005). Asthma is known to be caused by environmental and genetic factors (Martinez, 2007). These factors determine asthma severity and how easily it can be treated (Martinez, 2007). Many associations with asthma have been detected including exposure to cigarette smoke (Thomson et al., 2004), caesarean section birth relative to natural birth (Thavagnanam et al., 2008), early viral respiratory infections (Gold and Wright, 2005; Harju et al., 2006), early in life antibiotic use (Marra et al., 2006), and living in the US (Gold and Wright, 2005). One theory for the cause of the increase in asthma incidence is the hygiene hypothesis (Strachan, 1989), that the rise in prevalence of asthma is a direct consequence of the success of modern hygienic practices in preventing childhood infections. This hypothesis is supported by numerous studies that have shown that children coming from a less hygienic environment have less asthma and other allergenic diseases (Ball et al., 2000; Celedon et al., 1999; Jarvis et al., 1997). In addition, alterations in innate immune system genes have been shown to be associated with the inception and development of asthma. These genes include the toll-like receptors and other genes such as *MBL, MYLK, DEFB1, JUN, INF-α5,* and *NOS2A* reviewed in Wu and Chu (2009).

The Role of Infections in Asthma Exacerbations

Asthma exacerbations have long been associated with viral infections (Pattemore et al., 1992). More recently, the use of reverse transcriptase PCR has greatly facilitated the identification of the exacerbation-associated virus. Studies using this tool have suggested that 75–80% of asthma exacerbations are caused by virus infections (Wark et al., 2002).

Rhinovirus (RV) infections during early childhood are associated with the development of asthma, lower respiratory tract infections, and wheezing (Jackson et al., 2008; Lemanske et al., 2005). They are also associated with hospitalization for asthma in adults (Venarske et al., 2006). A separate study revealed that patients with allergic asthma infected with RV had increased admissions to hospitals and that dust mite allergen was the primary allergen when these patients were skin tested with a panel of aeroallergens (Green et al., 2002). Respiratory syncytial virus (RSV) in infants causes lower respiratory infection leading to pneumonia and bronchiolitis. RSV bronchiolitis is the leading cause of wheezing in infants and young children, and children infected with RSV resulting in bronchiolitis are more likely to develop wheezing and asthma later in childhood (Peebles, 2004). Similarly, the human metapneumovirus (hMPV) was first isolated from children in 2001 and has been found to be associated with asthma exacerbations in both children under 5 years of age and adults (Foulongne et al., 2006; Williams et al., 2004).

Mycoplasma pneumoniae and *Chlamydia pneumoniae* are bacteria that attach to airway epithelial cells and cause cell damage. Infections by these bacteria have been shown to be associated with asthma exacerbations (Johnston and Martin, 2005; Lieberman et al., 2003; Martin et al., 2001). Using a PCR assay, 31 of 55 patients with asthma were positive for either of these bacteria in lung tissue or BAL, suggesting that some level of colonization by these bacteria may be common in asthma patients (Martin et al., 2001). Studies in a mouse model suggest that preexisting allergic inflammation impairs the ability to upregulate TLR-2 and IL-6 in the lungs, leading to decreased clearance of *M. pneumoniae* and an increase in airway inflammation (Kraft et al., 2008).

The Role of Infections and Microbial Colonization in Asthma Development

Evidence is accumulating that infections are associated with the induction and development of asthma. First, long-term cohort studies on the development of asthma show that most childhood asthma begins in infancy. The first episode of wheezing begins before the age of 3 and is frequently associated with lower respiratory tract viral infections, usually RSV, but also RV (Gern et al., 2000; Sigurs et al., 2005). These infectious episodes and associated wheezing are strong predictors for the development of childhood asthma and atopy (Devulapalli et al., 2008; Kusel et al., 2007; Martinez et al., 1995; Singh et al., 2007). Second, many studies have associated viral infections with asthma prevalence in children (Devulapalli et al., 2008; Jackson et al., 2008; Kusel et al., 2007; Papadopoulos and Kalobatsou, 2007; Sigurs et al., 2005; Singh et al., 2007; Williams et al., 2004). Lastly, Wu et al. have provided evidence to suggest that viral infections have a causal role in asthma initiation and development where they show that viral infection during the first 4 months of age is strongly correlated with the development of asthma by age 5 (Wu et al., 2008). Only one-third of children with childhood wheezing and asthma, however, will develop persistent asthma symptoms in adulthood (Gerritsen, 2002; Taylor et al., 2005; Vonk et al., 2004). Management of the symptoms with corticosteroid therapy is effective but may not alter the asthma progression (Guilbert et al., 2006).

The role of infections in asthma induction and development will likely be shown to be mediated through the effect of these infections on the chronic inflammatory response in the airways of asthmatics. Microbial infections can generate either a Th2- or a Th1-biased response that could exacerbate or attenuate asthma, respectively. In asthmatics, a pro-inflammatory Th2 response persists even in the absence of allergens involving CD4+ Th2 cells, eosinophils, mast cells, and the Th2 cytokines IL-4, IL-5, IL-9, and IL-13 (Holgate, 2008).

Bacterial infections have been shown to contribute to asthma development. In a longitudinal prospective birth cohort study of 411 infants born to mothers with current or previous asthma, neonates colonized in the hypopharyngeal regions with *Streptococcus pneumoniae, Haemophilus influenzae*, or *Moraxella catarrhalis* or a combination of these organisms were found to be at increased risk for recurrent wheezing in early childhood and asthma at age 5 (Bisgaard et al., 2007). A protective role for some bacteria has been reported (Blaser et al., 2008). Several studies have found a protective effect of mycobacterial exposure on atopy and airway inflammation (Camporota et al., 2003; Shirakawa et al., 1997; Yang et al., 2002). These exposures include Bacillus Calmette-Guerin vaccination or heat-killed *Mycobacterium vaccae*. Early exposure to bacterial endotoxins may reduce future allergies or asthma (von Mutius et al., 2000), although endotoxins associated with house dust are associated with more asthma symptoms and worse lung function (Dales et al., 2006; Michel et al., 1996; Park et al., 2001).

Thus, the role of bacteria in asthma initiation and development appears to be complex. The causative interaction is likely to prove to be the interaction of bacteria and bacterial components in modulating the Th1 and Th2 innate immune system responses. The characterization of these interactions will be complicated by timing, dose, anatomical site, and duration of the bacterial exposure as well as the host genetic and environmental factors influencing the immune inflammatory response (Holt, 1996).

The Role of Fungi in Severe Asthma

It has recently been demonstrated that patients with severe asthma who are also atopic or sensitized to environmental fungi may benefit from treatment with the antifungal azole itraconazole (Denning et al., 2009). This observation has raised questions about the relationships among asthma severity, fungal sensitization, and fungal exposure. The issue is complicated by more than 1.5 million species of fungi that are thought to exist (Hawksworth and Rossman, 1997) and more than 80 species of fungi that have been associated with symptoms of airway allergy (Horner et al., 1995). For one species, *A. fumigatus*, 20 allergens are thought to participate in human airway allergies (www.allergome.org). Determining the clinical relevance of fungal allergens is confounded further by extensive cross-reactivity among fungal allergens (Crameri et al., 2009).

Fungal allergens can induce a number of different human bronchopulmonary disorders, each with a distinct immune pathogenesis. In allergic bronchopulmonary aspergillosis (ABPA), the respiratory system is chronically colonized typically with *A. fumigatus*. Evidence now suggests that severe asthmatics without ABPA are more likely to be atopic to fungi than patients with milder disease. The diagnostic label "severe asthma with fungal sensitization (SAFS)" has recently been applied to this group (Denning et al., 2006). In these patients, the fungal sensitization is most commonly *A. fumigatus, Candida albicans,* and *Penicillium notatum* (O'Driscoll et al., 2009).

The association between severe asthma and fungi has been identified in numerous studies. Atopy to environmental fungi has been associated with severe asthma (O'Driscoll et al., 2005). Many population studies have shown an association between local fungal spore counts and medical emergencies due to asthma exacerbations (Atkinson et al., 2006). Furthermore, studies have shown that fungus exposure in fungal-sensitized individuals induces asthma symptoms (Malling, 1986; Matheson et al., 2005; Pulimood et al., 2007; Salo et al., 2006; Woodcock et al., 2006). Treatment of patients with SAFS with antifungal drugs has generally led to improvement of asthma symptoms (Denning et al., 2006) concurrent with improvements in several markers of atopy such as reduced IgE values, reduced eosinophils counts, and reductions in the level of dose of oral and systemic steroids required (Pasqualotto et al., 2009).

These findings lead to the considerations of the fungal composition of the lung microbiome in asthmatic individuals and indeed in normal individuals. Environmental fungi colonize the lungs of otherwise healthy people (Lass-Florl et al., 1999; Okudaira et al., 1977). These studies were dependent on cultivation-based methods for the detection and identification of these fungi. As a cultivation-independent method, gas chromatography/mass spectroscopy on exhaled breath has revealed the presence of fungus specific biomarkers in patients with CF with and without fungal colonization by *A. fumigatus* (Syhre et al., 2008). This approach was limited to analyzing for known *A. fumigatus* markers. The application of sequencing-based approaches for studying the lung microbiome will be essential for revealing the role of fungi in the lung microbiome and the role of the lung microbiome on asthma.

The Lung Microbiome in Cystic Fibrosis (CF)

Introduction

CF is the most common inherited lung disease in the world. It is a severe autosomal recessive disease with an incidence of 1:2000 at birth in populations of northwestern European origin, with a mutant gene carrier frequency of 1:23 in these populations. The genetic defect occurs in the cystic fibrosis transmembrane regulator (CFTR) protein, which acts to transport chloride across cell membranes. Patients with CF are the archetype population with chronic bronchial colonization. Symptoms include permanent bacterial colonization of the lower airways, with a formation of a biofilm, fat maldigestion, male infertility, and elevated levels of chloride in the sweat (Knowles and Durie, 2002). The thick pulmonary system mucus in CF patients minimizes the effectiveness of the mucociliary elevator in clearing the lung of mucus-trapped microorganisms and other debris. As a consequence, microbes chronically colonize these patients' lungs and they suffer bouts of infection, requiring frequent hospital admission. Cultures reveal a wide range of bacteria, including *P. aeruginosa*, *Mycobacteria*, *A. fumigatus*, and sometimes viruses.

The Chronically Colonized CF Airways Represent a Surprising Complex and Diverse Ecosystem

The precise contributions of different microbes to patient morbidity, and the importance of inter-specific interactions remain largely unclear [reviewed in (Harrison, 2007)]. The complexity of this ecosystem is difficult to overstate. As an example of this complexity, the lungs of CF patients contain large numbers of neutrophils that migrate to this location in response to microbial colonization. These neutrophils secrete granule antimicrobial proteins called defensins that kill microbes. By analysis of CF sputum samples, the levels of extracellular defensins are sufficiently abundant that they may damage the airway epithelium (Soong et al., 1997). As another example of this inter-specific complexity, *P. aeruginosa* in CF lungs produces copious amounts of a tricyclic compound pyocyanin that kills competing microbes and eukaryotic cells. This compound was shown to specifically inactivate a human lung epithelial cell line vacuolar ATPase (Ran et al., 2003).

The Microbiome of the CF Lung Contains Microbial Diversity Not Evaluable by Standard Culture Techniques

A study of the microbiome of the lungs was conducted to explore the hypothesis that organisms not routinely identified by culture occur in the lungs of CF patient airways and may contribute to disease. To test this hypothesis, 16S rRNA sequence analysis was performed on BAL samples from 42 subjects, 28 CF patients, and 14 other disease controls (Harris et al., 2007). The findings of this analysis were that, for CF subjects, a single rRNA type was dominantly represented in the clone libraries prepared from lung microbiome genomic DNA. This was not found in the controls. Thirteen of the CF subjects' samples contained bacteria not routinely assessed by culture. Candidate pathogens were identified in four CF subjects. Candidate pathogens were also identified in the controls. This study documented the power of culture-independent molecular techniques to provide a broader view of the airway bacteria than standard clinical culture methods.

The CF viral metagenome was explored in a recent study using five CF individuals and five individuals without disease (Willner et al., 2009). In both cohorts, the overall viral diversity was low. The CF bacteriophage communities were highly similar to each other, whereas the non-CF individual had more distinct phage communities. CF eukaryotic viral communities were dominated by a few viruses, including human herpes viruses and retroviruses.

Fastidious or Noncultivatable Bacteria: An Emerging Concept in CF

The significance of fastidious or noncultivatable organisms in the airway of CF patients is beginning to be explored. Application of specific culture conditions to favor the growth of anaerobes coupled with molecular identification techniques have focused attention in CF on bacteria not routinely detected by standard culture and biochemical identification techniques (Harris et al., 2007; Tunney et al., 2008; Worlitzsch et al., 2009). Direct, culture-independent detection techniques have identified much larger numbers of bacterial species in CF airways and have demonstrated the ability to identify likely pathogenic bacteria occurring during exacerbations when routine cultures are negative (Harris et al., 2007). These molecular identification and detection methods have identified bacteria with different antibiotic susceptibilities relative to conventional pathogens and will undoubtedly lead to novel antimicrobial intervention trials in CF (Worlitzsch et al., 2009). Similar methodology to detect anaerobes or noncultivatable bacteria has not been applied systematically to patients with idiopathic bronchiectasis.

The Lung Microbiome in Idiopathic Bronchiectasis

Introduction

Bronchiectasis is characterized by chronic dilation and inflammation of the conducting airways associated with recurring infections (Barker, 2002). It is the pathologic manifestation of several genetic disorders, including CF and primary ciliary dyskinesia (PCD). However, many patients have no identifiable causes. Idiopathic bronchiectasis is estimated to affect approximately 110,000 US adults (Weycker et al., 2005). Symptoms include cough and chronic sputum production, recurring airway infection, dyspnea, wheezing, and chest pain (Barker, 2002). Microbial infections are central to the pathogenesis and progression of disease.

Bronchiectasis Phenotype and Microbial Flora

Much of the research characterizing the composition and significance of the lower airway microbial flora has been done in CF and relatively little is known about the microbial contribution to disease pathogenesis in idiopathic bronchiectasis. However, recent observations suggest a link between the lower airway microbial flora and host disease characteristics. For example, the prevalence of idiopathic bronchiectasis associated with nontuberculous mycobacteria (NTM) appears to be increasing (Billinger et al., 2009; Marras et al., 2007). Both familial clustering and a characteristic "tall asthenic" phenotype (scoliosis, pectus excavatum, mitral valve abnormalities) in postmenopausal women with bronchiectasis associated with NTM infection (Colombo et al., 2009; Kim et al., 2008) have been reported. It is unknown whether the age, female sex, and unique body morphotype associations are seen in idiopathic bronchiectasis unassociated with NTM. Correlating disease phenotype with microbial flora is dependent upon accurately categorizing the microbial status of the patients. For environmental organisms like NTM, it is important that this categorization include both accurate speciation and determination that the organism likely represents true infection rather than contamination or transient colonization. The American Thoracic Society and the Infectious Diseases Society of America (ATS/IDSA) microbiologic diagnostic criteria for pulmonary disease based on sputum specimens call for at least two positive sputum specimens for the same species (Kim et al., 2008).

Significance of Filamentous Fungi: A Novel Consideration in Bronchiectasis

Concomitant recovery of filamentous fungi from airway specimens is also common in bronchiectasis, but the pathophysiologic consequences are not known. A recent study in CF patients found that *A. fumigatus*, like NTM, was commonly present in older patients: 75% of patients aged 16–20 years and in 60% of patients over age 20 (Valenza et al., 2008). Amin and colleagues further noted that CF patients who were chronically infected with *A. fumigatus* (defined as two positive cultures in a given year) had significantly worse airway obstruction as evidenced by a lower forced expiratory volume in one second (FEV_1) and significantly higher risk of pulmonary exacerbations during subsequent follow-up than patients without *A. fumigatus* (Amin et al., 2009). This potential negative impact on the course of bronchiectasis and a possible benefit from antifungal treatment for chronic infection in CF have prompted initiation of a multicenter clinical trial of itraconazole in CF patients in Canada (Amin et al., 2009; Shoseyov et al., 2006). In non-CF bronchiectasis, a recent study suggested that *Aspergillus* is more common in patients infected with NTM than in those without NTM, and that it is commonly associated with fungal lung disease manifestations in NTM-infected patients (Kunst et al., 2006). However, outside the relatively small numbers of idiopathic bronchiectasis patients with allergic bronchopulmonary aspergillosis (ABPA) or chronic necrotizing aspergillosis, the pathologic significance of these fungi has not been systematically explored in large numbers of patients and very few data are available for *Aspergillus* species other than *fumigatus* or filamentous fungi other than *Aspergillus* (Kobashi et al., 2006; Kunst et al., 2006; Raju et al., 2008).

Serologic Assessment of Fungal Response

ABPA is well described in association with bronchiectasis occurring in asthmatics and patients with CF (Malde and Greenberger, 2004). The diagnostic criteria rely on an elevated total IgE as well as elevated *A. fumigatus*-specific IgE and IgG in the setting of episodic bronchial obstruction, pulmonary infiltrates, and central bronchiectasis. Kunst and colleagues assessed the prevalence of positive serologic markers for *A. fumigatus* [IgE by radioallergosorbent test (RAST) and precipitins] among idiopathic bronchiectasis patients and found these markers to be commonly present especially in the setting of concomitant NTM disease (Kunst et al., 2006). Patients with these serologic markers more commonly had radiographic manifestations suggesting *Aspergillus*-associated disease. While other filamentous fungi such as *Scedosporium* species have been commonly recovered from the airways of both CF and non-CF bronchiectasis patients, the role these fungi play in disease pathogenesis remains controversial (Cooley et al., 2007). While specific IgE antibody RAST and precipitin assays can be prepared using allergen prepared from the isolated species and correlated with the clinical presentation of allergic bronchopulmonary mycoses, these assays have not been commonly used to characterize the clinical significance of these fungal species recovered from the lower airway (Fedorova et al., 2008; Lake et al., 1991).

The Lung Microbiome in COPD

Introduction

COPD is the fourth leading cause of death in the US (Petty, 2000) and is expected to rank third in the world by 2020 (Lopez and Murray, 1998). Despite efforts aimed at smoking cessation, little impact has been made on COPD incidence, and current treatments are ineffective in slowing progression of the disease. COPD has been defined by the Global Initiative for Chronic Obstructive Lung Disease (GOLD) as "a disease state characterized by airflow limitation that is not fully reversible". The diagnosis of COPD can also encompass those with chronic obstructive bronchiolitis and emphysema. Tissue inflammation in COPD is characterized by a predominant neutrophil, CD8+ lymphocyte, and macrophage infiltration (Keatings et al., 1996; Lacoste et al., 1993; O'Shaughnessy et al., 1997; Saetta et al., 1998). It has been proposed that the mechanism of tissue damage involves the recruitment and activation of neutrophils, macrophages, and CD8+ T cells with concomitant upregulation of several cellular proteases and inflammatory cytokines.

Role of Infections in COPD

Although smoking is clearly the leading risk factor for COPD, not all smokers develop disease (Buist and Connett, 1993). While smoking can stimulate inflammation in the lungs, smokers with COPD have an increased inflammatory response than smokers without COPD, and inflammation can persist despite smoking cessation (Keatings et al., 1996; Lacoste et al., 1993; O'Shaughnessy et al., 1997; Saetta et al., 1998). These observations suggest that some other factor or factors contribute to development and perpetuation of the inflammatory response in COPD. Infection might be one such factor critical in triggering and perpetuating the inflammatory response in COPD. The mechanism by which infections might act to promote COPD progression has been termed the "vicious circle" hypothesis (Sethi, 2000a; Sethi and Murphy, 2008). In this scenario, smoking causes structural remodeling that renders smokers more likely to become colonized and/or less able to clear subclinical infection. Defects in mucociliary clearance and surfactant abnormalities caused by smoking also contribute to the tendency to develop chronic infection (Finley and Ladman, 1972; Honda et al., 1996; Raju et al., 2008; Vastag et al., 1985; Verra et al., 1995). Once colonization is established, the organism or organisms recruit white blood cells to the lungs, stimulating release of inflammatory cytokines and chemokines as well as proteases. Inability to clear the inciting organism perpetuates the cycle, ultimately resulting in tissue destruction, airway thickening, and clinical COPD.

Bacteria, Viruses, and Fungi Have All Been Linked to COPD

The most commonly implicated bacteria are *H. influenzae, M. catarrhalis, S. pneumoniae,* and *P. aeruginosa* (Sethi, 2004). Viruses that seem to be important in COPD include adenovirus, influenzae viruses, rhinovirus, respiratory syncytial virus, and human metapneumovirus (Mallia et al., 2006; Martinello et al., 2006; Retamales et al., 2001; Seemungal et al., 2001). These pathogens can be found in patients with COPD in the stable state and during exacerbations (Sethi, 2004). The colonization seen in patients with COPD is likely playing a role in disease and is not just an innocent bystander. For example, as bacterial load increases, FEV_1 falls, and colonization has been associated with greater sputum purulence, increased sputum neutrophils, and increased levels of interleukin (IL)-8, tumor necrosis factor (TNF)-α, and neutrophil elastase (Obrian et al., 2007; Patel et al., 2002; Sethi, 2000b; Stockley et al., 2000). Exacerbations associated with viruses are more severe and last longer than those without a viral trigger (Papi et al., 2006; Seemungal et al., 2001). In addition, exacerbations associated with both bacteria and viruses may be more severe than those associated with single organisms (Obrian et al., 2007), suggesting the usefulness of metagenomic techniques in this disease.

Colonization with the fungus *Pneumocystis jirovecii* (Pc, formerly *Pneumocystis carinii* f. sp. *hominis*) has recently been implicated in COPD pathogenesis. This organism generally causes acute *Pneumocystis* pneumonia (PCP) in patients with immunosuppression such as those infected with HIV, but colonization with the organism occurs in both HIV^+ and HIV^- individuals and may be important in COPD. Colonization with Pc is increased in HIV patients with COPD and correlates with disease severity (Morris et al., 2004b; Probst et al., 2000). Animal models also support the role of Pc colonization in COPD. Christensen and colleagues recently reported that in immunocompetent mice, exposure to cigarette smoke and Pc colonization resulted in pulmonary function deficits and airspace enlargement characteristic of emphysema (Christensen et al., 2008). In a model of Pc colonization in simian/human-immunodeficiency virus (SHIV)-infected nonhuman primates

(Norris et al., 2006), Pc-colonized animals developed airway obstruction and radiographic emphysema while animals infected with SHIV alone did not develop these changes (Shipley et al., 2010).

The Lung Microbiome and Human Immunodeficiency Virus Infections

Lung Diseases Remain a Leading Cause of Morbidity and Mortality in HIV Infection

Although pulmonary infections and neoplasms associated with HIV have decreased since the availability of highly active antiretroviral therapy (HAART) (Palella et al., 1998), some pulmonary conditions may actually be increasing in persons with HIV. Diseases such as COPD, asthma, and bronchiectasis were reported to be increased in those with HIV before the introduction of antiretroviral therapy, and a similar decrease in these conditions as seen in the opportunistic infections has not occurred after antiretroviral treatment of HIV. In fact, in a recent study of HIV^+ patients, almost 4% of deaths were due to obstructive airway disease in 1998, a three-fold increase from the pre-HAART era (Louie et al., 2002). Before the HAART era, HIV^+ subjects were noted to have an accelerated form of emphysema with significant emphysematous disease seen in subjects less than 40 years old (Diaz et al., 1992, 2000). Both emphysema and airflow obstruction have been reported in HIV infection. Unlike many of the acquired immunodeficiency syndrome (AIDS)-defining opportunistic infections, HIV-associated COPD may actually be more common in the current era of HIV as it is frequently reported in those without a history of AIDS-related pulmonary complications and the now aging HIV^+ population has a longer exposure to smoking and HIV.

Role of the Microbiome in Obstructive Lung Diseases in HIV

Given the immunological defects seen with HIV, it is quite possible that HIV^+ subjects, especially those who smoke, are more prone to develop subclinical pulmonary infections, even if successfully treated with HAART. The changes that occur in the lung microbiome have not been studied in HIV, but microbial colonization is a likely factor in the accelerated COPD seen in this population. The vicious circle hypothesis of COPD could be further worsened in HIV^+ patients by upregulation of HIV levels in the lung stimulated by pulmonary colonization. Several studies have shown that pulmonary infections increase lung levels of HIV. Koziel and colleagues reported that HIV RNA was detected in 62% of patients with active lung disease compared to 16% of asymptomatic subjects, independent of clinical stage of HIV and serum HIV RNA levels (Koziel et al., 1999). The lung appears to be an independent compartment for HIV replication as drug mutations found in BAL differ from those in blood (White et al., 2004). HIV in the lungs is associated with a lymphocytic alveolitis, particularly in those subjects with CD4 cell counts between 200 and 500 cells/μl, suggesting that the virus might act independently to stimulate pulmonary inflammation (Twigg et al., 1999). The relationship of HIV pulmonary viral levels, infections, inflammation, and COPD has not been examined.

Role of Pneumocystis *Colonization in HIV-Associated COPD*

Pneumocystis colonization is likely important in the pathogenesis of COPD in those with HIV as well as in the HIV^- population. In HIV^+ subjects, the prevalence of colonization is high, particularly if subjects smoke, and colonization is seen even in patients with high CD4 cell counts receiving HAART (Morris et al., 2004a). Anatomic emphysema is also more common in HIV^+ patients with Pc colonization (Morris, unpublished data). It has recently been shown that *Pneumocystis* colonization in HIV^+ subjects is associated with worse airway obstruction and an increased likelihood of clinical diagnosis of COPD, independent of smoking history and CD4 cell count (Morris et al., 2009). In addition, the SHIV-infected nonhuman primates described above serve as a model for the development of COPD in the setting of Pc colonization and HIV-like immunodeficiency (Norris et al., 2006; Shipley et al., in press).

Conclusions

Microorganisms including bacteria, fungi, and viruses play a central role in development, exacerbation, and progress of lung diseases. Even though normal lungs do not have a permanent resident microbiome, diseased lungs are acutely infected and/or chronically colonized. Standard laboratory practices have not properly reflected the entirety of the microbiomes, either in health or in disease, and thus newer sequence-based technologies have begun to reveal the true complexity of the lung microbiomes. Much more research still needs to be conducted in order to fully understand the microbial burden of the lungs, and how this burden relates to health and disease.

It is apparent that conditions that compromise the physical and immune system barriers to lung colonization by microbes result in chronic colonization and recurrent infections. These conditions include chronic inflammation as seen in HIV, asthma, and bronchiecstasis, or physical obstruction observed in CF and bronchiecstasis. The lung microbiomes in each of these conditions has not been properly explored to date, which limits our ability to make definitive conclusions on how to best manage these diseases. Our understanding of the fundamental role of viruses in the initial establishment and progress of asthma underscores how little knowledge exists on the role of viral infection in other chronic respiratory diseases. In addition, the fact that some bacterial populations and/or components seem to be protective against further and severe exacerbations in asthma opens the door to questions about the role of microorganisms in protecting against other diseases. The exact nature of this protection is not clearly understood, and great benefit would come from studies that further clarify these intriguing results. Furthermore, if some population structures aid in preventing disease progression, it is likely that other population structures may predispose episodes of acute acerbations and progression of the underlying disease condition. A comprehensive understanding of the dynamics of these microbiome interactions would likely result in better, more efficient therapies for these and other respiratory diseases.

Our developing microbiome analysis technology coupled with our increasing awareness of the potential positive and negative impact of lung population structure on respiratory system health and disease strongly supports the initiation of aggressive projects to characterize the human lung microbiome and its influence on health and disease.

References

Aly R, Maibach HE, Mandel A (1976) Bacterial flora in psoriasis. Br J Dermatol 95:603–606

Amin R, Dupuis A, Aaron SD, Ratjen F (2009) The effect of chronic infection with *Aspergillus fumigatus* on lung function and hospitalization in cystic fibrosis patients. Chest 137:171–176

Ammous Z, Hackett NR, Butler MW, Raman T, Dolgalev I, O'Connor TP, Harvey BG, Crystal RG (2008) Variability in small airway epithelial gene expression among normal smokers. Chest 133:1344–1353

Angly FE, Felts B, Breitbart M, Salamon P, Edwards RA, Carlson C, Chan AM, Haynes M, Kelley S, Liu H, Mahaffy JM, Mueller JE, Nulton J, Olson R, Parsons R, Rayhawk S, Suttle CA, Rohwer F (2006) The marine viromes of four oceanic regions. PLOS Biol 4:e368

Atkinson RW, Strachan DP, Anderson HR, Hajat S, Emberlin J (2006) Temporal associations between daily counts of fungal spores and asthma exacerbations. Occup Environ Med 63:580–590

Bahrani-Mougeot FK, Paster BJ, Coleman S, Barbuto S, Brennan MT, Noll J, Kennedy T, Fox PC, Lockhart PB (2007) Molecular analysis of oral and respiratory bacterial species associated with ventilator-associated pneumonia. J Clin Microbiol 45:1588–1593

Ball TM, Castro-Rodriguez JA, Griffith KA, Holberg CJ, Martinez FD, Wright AL (2000) Siblings, day-care attendance, and the risk of asthma and wheezing during childhood. N Engl J Med 343:538–543

Balzar S, Chu HW, Strand M, Wenzel S (2005) Relationship of small airway chymase-positive mast cells and lung function in severe asthma. Am J Respir Crit Care Med 171:431–439

Barker AF (2002) Bronchiectasis. N Engl J Med 346:1383–1393

Bickerman HA, Sproul EE, Barach AL (1958) An aerosol method of producing bronchial secretions in human subjects: a clinical technic for the detection of lung cancer. Dis Chest 33:347–362

Bik EM, Eckburg PB, Gill SR, Nelson KE, Purdom EA, Francois F, Perez-Perez G, Blaser MJ, Relman DA (2006) Molecular analysis of the bacterial microbiota in the human stomach. Proc Natl Acad Sci USA 103:732–737

Billinger ME, Olivier KN, Viboud C, de Oca RM, Steiner C, Holland SM, Prevots DR (2009) Nontuberculous mycobacteria-associated lung disease in hospitalized persons, United States, 1998–2005. Emerg Infect Dis 15:1562–1569

Bisgaard H, Hermansen MN, Buchvald F, Loland L, Halkjaer LB, Bønnelykke K, Brasholt M, Heltberg A, Vissing NH, Thorsen SV, Stage M, Pipper CB (2007) Childhood asthma after bacterial colonization of the airway in neonates. N Engl J Med 357:1487–1495

Bittar F, Richet H, Dubus JC, Reynaud-Gaubert M, Stremler N, Sarles J, Raoult D, Rolain JM (2008). Molecular detection of multiple emerging pathogens in sputa from cystic fibrosis patients. PLOS One 3:e2908

Blaser MJ, Chen Y, Reibman J (2008) Does *Helicobacter pylori* protect against asthma and allergy? Br Med J 57:561

Bozza S, Gaziano R, Spreca A, Bacci A, Montagnoli C, di Francesco P, Romani L (2002) Dendritic cells transport conidia and hyphae of *Aspergillus fumigatus* from the airways to the draining lymph nodes and initiate disparate Th responses to the fungus. J Immunol 168:1362

Buist AS, Connett JE (1993) The lung health study. Baseline characteristics of randomized participants. Chest 103:1644

Busse WW, Lemanske RF Jr (2001) Asthma. N Engl J Med 344:350–362

Camporota L, Corkhill A, Long H et al (2003) The effects of *Mycobacterium vaccae* on allergen-induced airway responses in atopic asthma. Eur Respir J 21:287–293

Celedon JC, Litonjua AA, Weiss ST, Gold DR (1999) Day-care attendance in the first year of life and illnesses of the upper and lower respiratory tract in children with a familial history of atopy. Pediatrics 104:495–500

Chakravorty S, Helb D, Burday M, Connell N, Alland D (2007) A detailed analysis of 16S ribosomal RNA gene segments for the diagnosis of pathogenic bacteria. J Microbiol Methods 69:330–339

Christensen PJ, Preston AM, Ling T, Du M, Fields WB, Curtis JL, Beck JM (2008) Pneumocystis murina infection and cigarette smoke exposure interact to cause increased organism burden, development of airspace enlargement, and pulmonary inflammation in mice. Infect Immun 76:3481–3490

Colombo RE, Hill SC, Claypool RJ, Holland SM, Olivier KN (2009) Familial clustering of pulmonary nontuberculous mycobacterial disease. Chest 137:629–634

Cooley L, Spelman D, Thursky K, Slavin M (2007) Infection with *Scedosporium apiospermum and S. prolificans*, Australia. Emerg Infect Dis 13:1170–1177

Crameri R, Zeller S, Glaser AG, Vilhelmsson M, Rhyner C (2009) Cross-reactivity among fungal allergens: a clinically relevant phenomenon? Mycoses 52:99–106

Dales R, Miller D, Ruest K, Guay M, Judek S (2006) Airborne endotoxin is associated with respiratory illness in the first 2 years of life. Environ Health Perspect 114:610–614

Dekio I, Hayashi H, Sakamoto M, Kitahara M, Nishikawa T, Suematsu M, Benno Y (2005) Detection of potentially novel bacterial components of the human skin microbiota using culture-independent molecular profiling. J Med Microbiol 54:1231–1238

Denning DW, O'Driscoll BR, Hogaboam CM, Bowyer P, Niven RM (2006) The link between fungi and severe asthma: a summary of the evidence. Eur Respir J 27:615–626

Denning DW, O'Driscoll BR, Powell G, Chew F, Atherton GT, Vyas A, Miles J, Morris J, Niven RM (2009) Randomized controlled trial of oral antifungal treatment for severe asthma with fungal sensitization: the fungal asthma sensitization trial (FAST) study. Am J Respir Crit Care Med 179:11–18

Devulapalli CS, Carlsen KC, Haland G, Munthe-Kaas MC, Pettersen M, Mowinckel P, Carlsen KH (2008) Severity of obstructive airways disease by age 2 years predicts asthma at 10 years of age. Thorax 63:8–13

Diamond G, Legarda D, Ryan LK (2000) The innate immune response of the respiratory epithelium. Immunol Rev 173:27–38

Diaz PT, Clanton TL, Pacht ER (1992) Emphysema-like pulmonary disease associated with human immunodeficiency virus infection. Ann Intern Med 116:124–128

Diaz PT, King MA, Pacht ER, Wewers MD, Gadek JE, Nagaraja HN, Drake J, Clanton TL (2000) Increased susceptibility to pulmonary emphysema among HIV-seropositive smokers. Ann Intern Med 132:369–372

Djikeng A, Halpin R, Kuzmickas R, DePasse J, Feldblyum J, Sengamalay N, Afonso C, Zhang X, Anderson NG, Ghedin E, Spiro DJ (2008) Viral genome sequencing by random priming methods. BMC Genomics 9:5

Eckburg PB, Bik EM, Bernstein CN, Purdom E, Dethlefsen L, Sargent M, Gill SR, Nelson KE, Relman DA (2005) Diversity of the human intestinal microbial flora. Science 308:1635–1638
Fedorova ND, Khaldi N, Joardar VS, Maiti R, Amedeo P, Anderson MJ, Crabtree J, Silva JC, Badger JH, Albarraq A, Angiuoli S, Bussey H, Bowyer P, Cotty PJ, Dyer PS, Egan A, Galens K, Fraser-Liggett CM, Haas BJ, Inman JM, Kent R, Lemieux S, Malavazi L, Orvis J, Roemer T, Ronning CM, Sundaram JP, Sutton G, Turner G, Venter JC, White OR, Whitty BR, Youngman P, Wolfe KH, Goldman GH, Wortman JR, Jiang B, Denning DW, Nierman WC (2008) Genomic islands in the pathogenic filamentous fungus *Aspergillus fumigatus*. PLOS Genet 4:e1000046
Fennessy JJ (1967) Transbronchial biopsy of peripheral lung lesions. Radiology 88:878–882
Finley TN, Ladman AJ (1972) Low yield of pulmonary surfactant in cigarette smokers. N Engl J Med 286:223–227
Flanagan JL, Brodie EL, Weng L, Lynch SV, Wiener-Kronish JP, Bristow J (2007) Loss of bacterial diversity during antibiotic treatment of intubated patients colonized with *Pseudomonas aeruginosa*. J Clin Microbiol 45:1954–1962
Foulongne V, Guyon G, Rodiere M, Segondy M (2006) Human metapneumovirus infection in young children hospitalized with respiratory tract disease. Pediatr Infect Dis J 25:354–359
Frank DN, St Amand AL, Feldman RA, Boedeker EC, Harpaz N, Pace NR (2007) Molecular-phylogenetic characterization of microbial community imbalances in human inflammatory bowel diseases. Proc Natl Acad Sci USA 104:13780–13785
Fraser RS (2005) Histology and gross anatomy of the respiratory tract. In: Hamid Q, Shannon J, Martin J (eds) Physiologic basis of respiratory disease. BK Decker, Hamilton, ON, pp 133–141
Fujita SI, Senda Y, Nakaguchi S, Hashimoto T (2001) Multiplex PCR using internal transcribed spacer 1 and 2 regions for rapid detection and identification of yeast strains. J Clin Microbiol 39:3617–3622
Gao Z, Tseng CH, Pei Z, Blaser MJ (2007) Molecular analysis of human forearm superficial skin bacterial biota. Proc Natl Acad Sci USA 104:2927–2932
Gao Z, Tseng CH, Strober BE, Pei Z, Blaser MJ (2008) Substantial alterations of the cutaneous bacterial biota in psoriatic lesions. PLOS One 3:e2719
Gern JE, Vrtis R, Grindle KA, Swenson C, Busse WW (2000) Relationship of upper and lower airway cytokines to outcome of experimental rhinovirus infection. Am J Respir Crit Care Med 162:2226–2231
Gerritsen J (2000) Host defence mechanisms of the respiratory system. Paediatr Respir Rev 1: 128–134
Gerritsen J (2002) Follow-up studies of asthma from childhood to adulthood. Paediatr Respir Rev 3:184–192
Gill SR, Pop M, Deboy RT, Eckburg PB, Turnbaugh PJ, Samuel BS, Gordon JI, Relman DA, Fraser-Liggett CM, Nelson KE (2006) Metagenomic analysis of the human distal gut microbiome. Science NY 312:1355–1359
Gold DR, Wright R (2005) Population disparities in asthma. Annu Rev Public Health 26:89–113
Gophna U, Sommerfeld K, Gophna S, Doolittle WF, Veldhuyzen van Zanten SJ (2006) Differences between tissue-associated intestinal microfloras of patients with Crohn's disease and ulcerative colitis. J Clin Microbiol 44:4136–4141
Green RM, Custovic A, Sanderson G, Hunter J, Johnston SL, Woodcock A (2002) Synergism between allergens and viruses and risk of hospital admission with asthma: case-control study. BMJ 324:763
Guilbert TW, Morgan WJ, Zeiger RS et al (2006) Long-term inhaled corticosteroids in preschool children at high risk for asthma. N Engl J Med 354:1985–1997
Handelsman J, Rondon MR, Brady SF, Clardy J, Goodman RM (1998) Molecular biological access to the chemistry of unknown soil microbes: a new frontier for natural products. Chem Biol 5:R245–249
Harju TH, Leinonen M, Nokso-Koivisto J et al (2006). Pathogenic bacteria and viruses in induced sputum or pharyngeal secretions of adults with stable asthma. Thorax 61:579–584

Harris JK, De Groote MA, Sagel SD et al (2007) Molecular identification of bacteria in bronchoalveolar lavage fluid from children with cystic fibrosis. Proc Natl Acad Sci USA 104:20529–20533

Harrison F (2007) Microbial ecology of the cystic fibrosis lung. Microbiology 153: 917–923

Hawksworth DL, Rossman AY (1997) Where are all the undescribed fungi? Phytopathology 87:888–891

Henderson AJ (1994) Bronchoalveolar lavage. Arch Dis Child 70:167–169

Hilty M, Burke C, Pedro H et al (2010) Disordered microbial communities in asthmatic airways. PLOS One 5:e8578

Holgate ST (2008) Pathogenesis of asthma. Clin Exp Allergy 38:872–897

Holt PG (1996) Infections and the development of allergy. Toxicol Lett 86:205–210

Honda Y, Takahashi H, Kuroki Y, Akino T, Abe S (1996) Decreased contents of surfactant proteins A and D in BAL fluids of healthy smokers. Chest 109:1006–1009

Horner WE, Helbling A, Salvaggio JE, Lehrer SB (1995) Fungal allergens. Clin Microbiol Rev 8:161–179

Hugenholtz P, Goebel BM, Pace NR (1998) Impact of culture-independent studies on the emerging phylogenetic view of bacterial diversity. J Bacteriol 180:4765–4774

Hyman RW, Fukushima M, Diamond L, Kumm J, Giudice LC, Davis RW (2005) Microbes on the human vaginal epithelium. Proc Natl Acad Sci USA 102:7952–7957

Jackson DJ, Gangnon RE, Evans MD et al (2008) Wheezing rhinovirus illnesses in early life predict asthma development in high-risk children. Am J Respir Crit Care Med 178:667–672

Jarvis D, Chinn S, Luczynska C, Burney P (1997) The association of family size with atopy and atopic disease. Clin Exp Allergy 27:240–245

Johnston SL, Martin RJ (2005) *Chlamydophila pneumoniae* and *Mycoplasma pneumoniae*: a role in asthma pathogenesis? Am J Respir Crit Care Med 172:1078–1089

Kazor CE, Mitchell PM, Lee AM, Stokes LN, Loesche WJ, Dewhirst FE, Paster BJ (2003) Diversity of bacterial populations on the tongue dorsa of patients with halitosis and healthy patients. J Clin Microbiol 41:558–563

Keatings VM, Collins PD, Scott DM, Barnes PJ (1996) Differences in interleukin-8 and tumor necrosis factor-alpha in induced sputum from patients with chronic obstructive pulmonary disease or asthma. Am J Respir Crit Care Med 153:530–534

Kirby AC, Coles MC, Kaye PM (2009) Alveolar macrophages transport pathogens to lung draining lymph nodes. J Immunol 183:1983

Kistler A, Avila PC, Rouskin S et al (2007) Pan-viral screening of respiratory tract infections in adults with and without asthma reveals unexpected human coronavirus and human rhinovirus diversity. J Infect Dis 196:817–825

Knowles MR, Durie PR (2002) What is cystic fibrosis? N Engl J Med 347:439–442

Kobashi Y, Fukuda M, Yoshida K, Miyashita N, Niki Y, Oka M (2006) Chronic necrotizing pulmonary aspergillosis as a complication of pulmonary *Mycobacterium avium* complex disease. Respirology 11:809–813

Korting HC, Lukacs A, Braun-Falco O (1988) Microbial flora and odor of the healthy human skin. Hautarzt 39:564–568

Koziel H, Kim S, Reardon C, Li X, Garland R, Pinkston P, Kornfeld H (1999) Enhanced in vivo human immunodeficiency virus-1 replication in the lungs of human immunodeficiency virus-infected persons with *Pneumocystis carinii* pneumonia. Am J Respir Crit Care Med 160: 2048–2055

Kraft M, Adler KB, Ingram JL, Crews AL, Atkinson TP, Cairns CB, Krause DC, Chu HW (2008) *Mycoplasma pneumoniae* induces airway epithelial cell expression of MUC5AC in asthma. Eur Respir J 31:43–46

Krishnan V, Diette GB, Rand CS, Bilderback AL, Merriman B, Hansel NN, Krishnan JA (2006) Mortality in patients hospitalized for asthma exacerbations in the United States. Am J Respir Crit Care Med 174:633–638

Kroes I, Lepp PW, Relman DA (1999) Bacterial diversity within the human subgingival crevice. Proc Natl Acad Sci USA 96:14547–14552
Kunst H, Wickremasinghe M, Wells A, Wilson R (2006) Nontuberculous mycobacterial disease and aspergillus-related lung disease in bronchiectasis. Eur Respir J 28:352–357
Kurokawa K, Itoh T, Kuwahara T et al (2007) Comparative metagenomics revealed commonly enriched gene sets in human gut microbiomes. DNA Res 14:169–181
Kusel MM, de Klerk NH, Kebadze T, Vohma V, Holt PG, Johnston SL, Sly PD (2007) Early-life respiratory viral infections, atopic sensitization, and risk of subsequent development of persistent asthma. J Allergy Clin Immunol 119:1105–1110
Lacoste JY, Bousquet J, Chanez P et al (1993) Eosinophilic and neutrophilic inflammation in asthma, chronic bronchitis, and chronic obstructive pulmonary disease. J Allergy Clin Immunol 92:537–548
Lake FR, Froudist JH, McAleer R, Gillon RL, Tribe AE, Thompson PJ (1991) Allergic bronchopulmonary fungal disease caused by bipolaris and curvularia. Aust N Z J Med 21: 871–874
Lass-Florl C, Salzer GM, Schmid T, Rabl W, Ulmer H, Dierichi, MP (1999) Pulmonary aspergillus colonization in humans and its impact on management of critically ill patients. Br J Haematol 104:745–747
Lemanske RF Jr, Jackson DJ, Gangnon RE et al (2005) Rhinovirus illnesses during infancy predict subsequent childhood wheezing. J Allergy Clin Immunol 116:571–577
Lieberman D, Printz S, Ben-Yaakov M, Lazarovich Z, Ohana B, Friedman MG, Dvoskin B, Leinonen M, Boldur, I (2003) Atypical pathogen infection in adults with acute exacerbation of bronchial asthma. Am J Respir Crit Care Med 167:406–410
Lilly CM (2005) Diversity of asthma: evolving concepts of pathophysiology and lessons from genetics. J Allergy Clin Immunol 115:S526–531
Lopez AD, Murray CC (1998) The global burden of disease, 1990–2020. Nat Med 4:1241–1243
Louie JK, Hsu LC, Osmond DH, Katz MH, Schwarcz SK (2002) Trends in causes of death among persons with acquired immunodeficiency syndrome in the era of highly active antiretroviral therapy, San Francisco, 1994–1998. J Infect Dis 186:1023–1027
Makimura K (2001) Species identification system for dermatophytes based on the DNA sequences of nuclear ribosomal internal transcribed spacer 1. Nihon Ishinkin Gakkai zasshi = Jpn J Med Mycol 42:61–67
Malde B, Greenberger PA (2004) Allergic bronchopulmonary aspergillosis. Allergy Asthma Proc 25:S38–39
Mallia P, Message SD, Kebadze T, Parker HL, Kon OM, Johnston SL (2006) An experimental model of rhinovirus induced chronic obstructive pulmonary disease exacerbations: a pilot study. Respir Res 7:116
Malling HJ (1986) Diagnosis and immunotherapy of mould allergy. IV. Relation between asthma symptoms, spore counts and diagnostic tests. Allergy 41:342–350
Manichanh C, Rigottier-Gois L, Bonnaud E et al (2006) Reduced diversity of faecal microbiota in Crohn's disease revealed by a metagenomic approach. Gut 55:205–211
Marra F, Lynd L, Coombes M, Richardson K, Legal M, Fitzgerald JM, Marra CA (2006) Does antibiotic exposure during infancy lead to development of asthma?: a systematic review and metaanalysis. Chest 129:610–618
Marras TK, Chedore P, Ying AM, Jamieson F (2007) Isolation prevalence of pulmonary nontuberculous mycobacteria in Ontario, 1997–2003. Thorax 62:661–666
Martin RJ, Kraft M, Chu HW, Berns EA, Cassell GH (2001) A link between chronic asthma and chronic infection. J Allergy Clin Immunol 107:595–601
Martinello RA, Esper F, Weibel C, Ferguson D, Landry ML, Kahn JS (2006) Human metapneumovirus and exacerbations of chronic obstructive pulmonary disease. J Infect 53:248–254
Martinez FD, Wright AL, Taussig LM, Holberg CJ, Halonen M, Morgan WJ (1995) Asthma and wheezing in the first six years of life. The group health medical associates. N Engl J Med 332:133–138

Martinez FD (2007) Genes, environments, development and asthma: a reappraisal. Eur Respir J 29:179–184

Matheson MC, Abramson MJ, Dharmage SC, Forbes AB, Raven JM, Thien FC, Walters, EH (2005) Changes in indoor allergen and fungal levels predict changes in asthma activity among young adults. Clin Exp Allergy 35:907–913

Michel O, Kips J, Duchateau J, Vertongen F, Robert L, Collet H, Pauwels R, Sergysels R (1996) Severity of asthma is related to endotoxin in house dust. Am J Respir Crit Care Med 154: 1641–1646

Morris A, Kingsley LA, Groner G, Lebedeva IP, Beard CB, Norris KA (2004a) Prevalence and clinical predictors of pneumocystis colonization among HIV-infected men. AIDS 18: 793–798

Morris A, Sciurba FC, Lebedeva IP, Githaiga A, Elliott WM, Hogg JC, Huang L, Norris KA (2004b) Association of chronic obstructive pulmonary disease severity and pneumocystis colonization. Am J Respir Crit Care Med 170:408–413

Morris A, Alexander T, Radhi S, Lucht L, Sciurba FC, Kolls JK, Srivastava R, Steele C, Norris KA (2009) Airway obstruction is increased in pneumocystis-colonized human immunodeficiency virus-infected outpatients. J Clin Microbiol 47:3773–3776

Nierman WC, Pain A, Anderson MJ et al (2005) Genomic sequence of the pathogenic and allergenic filamentous fungus *Aspergillus fumigatus*. Nature 438:1151–1156

Norris KA, Morris A, Patil S, Fernandes E (2006) *Pneumocystis* colonization, airway inflammation, and pulmonary function decline in acquired immunodeficiency syndrome. Immunol Res 36:175–187

O'Driscoll BR, Hopkinson LC, Denning DW (2005) Mold sensitization is common amongst patients with severe asthma requiring multiple hospital admissions. BMC Pulm Med 5:4

O'Driscoll BR, Powell G, Chew F, Niven RM, Miles JF, Vyas A, Denning DW (2009) Comparison of skin prick tests with specific serum immunoglobulin E in the diagnosis of fungal sensitization in patients with severe asthma. Clin Exp Allergy 39:1677–1683

O'Shaughnessy TC, Ansari TW, Barnes NC, Jeffery PK (1997) Inflammation in bronchial biopsies of subjects with chronic bronchitis: inverse relationship of CD8+ T lymphocytes with FEV1. Am J Respir Crit Care Med 155:852–857

Obrian GR, Georgianna DR, Wilkinson JR, Yu J, Abbas HK, Bhatnagar D, Cleveland TE, Nierman W, Payne GA (2007) The effect of elevated temperature on gene transcription and aflatoxin biosynthesis. Mycologia 99:232

Okudaira M, Kurata H, Sakabe F (1977) Studies on the fungal flora in the lung of human necropsy cases. A critical survey in connection with the pathogenesis of opportunistic fungus infections. Mycopathologia 61:3–18

Palella FJ Jr, Delaney KM, Moorman AC, Loveless MO, Fuhrer J, Satten GA, Aschman DJ, Holmberg SD (1998) Declining morbidity and mortality among patients with advanced human immunodeficiency virus infection. HIV outpatient study Investigators. N Engl J Med 338:853–860

Papadopoulos NG, Kalobatsou A (2007) Respiratory viruses in childhood asthma. Curr Opin Allergy Clin Immunol 7:91–95

Papi A, Bellettato CM, Braccioni F, Romagnoli M, Casolari P, Caramori G, Fabbri LM, Johnston SL (2006) Infections and airway inflammation in chronic obstructive pulmonary disease severe exacerbations. Am J Respir Crit Care Med 173:1114–1121

Park JH, Gold DR, Spiegelman DL, Burge HA, Milton DK (2001) House dust endotoxin and wheeze in the first year of life. Am J Respir Crit Care Med 163:322–328

Pasqualotto AC, Powell G, Niven R, Denning DW (2009) The effects of antifungal therapy on severe asthma with fungal sensitization and allergic bronchopulmonary aspergillosis. Respirology 14:1121–1127

Paster BJ, Boches SK, Galvin JL, Ericson RE, Lau CN, Levanos VA, Sahasrabudhe A, Dewhirst FE (2001) Bacterial diversity in human subgingival plaque. J Bacteriol 183: 3770–3783

Patel IS, Seemungal TA, Wilks M, Lloyd-Owen SJ, Donaldson GC, Wedzicha JA (2002) Relationship between bacterial colonisation and the frequency, character, and severity of COPD exacerbations. Thorax 57:759–764

Pattemore PK, Johnston SL, Bardin PG (1992) Viruses as precipitants of asthma symptoms. I. Epidemiology. Clin Exp Allergy 22:325–336

Peebles RS Jr (2004) Viral infections, atopy, and asthma: is there a causal relationship? J Allergy Clin Immunol 113:S15–18

Petty TL (2000) Scope of the COPD problem in North America: early studies of prevalence and NHANESIII data: basis for early identification and intervention. Chest 117:326S–331S

Probst M, Ries H, Schmidt-Wieland T, Serr A (2000) Detection of *Pneumocystis carinii* DNA in patients with chronic lung diseases. Eur J Clin Microbiol Infect Dis 19:644–645

Pulimood TB, Corden JM, Bryden C, Sharples L, Nasser SM (2007) Epidemic asthma and the role of the fungal mold *Alternaria alternata*. J Allergy Clin Immunol 120:610–617

Raju B, Hoshino Y, Belitskaya-Lévy I et al (2008) Gene expression profiles of bronchoalveolar cells in pulmonary TB. Tuberculosis 88:39–51

Ran H, Hassett DJ, Lau GW (2003) Human targets of *Pseudomonas aeruginosa* pyocyanin. Proc Natl Acad Sci USA 100:1315–14320

Relman DA, Falkow S, LeBoit PE, Perkocha LA, Min KW, Welch DF, Slater LN (1991).The organism causing bacillary angiomatosis, peliosis hepatis, and fever and bacteremia in immunocompromised patients. N Engl J Med 324:1514

Retamales I, Elliott WM, Meshi B, Coxson HO, Pare PD, Sciurba FC, Rogers RM, Hayashi S, Hogg JC (2001) Amplification of inflammation in emphysema and its association with latent adenoviral infection. Am J Respir Crit Care Med 164:469–473

Reynolds HY, Chretien J (1984) Respiratory tract fluids: analysis of content and contemporary use in understanding lung diseases. Dis Mon 30:1–103

Rokas A, Payne G, Fedorova ND et al (2007) What can comparative genomics tell us about species concepts in the genus *Aspergillus*? Stud Mycol 59:11

Saetta M, Di Stefano A, Turato G, Facchini FM, Corbino L, Mapp CE, Maestrelli P, Ciaccia A, Fabbri LM (1998) CD8+ T lymphocytes in peripheral airways of smokers with chronic obstructive pulmonary disease. Am J Respir Crit Care Med 157:822–826

Salo PM, Arbes SJ Jr, Sever M, Jaramillo R, Cohn RD, London SJ, Zeldin DC (2006) Exposure to *Alternaria alternata* in US homes is associated with asthma symptoms. J Allergy Clin Immunol 118:892–898

Scott JP, Fradet G, Smyth RL, Mullins P, Pratt A, Clelland CA, Higenbottam T, Wallwork J (1991) Prospective study of transbronchial biopsies in the management of heart–lung and single lung transplant patients. J Heart Lung Transplant 10:626–636;Discussion 636–627

Seemungal T, Harper-Owen R, Bhowmik A et al (2001) Respiratory viruses, symptoms, and inflammatory markers in acute exacerbations and stable chronic obstructive pulmonary disease. Am J Respir Crit Care Med 164:1618–1623

Segreti J, House HR, Siegel RE (2005) Principles of antibiotic treatment of community-acquired pneumonia in the outpatient setting. Am J Med 118:21–28

Sethi S (2000a) Bacterial infection and the pathogenesis of COPD. Chest 117:286S–291S

Sethi S (2000b) Infectious etiology of acute exacerbations of chronic bronchitis. Chest 117: 380S 385S

Sethi S (2004) Bacteria in exacerbations of chronic obstructive pulmonary disease: phenomenon or epiphenomenon? Proc Am Thorac Soc 1:109–114

Sethi S, Murphy TF (2008) Infection in the pathogenesis and course of chronic obstructive pulmonary disease. N Engl J Med 359:2355–2365

Shirakawa T, Enomoto T, Shimazu S, Hopkin JM (1997) The inverse association between tuberculin responses and atopic disorder. Science 275:77–79

Shipley TW, Kling HM, Morris A, Patil S, Kristoff J, Guyach SE, Murphy JM, Shao X, Sciuraba FC, Rodgers RM, Richards T, Thompson P, Montelaro RC, Coxson HO, Higg JC, Norris KA (2010) Persistent pneumocystis colonization leads to development of chronic obstructive pulmonary disease in a non-human primate model of AIDS. J Inf Dis 202:302–312

Shoseyov D, Brownlee KG, Conway SP, Kerem E (2006) Aspergillus bronchitis in cystic fibrosis. Chest 130:222–226
Sigurs N, Gustafsson PM, Bjarnason R, Lundberg F, Schmidt S, Sigurbergsson F, Kjellman B (2005) Severe respiratory syncytial virus bronchiolitis in infancy and asthma and allergy at age 13. Am J Respir Crit Care Med 171:137–141
Singh AM, Moore PE, Gern JE, Lemanske RF Jr, Hartert TV (2007) Bronchiolitis to asthma: a review and call for studies of gene–virus interactions in asthma causation. Am J Respir Crit Care Med 175:108–119
Soong LB, Ganz T, Ellison A, Caughey GH (1997) Purification and characterization of defensins from cystic fibrosis sputum. Inflamm Res 46:98–102
Speert DP (2006) Bacterial infections of the lung in normal and immunodeficient patients. Novartis Found Symp 279:42–51
Stockley RA, Hill AT, Hill SL, Campbell EJ (2000). Bronchial inflammation: its relationship to colonizing microbial load and alpha(1)-antitrypsin deficiency. Chest 117: 291S–293S
Stone KC, Mercer RR, Freeman BA, Chang LY, Crapo JD (1992) Distribution of lung cell numbers and volumes between alveolar and nonalveolar tissue. Am Rev Respir Dis 146:454
Strachan DP (1989) Hay fever, hygiene, and household size. Br Med J 299:1259–1260
Syhre M, Scotter JM, Chambers ST (2008) Investigation into the production of 2-Pentylfuran by *Aspergillus fumigatus* and other respiratory pathogens in vitro and human breath samples. Med Mycol 46:209–215
Taylor DR, Cowan JO, Greene JM, Willan AR, Sears MR (2005) Asthma in remission: can relapse in early adulthood be predicted at 18 years of age? Chest 127:845–850
Thavagnanam S, Fleming J, Bromley A, Shields MD, Cardwell CR (2008) A meta-analysis of the association between caesarean section and childhood asthma. Clin Exp Allergy 38:629–633
Thomson NC, Chaudhuri R, Livingston E (2004) Asthma and cigarette smoking. Eur Respir J 24:822–833
Trulock EP, Ettinger NA, Brunt EM, Pasque MK, Kaiser LR, Cooper JD (1992) The role of transbronchial lung biopsy in the treatment of lung transplant recipients. An analysis of 200 consecutive procedures. Chest 102:1049–1054
Tunney MM, Field TR, Moriarty TF et al (2008) Detection of anaerobic bacteria in high numbers in sputum from patients with cystic fibrosis. Am J Respir Crit Care Med 177: 995–1001
Turnbaugh PJ, Ley RE, Mahowald MA, Magrini V, Mardis ER, Gordon JI (2006) An obesity-associated gut microbiome with increased capacity for energy harvest. Nature 444: 1027–1031
Twigg HL, Soliman DM, Day RB, Knox KS, Anderson RJ, Wilkes DS, Schnizlein-Bick CT (1999) Lymphocytic alveolitis, bronchoalveolar lavage viral load, and outcome in human immunodeficiency virus infection. Am J Respir Crit Care Med 159:1439–1444
Valenza G, Tappe D, Turnwald D, Frosch M, Konig C, Hebestreit H, Abele-Horn M (2008) Prevalence and antimicrobial susceptibility of microorganisms isolated from sputa of patients with cystic fibrosis. J Cyst Fibros 7:123–127
Vastag E, Matthys H, Kohler D, Gronbeck L, Daikeler G (1985) Mucociliary clearance and airways obstruction in smokers, ex-smokers and normal subjects who never smoked. Eur J Respir Dis Suppl 139:93–100
Venarske DL, Busse WW, Griffin MR et al (2006). The relationship of rhinovirus-associated asthma hospitalizations with inhaled corticosteroids and smoking. J Infect Dis 193:1536–1543
Verra F, Escudier E, Lebargy F, Bernaudin JF, De Cremoux H, Bignon J (1995) Ciliary abnormalities in bronchial epithelium of smokers, ex-smokers, and nonsmokers. Am J Respir Crit Care Med 151:630–634
von Mutius E, Braun-Fahrlander C, Schierl R, Riedler J, Ehlermann S, Maisch S, Waser M, Nowak D (2000) Exposure to endotoxin or other bacterial components might protect against the development of atopy. Clin Exp Allergy 30:1230–1234

Vonk JM, Postma DS, Boezen HM, Grol MH, Schouten JP, Koeter GH, Gerritsen J (2004) Childhood factors associated with asthma remission after 30 year follow up. Thorax 59: 925–929

Wark PA, Johnston SL, Moric I, Simpson JL, Hensley MJ, Gibson PG (2002) Neutrophil degranulation and cell lysis is associated with clinical severity in virus-induced asthma. Eur Respir J 19:68–75

Weycker D, Edelsberg J, Oster G, Tino G (2005) Prevalence and economic burden of bronchiectasis. Clin Pulm Med 12:205–209

White NC, Israel-Biet D, Coker RJ, Mitchell DM, Weber JN, Clarke JR (2004) Different resistance mutations can be detected simultaneously in the blood and the lung of HIV-1 infected individuals on antiretroviral therapy. J Med Virol 72:352–357

Williams JV, Harris PA, Tollefson SJ, Halburnt-Rush LL, Pingsterhaus JM, Edwards KM, Wright PF, Crowe JE Jr (2004) Human metapneumovirus and lower respiratory tract disease in otherwise healthy infants and children. N Engl J Med 350:443–450

Willner D, Furlan M, Haynes M et al (2009) Metagenomic analysis of respiratory tract DNA viral communities in cystic fibrosis and non-cystic fibrosis individuals. PLOS One 4:e7370

Woodcock AA, Steel N, Moore CB, Howard SJ, Custovic A, Denning DW (2006) Fungal contamination of bedding. Allergy 61:140–142

Worlitzsch D, Rintelen C, Bohm K, Wollschlager B, Merkel N, Borneff-Lipp M, Doring G (2009) Antibiotic-resistant obligate anaerobes during exacerbations of cystic fibrosis patients. Clin Microbiol Infect 15:454–460

Wu P, Dupont WD, Griffin MR, Carroll KN. Mitchel EF, Gebretsadik T, Hartert TV (2008) Evidence of a causal role of winter virus infection during infancy in early childhood asthma. Am J Respir Crit Care Med 178:1123–1129

Wu Q, Chu HW (2009) Role of infections in the induction and development of asthma: genetic and inflammatory drivers. Expert Rev Clin Immunol 5:97–109

Yang X, Fan Y, Wang S, Han X, Yang J, Bilenki L, Chen L (2002) Mycobacterial infection inhibits established allergic inflammatory responses via alteration of cytokine production and vascular cell adhesion molecule-1 expression. Immunology 105:336–343

Zavala DC, Richardson RH, Mukerjee PK, Rossi NP, Bedell GN (1973) Use of the bronchofiberscope for bronchial brush biopsy. Diagnostic results and comparison with other brushing techniques. Chest 63:889–892

Zhang H, Su YA, Hu P, Yang J, Zheng B, Wu P, Peng J, Tang Y, Zhang L (2006) Signature patterns revealed by microarray analyses of mice infected with influenza virus A and *Streptococcus pneumoniae*. Microbes Infect 8:2172–2185

Zhou X, Bent SJ, Schneider MG, Davis CC, Islam MR, Forney LJ (2004) Characterization of vaginal microbial communities in adult healthy women using cultivation-independent methods. Microbiology 150:2565–2573

Chapter 8
The Human Skin Microbiome in Health and Skin Diseases

Huiying Li

Human Skin Microbiome

The human skin, as the largest organ of the human body, protects the underlying tissues and plays an important role as a front-line defense system against external environmental changes and invading pathogens. It is colonized by a unique and complex microbial ecosystem, including bacteria, fungi, and bacteriophages, some of which could become pathogenic under certain circumstances. The skin microbiota is complex. Several hundred different microbial species reside on the skin. Its composition and distribution are uniquely different from the flora of other organs.

In the past, microorganisms on the skin have been identified mainly by culture-based methods. These methods have limitations and are biased toward the "cultivatable" microorganisms under the specific culturing conditions. Many microbes that are fastidious and rare are difficult to identify from a mixed microbial cultures, and some species cannot grow under known culturing conditions. Therefore, the composition and distribution of the microbes on the human skin were not extensively described until culture-independent molecular methods have been used.

In recent years, molecular characterizations of the human external skin microbiota based on 16S ribosomal RNA (rRNA) analysis have been carried out on a large scale (Costello et al., 2009; Gao et al., 2007; Grice et al., 2009, 2008). These studies have revealed the human skin microbiota to be much more diverse than previously demonstrated using culture-based methods. Although only a few phyla are represented, the skin microbiota is highly diverse at the species level – several hundreds of species have been identified.

The human skin microbiota can be affected by many environmental factors, such as temperature, humidity, pH, exposure to air and light, and host factors, such as genetic background, body locations, gender, immune response, hygiene habit, use of antibiotics and antimicrobial detergent, and cosmetic use. Nearly 40 years ago, Marples (1969) pointed out that no area of the skin can be taken as representative

H. Li (✉)
Crump Institute for Molecular Imaging, California NanoSystems Institute, Room 4339, 570 Westwood Plaza, Los Angeles, CA 90095-1770, USA
e-mail: huiying@mednet.ucla.edu

K.E. Nelson (ed.), *Metagenomics of the Human Body*,
DOI 10.1007/978-1-4419-7089-3_8, © Springer Science+Business Media, LLC 2011

of the whole surface. Great differences always exist between different people and between different areas of the skin on the same person (Holt, 1971; Noble, 1969; Shaw et al., 1970). Recent 16S rRNA-based methods also revealed that significant variations exist between individuals, and between different sampling sites within the same individual (Costello et al., 2009; Fierer et al., 2008; Gao et al., 2007; Grice et al., 2009, 2008). The skin microbiome of the same site from the same individual appears to be less variable within a few months.

The main techniques in sampling the skin microbiota include swabbing, scraping, punch biopsy, and taping using adhesive tapes. Both swabbing and taping are simple, quick, and noninvasive. Scraping usually comprises significant amount of human skin cells. Punch biopsy is invasive, but can cover all layers of the skin microbiota. Grice et al. (2008) have reported the skin microbiota composition using three different sampling techniques, swabbing, scraping, and punch biopsy, to survey the skin microbiota at all depths of the skin. They concluded that the dominant operational taxonomic units (OTUs) were captured by all three methods, although the rare OTUs are different.

Bacterial Composition of the Skin Microbiota

Commensal bacteria play a crucial role in the development of the immune system in humans. They are exposed to the external environment and constantly signal and interact with the host by activating innate and adaptive immunity. Reduced exposure to microorganisms in early childhood may result in an abnormal immune response to commensal bacteria, which are otherwise innocuous. Commensal bacteria may be protective of the host to other pathogenic species and could trigger diseases under specific circumstances. It has been suggested that targeted therapies to maintain healthy skin may require not only eliminating pathogenic bacteria, but also promoting the growth of commensal microorganisms (Grice et al., 2009).

It has been estimated that the human skin harbors hundreds of bacterial species and that the number of bacterial cells of the skin is on the order of 10^{12}. Recent progress in the study of the human skin microbiome using large-scale sequencing of the 16S rRNA has revealed the diversity and variation of the skin microbiome (Costello et al., 2009; Fierer et al., 2008; Gao et al., 2007; Grice et al., 2009, 2008). These studies have shown that highly diverse microorganisms from hundreds of different species make up the skin microbiota. Among them, three major bacterial phyla, Actinobacteria, Firmicutes, and Proteobacteria, account for more than 90% of the microbiota (Gao et al., 2007). The bacterial distribution also varies among different skin sites, such as dry, moist, or lipid-rich locations (Grice et al., 2009). The balance between various commensal bacterial organisms of the human skin may function as part of the skin defense system and protect the skin from pathogen invasions.

In studying the bacterial composition of the volar forearm of six healthy subjects using 16S rRNA sequencing, Gao et al. (2007) have found bacteria from eight phyla, 182 species-level OTUs (SLOTUs), which was estimated as representing 74% of the

SLOTUs in this ecosystem. Actinobacteria, Firmicutes, and Proteobacteria were the top three dominant phyla representing 94.6% of all the sequences. The most abundant genera among the six individuals are *Propionibacterium*, *Corynebacterium*, *Staphylococcus*, *Streptococcus*, *Acinetobacter*, and *Finegoldia.*

In sampling the inner elbow of five healthy subjects, Grice et al. (2008) identified bacteria from six phyla and 113 SLOTUs. The authors suggested that it was not a complete sampling of the species and the OTU richness will increase with additional sequencing. It was estimated that at least another 15–20 more OTUs per arm per subject have yet to be revealed. Proteobacteria dominated the microbiota of the sampling site. When comparing this study with the study of Gao et al. (2007), many OTUs overlap. However, the most abundant OTUs from Proteobacteria in this study were not present in the study of the forearm. This could reflect the biological differences between the two sites, and could also be due to technical differences. In summary, these two studies suggest that the human skin harbors possibly several hundreds of bacterial species and different body sites may have different microbial composition, as will be discussed in the next section.

It has been suggested that the skin microbiota consists of long-term resident microbes and transient, short-term residents. The long-term resident microbes are the ones that are consistently detected from the human skin (Holland and Bojar, 2002); *Staphylococcus*, *Propionibacteria*, *Micrococcus*, *Corynebacterium*, *Brevibacteria*, and *Acinetobacter* species, along with others, are commonly cultivated from normal skin. Two commensal bacterial species are thought to be the dominant species – *Staphylococcus epidermidis* and *Propionibacterium acnes*. Both species have been cultivated and their genomes have been fully sequenced (Bruggemann et al., 2004; Zhang et al., 2003).

Staphylococcus epidermidis is a Gram-positive, facultative anaerobe. It is one of the most common clinical isolates of the skin microbiota. It is thought to comprise >90% of the aerobic resident flora of the skin (Cogen et al., 2008). *S. epidermidis* has been proposed to function in preventing colonization of other pathogens, such as *Staphylococcus aureus* (Lina et al., 2003). However, the organism is also an important cause of intravenous catheter-related infections. The bacterium also can form biofilms.

P. acnes is an aerotolerant anaerobic Gram-positive bacterium that was first isolated in 1915 by Harris and Wade from an acne lesion (Harris and Wade, 1915). Despite being one of the most prominent members of the normal human skin commensals, there is very limited knowledge regarding the genetic diversity of *P. acnes*. Methods including phage-typing, serotyping, and mass spectrometry have been used in an effort to define the extent of diversity among *P. acnes* isolates (Cummins, 1976; Goodacre et al., 1996; Webster and Cummins, 1978). *P. acnes* has been divided into two serotypes based on Cummings (1976). Despite hundreds of clinical strains being isolated, to this date, only one strain has been completely sequenced (Bruggemann et al., 2004), although several are known to be in progress as a result of the US National Institutes of Health Human Microbiome Project (HMP: http://nihroadmap.nih.gov/hmp/) and there are no high-resolution genotyping data such as multilocus sequence typing (MLST). As the HMP progresses, we expect to gain more insights on the strain diversity and the pan-genome structure of this important commensal bacterium by sequencing and comparing the genomes of multiple *P. acnes* isolates.

Skin Microbiome Variation Among Different Body Sites

Human skin is a habitat of multiple microbial communities. Skin commensal microbes co-evolve with the host and adapt to the unique environment of the specific body sites and skin structures where they reside. The skin microbiota can differ significantly among different anatomical sites. Bacterial density on the skin can vary from a few hundred per cm^2 on dry surfaces to $10{,}000/cm^2$ in the moist areas (Ruocco et al., 2007).

Grice et al. (2009) have reported the topographical diversity of the human skin microbiota. They have surveyed bacterial composition from 20 diverse skin sites on 10 healthy human individuals based on nearly full-length 16S rRNA gene sequences. They mainly compared three microenvironment types: sebaceous, moist, and dry locations. The 20 skin sites include glabella, alar crease, external auditory canal, occiput, manubrium, back, nare, axillary vault, antecubital fossa, interdigital web space, inguinal crease, gluteal crease, popliteal fossa, plantar heel, umbilicus, volar forearm, hypothenar palm, and buttock. In general, the bacterial composition at sebaceous sites is less diverse, less even, and less rich than moist and dry sites. In other words, sebaceous sites are dominated by only a few major OTUs. The most abundant species found in sebaceous sites include *Propionibacteria* spp., *Corynebacteria* spp., other actinobacteriales, and *Staphylococci* spp. This conclusion is consistent with the findings by our group in the study of the pilosebaceous units on the face, which is rich in lipids. We found that ~90% of all the 16S rRNA clones are from one species, *P. acnes* (unpublished data). This suggests that the microbiota in the highly specialized microcompartment has evolved and adapted to the unique environment. In moist sites, *Corynebacteria* species are more abundant. The dry sites have increased β-Proteobacteria and Flavobacteriales with a mixed population of bacteria (Grice et al., 2009).

Costello et al. (2009) studied the bacterial distribution of more than 10 skin locations and a few other body sites (oral cavity and gut) by sequencing the variable region 2 (V2) of the 16S rRNA gene using 454/Roche high-throughput sequencing platform. The skin sites include forehead, hair, external nose, external ear, external auditory canal, nostril, navel, palm, index finger, forearm, armpit, back of knee, sole of foot, glans penis, and labia minora. Some of the sites overlap with the experiments performed by Grice et al. (2009). Based on the sequence information, the authors found that the communities were grouped strongly by body site, then by host individual, and lastly by the sampling time point (Costello et al., 2009). This conclusion again shows that each body site has its unique microenvironment, thus unique bacterial communities have evolved with the environment of specific body sites, such as lipid-rich, moist, or dry sites.

When considering the data, one issue to point out is that pH of the skin could play a role in the microbiota composition in addition to the factors mentioned above, since different dry or moist sites may have different pH. Matousek and Campbell (2002) summarized the cutaneous pH variations among different anatomical locations. Areas that are occluded (axillae, genitoanal region, intertriginous, and interdigital) have a relatively higher pH. This could be due to the fact that occlusion of skin areas limits evaporation of eccrine perspiration (Goodman, 1943).

Left and Right Symmetry

The human body has a twofold symmetry. Is the skin microbiota on the left and right sides symmetric?

Studies by Costello et al. (2009), Fierer et al. (2008), Gao et al. (2007), and Grice et al. (2009) sampled symmetric skin sites, and concluded that there is a symmetry between the microbiota on the left and right sides.

Gao et al. (2007) observed a high level of conservation between the two symmetric body sites. Although the microbiome changes significantly over time, the conservation between the two sides was relatively stable. This indicates that although the bacterial biota is dynamic over time, it changes in both sides similarly. Grice et al. (2009) analyzed three paired symmetric sites (left and right antecubital fossae, axillae, and forearms) and showed that the left and right sites are more similar to each other than to other subjects. Costello et al. (2009) showed that the microbiome of the left and right sides of most of the same skin sites are grouped together, except the palm and index finger.

To assess whether the biogeographic patterns on different skin sites arise due to the microenvironmental factors or previous exposures to microbes or both, Costello et al. (2009) performed "inoculation" experiments. They disinfected the foreheads and left volar forearms of volunteers and then transplanted with their tongue bacteria. Forehead and forearm gave different results at 8 h post-transplantation. Forearm communities became more similar to tongue communities than the native communities in composition and diversity. However, forehead became more similar to original forehead community over time. Although it is an experiment with a short observation time and the conclusions might change as the post-transplant time extends, it suggests that the microenvironment of the skin sites plays an important role in selecting the microbes and shaping the bacterial communities.

Skin Microbiome Variation Among Individuals

The bacterial biota in normal skin is highly diverse. There is low-level interpersonal consensus. Gao et al. (2007) observed that the microbiome differed significantly among the six individuals that they studied. Only 2.2% of the SLOTUs and 6.6% of the 119 genera (four major ones – *Propionibacteria*, *Corynebacteria*, *Staphylococcus*, and *Acinetobacter*) were shared among the six individuals. However, 71.4% of all the SLOTUs and 68.1% of the genera were found only from a single individual.

Grice et al. (2009) compared the interpersonal variation among different skin sites. The most variable sites among different individuals are interdigital web spaces, toe webs, axillae, and umbilici. The least variable sites include alar creasese, nares, and backs, where sebaceous glands are abundant. Intrapersonal variation is less than interpersonal variation.

Temporal Variation of the Skin Microbiome

The human skin microbiota varies with time as the environmental factors or host factors change.

In the study of the skin microbiome of the forearm, Gao et al. (2007) resampled four of the six subjects 8–10 months later. Two additional phyla and 65 more SLOTUs were revealed in the second sampling. When the similarity of the microbiota among different samples was analyzed, the authors showed that the microbiota of the same subject at different time points was not more similar to each other than to samples from other subjects. This study suggested that the skin microbiota from the same individual could have significant temporal variation.

Grice et al. (2009) collected follow-up samples from five of the ten subjects 4–6 months later. Although some sites (external auditory canal, inguinal crease, alar crease, and nare) were relatively stable, significant variation was seen in the popliteal fossa, volar forearm, and buttock. This suggests that the longitudinal stability of the microbiota is also site dependent. The authors noted that overall, most of the subjects were significantly more similar to themselves over time than to other subjects. This conclusion is different from the one by Gao et al. (2007). This inconsistency could be due to the differences in resampling time interval, the sampling sites, and the sequencing depth.

Costello et al. (2009) resampled subjects 24 h and 3 months later, respectively. The microbiota was found to vary most significantly among different body sites, then by individuals, and finally by time. Variation of the microbiota was significantly less over 24 h than over 3 months. The authors suggested that the microbiota of each individual stays relatively stable over time.

Eukaryotes of the Skin Microbiota

Human skin is colonized not only by bacteria, but also by small eukaryotes, mainly fungi. Most of the fungal organisms in the normal healthy human forearm skin belong to the genus *Malassezia*. *Malassezia* species are members of the normal skin microbiota and have been associated with various skin diseases such as seborrheic dermatitis, dandruff, and atopic dermatitis.

Currently there are 13 species classified in the genus *Malassezia*. Most of the species are lipid dependent. Although culture-based methods have been used to examine the distribution of the *Malassezia* species in different skin locations, the conclusions could be biased by the limitation in culturing conditions.

Paulino et al. (2006) used broad-range primers (modified primer EF4 and primer 1,536) to amplify fungal 18S rRNA from the skin of the forearms of normal individuals and patients with psoriasis. More than 95% of the sequences were classified into two unknown phylotypes, which are related to *Malassezia furfur* (<97% identity). When using the primers that specifically amplify the 5.8S rRNA gene and ITS2 from *Malassezia* species, five different *Malassezia* species were identified from five healthy subjects, including *M. restricta*, *M. globosa*, *M. sympodialis*, *M. pachydermatis*, and *M. furfur*. *M. restricta* and *M. globosa* were the most abundant of the *Malassezia* species identified. These two species were previously identified by molecular and culture-dependent methods. In addition, four unknown phylotypes were also found. Three of them are closely related to *M. restricta* and one is related to *M. sympodialis*. These unknown phylotypes may represent new *Malassezia* species.

The *Malassezia* species identified from the samples collected from the same body site from the same donor obtained 6–10 months later were similar to the first sampling in four of the five subjects. This suggests that the *Malassezia* microbiota is relatively stable over the time period (Paulino et al., 2006).

Paulino et al. (2008) further analyzed the distribution of the *Malassezia* species on multiple body locations using multiplex real-time PCR. The microbiota from left and right forearms, forehead, scalp, upper back, and lower back was analyzed to detect and quantify the *Malassezia* species. Although only one normal subject and one patient with psoriasis were analyzed, this study revealed that, similar to the *Malassezia* biota in the forearm, *M. restricta* was the most abundant species, accounting for 57–100% and 7–99% of all the *Malassezia* organisms in the healthy subject and the patient, respectively. Based on the species composition, different body sites from the same host clustered together suggesting that, in this case, host factor plays a more important role in determining the *Malassezia* species distribution. There was no significant difference between samples from healthy skin and psoriatic lesions.

Skin Microbiome and Diseases

The microbial composition and distribution of the human skin microbiota have been associated with multiple skin diseases, including acne vulgaris, atopic dermatitis, psoriasis, rosacea, dandruff, seborrheic dermatitis, and folliculitis.

Acne Vulgaris

Acne vulgaris (commonly known as acne) is the most common skin disorder affecting approximately 50 million people just in the US, and much more throughout the world. It is a global disease that has no predilection for a specific race or gender. More than 80% of the population suffers from acne at some point in their life. Historical record suggests that several ancient Egyptian pharaohs had acne, and the ancient Roman writer Celsus described the skin condition and its treatment in *De Medicina*. The English word "acne" itself likely was a corruption of the Greek term "acme", used by the Byzantine physician and medical writer Aetius Amidenus to describe skin eruptions. The disease has significant morbidity and profoundly affects patients' self-esteem, especially in adolescent population. Acne can be extremely painful and cause lasting marks or scars as well as lead to psychosocial suffering. Even mild-to-moderate disease can be associated with significant depression and suicidal ideation. Economically, it is estimated that US consumers spend more than 1.2 billion dollars each year for the treatment of acne (Lee et al., 2003).

Pathogenesis, Clinical Presentation, and Treatment of Acne Vulgaris

Despite the medical and economical impacts of the disease, pathogenesis and treatment of acne remain incompletely defined. Acne is a multifactorial disease. Four basic mechanisms contributing to acne are hormones, increased sebum production, changes inside hair follicles, and bacteria. The most commonly cited theory regarding the pathogenesis of acne states that increased sebum production, usually associated with increase in androgen hormone with onset of puberty, leads to alterations in the lipid composition of hair follicles (Cunliffe, 2002). The altered microenvironment is then hypothesized to promote propagation of an acnegenic microbial community, which leads to further pathology, including initiation of host inflammatory response and eventual progression to advanced acne (Bojar and Holland, 2004). One species of bacteria that has been frequently implicated as the key player of the acnegenic microbes is *P. acnes* (Bojar and Holland, 2004). In inflammatory acne, *P. acnes* plays a key role in eliciting a host inflammatory response that is thought to be important for the pathogenesis of the disease (Webster, 1995).

Clinically, acne lesions are limited to skin with a high density of pilosebaceous units (consisting of hair follicle and its associated sebaceous gland) and are characterized by a continuum of lesions from noninflammed comedones and follicular papules to inflammatory papules, pustules, and nodules (Strauss and Thiboutot, 1999). Acne lesions can be graded on a severity scale starting with microcomedones, which are not readily visible to the naked eye, and consist of bacteria, keratinocytes, and sebum forming a plug. The microcomedone is usually considered the preclinical stage of acne. However, the experiments of Strauss and Kligman (1960) were pivotal in showing that the microcomedone is one pathogenetic feature of acne. The further studies of Plewig et al. (1971) and Holmes et al. (1972) showed that even normal-looking skin from acne patients may already be developing very early microcomedones; this was further supported by the work of Cunliffe et al. (2000) using in situ hybridization techniques. As microcomedones enlarge, infundibulum dilates and follicular walls thin, they become visible to the naked eye and are called comedones. Comedones can be classified as closed (whitehead) or open (blackhead) and are considered stage I of acne. Both types of lesions are associated with mild localized inflammation in the dermis. In some cases, an intense inflammatory reaction is initiated within the follicles when they rupture and leads to formations of undrained papules and pustules. In rare instances, the lesions proceed to further enlarge and become nodules and cysts. Scars are formed often when the follicular damage is severe.

Current therapy for acne includes topical and oral antibiotics as well as benzoyl peroxide that are active against *P. acnes* (Katsambas and Dessinioti, 2008). Topical and systemic retinoids (derivatives of vitamin A) are also widely prescribed as a treatment for moderate-to-severe acne (Katsambas and Dessinioti, 2008). The efficacy of retinoids has been attributed to decreased sebum production, decreased desmosomal attachment of keratinocytes lining the pilosebaceous unit, inhibition of Toll-like receptor-mediated inflammatory events (Liu et al., 2006), and enhanced macrophage microbicidal activity (Liu et al., 2008a). Severe side effects including birth defect and depression have been reported in patients treated with retinoids. Intralesional and oral steroids are occasionally used to reduce inflammation in severe cases of acne (Olsen, 1982).

Propionibacterium acnes and Acne

P. acnes has long been implicated as a microbial factor in the pathogenesis of acne. However, the causal link between *P. acnes* and the development of acne has never been proven. *P. acnes* can be recovered from skin surface as well as follicles on both normal and acne skin (Bojar and Holland, 2004). There is also no correlation between the numbers of *P. acnes* bacteria within a lesion and the clinical stage of acne (Leyden et al., 1998). Preclinical lesions (microcomedones) have been shown to contain up to 10^6 organisms whereas Leeming et al. (1988) found no viable *P. acnes* in 20% of inflammatory acne lesions. On the other hand, a microbial etiology for acne development is strongly suggested by the fact that one of the most common and effective acne treatments is either topical or oral antibiotics, such as erythromycin and clindamycin with activity against *P. acnes* (Katsambas and Dessinioti, 2008). It is also supported in part by the reduced efficacy of these treatments when antibiotic-resistant *P. acnes* strains are present (Eady et al., 1989).

Even though the precise role of *P. acnes* in acne pathogenesis is far from being clearly understood, multiple studies have focused on putative "acnegenic components" of *P. acnes* as potential "virulence factors." One focus has been on the capacity for lipid metabolism of *P. acnes*. The sequenced genome of *P. acnes* strain KP171202 encodes multiple lipases/esterase, including GehA, an extracellular triacylglycerol lipase (Bruggemann et al., 2004). These lipases are hypothesized to alter lipid composition within the sebaceous gland environment, resulting in a host response that could lead to acne formation (Higaki et al., 2000). In particular, *P. acnes* is able to metabolize triglycerides into free fatty acids and glycerol. The free fatty acids have been shown to be able to act directly as an immunological stimulant (Strauss et al., 1976) or as a cytotoxic agent that leads to breakage in the follicular epithelium, allowing additional cellular components of *P. acnes* to induce inflammation (Higaki et al., 2000). Other proposed *P. acnes* virulence factors include enzymes that are involved in adherence and colonization of the follicle (Cogen et al., 2008). One example is hyaluronate lyase, which degrades hyaluronan in the extracellular matrix, potentially contributing to adherence and invasion (Steiner et al., 1997). The *P. acnes* genome encodes several sialidases and endoglycoceramidases, which may be involved in host tissue degradation (Bruggemann et al., 2004).

Follicular Microbiota and Acne

In addition to *P. acnes*, another bacterial species that is recovered frequently from sampling of pilosebaceous units is *S. epidermidis*. In fact, the earliest study of bacterial flora in acne done by Unna in 1891 reported two distinct microorganisms in acne: a microbacillus (*P. acnes*) that predominated in comedone and early lesions and a micrococcus that was more predominant in pustular lesions and later identified as *S. epidermidis* (Rosenberg, 1969).

Other *Propionibacterium* species have also been isolated from acne lesions, including *P. granulosum* (Gloor and Franke, 1978), which was also found in our study by 16S rRNA analysis. It is often significantly less abundant than *P. acnes* in acne lesions. Nevertheless, it has been shown to metabolize triglycerides, and in at least one study, its abundance appears to correlate with clinical acne (Gloor and Franke, 1978; Leyden et al., 1998). More severe acne has also been associated with higher number of *P. granulosum* bacteria (Marples et al., 1973).

In addition to bacterial species, another microbial group that has been linked to acne is the fungal genus *Malassezia*, of which *M. furfur* is the most commonly cultured species from human skin due to its role as the causative agent of pityriasis versicolor (Crespo-Erchiga and Florencio, 2006). Additional members including *M. globosa* and *M. sympodialis* can also cause pityriasis versicolor. As a whole, the genus is lipophilic and encodes multiple lipases that can degrade components of sebum (Dawson, 2007). Although members of the *Malassezia* genus have been implicated in the pathogenesis of seborrheic dermatitis and dandruff (Ro and Dawson, 2005), their link to acne remains far more speculative.

Propionibacterium acnes Bacteriophages

A surprisingly common component of the skin microbiota associated with acne that is starting to be appreciated is bacteriophage. Zierdt et al. (1968) isolated a phage, phage 174, from spontaneous plaques of a *P. acnes* isolate (at that time known as *Corynebacterium acnes*). Phage 174 was able to lyse nearly all *P. acnes* strains tested in the study (Zierdt et al., 1968). Subsequently, seven different phage types have been identified on the basis of their patterns of bacterial lysis (Bataille et al., 2002; Cunliffe, 2002). However, the study of *P. acnes* bacteriophages has until recently been limited to the development of phage typing systems to distinguish the different serotypes of *P. acnes* (Jong et al., 1975; Webster and Cummins, 1978). Molecular characterization of *P. acnes* phage has been lacking in the past decades. The first genomic sequence of a *P. acnes* bacteriophage, PA6, published in 2007, has provided new insight into *P. acnes* phage biology (Farrar et al., 2007). PA6 shows a high degree of similarity to mycobacteriophages both morphologically and genetically. The 29,739-bp genome codes for 48 genes and the organization of the PA6 genome is similar to, although smaller than, that of the temperate mycobacteriophages. Electron microscopy demonstrated an isometric head and a long flexible tail, similar to previous characterizations (Farrar et al., 2007). Characteristic genes were found to encode phage replication and assembly functions. A striking feature of the genome is the lack of known genes involved in lysogeny such as those conferring integrase functions. Even so, the turbid appearance of PA6 plaques suggests that it may be capable of lysogeny (Farrar et al., 2007).

Much remains unknown regarding the potential role, if any, of *P. acnes* phage in acne pathogenesis. For example, does it harbor an acnegenic determinant that is transmitted among *P. acnes* strains? Does lytic replication of *P. acnes* phage affect the population dynamics of *P. acnes* and potentially affect its acnegenic potential? With our efforts in sequencing multiple *P. acnes* phages and metagenomic sequencing of the acne microbiome comes the potential to genetically manipulate the host bacterium using phages and to answer these questions. Further studies of *P. acnes* bacteriophages may also lead to the development of phage therapy for acne. Such therapy may overcome the current problems surrounding long-term use of antibiotics and the emergence of resistance in *P. acnes* and other skin commensals (Farrar et al., 2007).

Host Factors in Acne Pathogenesis

A host response pattern combined with bacterial triggering is generally accepted as being important for acne pathogenesis.

In several twin studies, the development of acne has a high concordance rate of 80% or higher among monozygotic twins and a heritability estimate of greater than 50%, suggesting a heritable host component to the disease (Bataille et al., 2002). Several host factors have been implicated in the development of acne, particularly sebum production and host immune response. One of the earliest host events in acne formation is thought to be the narrowing at the top of the hair follicle (follicular infundibulum), caused by hyperkeratinization. The hyperkeratinization might be induced by neurologic or hormonal stimuli. In vitro, the same hyperkeratinization could be induced by chemokines such as interleukin-1α (Bataille et al., 2002).

Sebum is thought to be a key component of the milieu of sebaceous units that allow the growth and proliferation of the acne microbiota. Sebum production has long been linked to acne formation (Cunliffe, 2002). One of the reasons often cited for the prevalence of acne during puberty is the stimulatory effect of androgens on sebum production. However, no direct relationship between sebum production and severity of acne has ever been demonstrated, nor is there a correlation between androgen level and mild-to-moderate acne (Cibula et al., 2000; Darley et al., 1984).

It is not clear if the host inflammatory response is pathologic, although high-grade acne shows evidence of inflammation. Clinical and experimental evidence indicates that the immune response in acne includes both innate and adaptive immunity. Patients with acne have been shown to have elevated antibody titers to *P. acnes* (Gowland et al., 1978), increased lymphocyte transformation to *P. acnes* (Puhvel et al., 1977), delayed skin test reactivity to *P. acnes* (Kersey et al., 1980), an activated helper T-cell population (Puhvel et al., 1977), and complement deposition in the walls of dermal blood vessels (Dahl and McGibbon, 1979). Investigation of the inflammatory infiltrates in acne lesions demonstrates dominance of $CD4^+$ T cells in early acne lesions (6-h lesions), whereas neutrophils and macrophages are found in later lesions (24- to 72-h lesions) (Norris and Cunliffe, 1988). It is not known if all the immune responses are the effect or cause of severe acne. Furthermore, any link between host immune response in acne lesions to its microbiota will have to account for the fact that only a subset of all sebaceous units on a patient's skin develop into acne lesions at any given time.

As part of the HMP, the skin microbiome associated with acne will be analyzed in detail using metagenomic approaches. By elucidating the microbial component in acne, we hope to better understand the disease pathogenesis, to improve currently available therapy, and to develop effective preventive strategies for this significant cause of morbidity among adolescents.

Atopic Dermatitis

Atopic dermatitis (AD) is a chronic inflammatory skin disease. In industrialized countries, 10–20% of children and 1–3% of adults are affected by the disease (Bieber, 2008; Leung et al., 2004). More than 60% of the AD begins during the first year of life. The estimated prevalence of AD in the United States is 9–30%, and the prevalence has increased significantly in the past 30 years (Terui, 2009).

Clinical Presentation and Pathogenesis of Atopic Dermatitis

AD is characterized by itchy patches on the scalp, forehead, and face. Other common sites for these patches are hands, wrists, neck, upper chest, feet, and ankles. Other manifestations of AD include defects in the skin barrier function, causing dry skin, and IgE-mediated sensitization to food and environmental allergens (Akdis et al., 2006; Terui, 2009).

AD is associated with defective skin barrier function and abnormalities in cutaneous immunity, including decreased expression of antimicrobial peptides, increased Th2 cytokine expression, and low numbers of plasmacytoid dendritic cells. The disease had largely been thought to have a primarily immunologic etiopathogenesis. Recent findings of mutations in the filaggrin gene, encoding a structural protein in the skin, provided new insight on the mechanisms of the disease (O'Regan et al., 2009). The filaggrin gene mutations have been identified as a major risk factor for AD. More than 10 different mutations in filaggrin have been reported; however, two most common mutations (R501X and 2282del4) account for the majority of cases (Terui, 2009). In total, 18–48% of individuals with AD carry filaggrin gene null alleles (Cookson, 2004). Individuals with a deficiency of filaggrin show certain immune abnormalities, including increased sensitization to allergens, resulting in IgE antibodies to these allergens. Thus, it has been hypothesized that the heritable epithelial barrier defect due to reduced levels of filaggrin results in reduced epidermal defense mechanisms to allergens and microbes, which is followed by polarized Th2 lymphocyte responses with resultant chronic inflammation (O'Regan et al., 2009).

Skin Microbiota and Atopic Dermatitis

The skin of AD patients is highly susceptible to infection by viruses, bacteria, and fungi. These microorganisms interact with the human immune system to trigger the onset of or exacerbate the disease (Baker, 2006). Unlike other inflammatory skin diseases, AD patients tend to develop disseminated viral infections, which complicate the disease (Leung and Bieber, 2003; Wollenberg et al., 2003). Fungi are also major infectious agents of AD skin. In addition, it has been shown that several bacterial species, especially *S. aureus*, can be detected on the skin of more than 90% of patients with AD (Leung, 2003). Traditional culture analysis showed that *Staphylococcus* species are densely colonized on the skin in AD patients, even at sites that appear normal (Gloor et al., 1982). These observations are consistent with the understanding that patients with AD have chronic inflammation and defects in innate and adaptive immune responses, and thus have a general defect in the control of skin infections. It has been shown that *Staphylococcus* colonization occurs due to reduced secretion of antibiotic peptides in sweat (Rieg et al., 2005).

S. aureus is a Gram-positive coccus and also a major human pathogen. It is facultatively anaerobic, hemolytic, and coagulase-positive. It forms large, round, golden-yellow colonies. It can often be found as part of the microbiota in the nose and on the skin. Approximately 20% of the population are long-term carriers of *S. aureus* (Kluytmans et al., 1997). The colonization of *S. aureus* is associated with increased skin inflammation and exacerbations of AD. It has been suggested that products of *S. aureus*, such as enterotoxin A and enterotoxin B, which may be present in lesions, can function as superantigens (Cogen et al., 2008). This may lead to T-cell activation, inflammation cascade, and IgE antibody responses in patients with AD. The severity of AD may also correlate with the expression of superantigens.

Besides *S. aureus*, no other bacteria have been directly linked to AD. Dekio et al. (2007) applied T-RFLP, a bacterial 16S rRNA gene-based terminal RFLP, to analyze swab-scrubbed skin samples of forehead from AD patients and compared its microbiota with the ones from normal subjects. They suggested that bacteria other than *Staphylococcus* species and *P. acnes* should also be analyzed in AD patients.

In AD patients, the antecubital and popliteal fossae are sites that are preferentially involved in the disease. The bacterial communities at these sites in normal individuals are highly diverse and rich in species composition compared to other skin sites, harboring similar ranges of organisms (Grice et al., 2009). A hygiene hypothesis – reduced exposure to infections in the clean environment during early life – has been proposed to explain why AD has become increasingly prevalent and severe over the past 30–40 years in developed countries (Strachan, 1989). If the hypothesis is correct, we would expect to see that the species distribution of the commensal microbes in AD patients is different from the one on the skin of normal individuals where the innate and adaptive immune responses are adequate. A potentially "skewed" microbial community structure on the skin of AD patients may influence the colonization and infection by various microorganisms. Studying the skin microbiome in AD patients can help us better understand the disease pathogenesis and may provide insight on the treatment of the disease.

Psoriasis

Psoriasis is a chronic inflammatory disease mainly involving the skin, and sometimes affecting multiple organs. It is not contagious but often recurs. It varies in severity and area of the skin affected. The disease has significant impact on the quality of life of the patients (Choi and Koo, 2003; Gupta et al., 1993). Approximately 2% of the world's population has been affected by the disease. Psoriasis affects both genders and can occur at any age. The cost for treating the disease is over 3 billion dollars a year (Sander et al., 1993).

Clinical Presentation and Pathogenesis of Psoriasis

Psoriasis often causes red, scaly patches on the skin. The manifestation of the disease includes hyperkeratosis, hyperproliferation of keratinocytes, infiltration of skin by immune cells, and angiogenesis. Most commonly affected skin areas are elbows and knees. In contrast to AD, psoriasis is more often to be found on the extensor of the joints. Psoriasis can also cause inflammation of the joints, which is known as psoriatic arthritis. Ten to fifteen percent of the people with psoriasis have psoriatic arthritis.

The cause of psoriasis is poorly understood and currently is thought to be autoimmune in origin. About one-third of people with psoriasis report a family history of the disease. Researchers have identified genetic loci associated with the condition, including *HLA-C*, *IL12B*, *IL23R*, *IL23A*, *IL4/IL13*, *TNFAIP3*, and *TNIP1* (Capon et al., 2008, 2007; Cargill et al., 2007; Elder et al., 2010; Liu et al., 2008b; Nair et al., 2008; Tsunemi et al., 2002; Zhang et al., 2009). Studies of monozygotic twins suggest a 70% chance of a twin developing psoriasis if the other twin has psoriasis. The concordance is around 20% for dizygotic twins. These findings suggest that psoriasis is multifactorial, affected by both genetic predisposition of multiple genes and environmental factors including stress, trauma, and infections (Krueger and Ellis, 2005; Watson et al., 1972).

Skin Microbiota and Psoriasis

Several bacterial species, including *S. aureus* and *Streptococcus pyogenes*, have been suggested to play a role in the pathogenesis of psoriasis. Fungal organisms, including *M. furfur* and *Candida albicans*, have also been linked with the development of psoriatic skin lesions. With the advancement in the human microbiome research, it is now feasible to analyze whether the skin microbiome plays a role in the pathogenesis of psoriasis.

Paulino et al. (2006) compared the composition of fungal species and the distribution of *Malassezia* species in healthy skin and psoritic lesions and have not identified significant differences between the two groups.

Gao et al. (2008) used broad-range 16S rRNA PCR to analyze the bacterial biota in psoriasis. At the phylum level, *Actinobacteria* is the most prevalent phylum in normal skin (47.6%); however, it was detected less frequently from psoriatic lesions (37.3%). By contrast, Firmicutes became the most dominant phylum in diseased skin (46.2%), significantly overrepresented than the uninvolved skin from the same patients (39.0%) and from healthy subjects (24.4%). The decrease in the *Propionibacterium* species and increase in the genus *Streptococcus* can explain the main difference between the normal skin and psoriatic lesions. The ratio of *Streptococcus* to *Propionibacterium* is 0.4 in normal skin, but significantly increased to 5.0 in psoriatic lesion samples. *P. acnes* was the most prevalent species in the healthy subjects (20.2 ± 16.0%), but less represented in the normal skin samples from the psoriasis patients (11.7 ± 20.4%) and from the lesions (2.6 ± 5.1%). This suggests that the skin microbiome in psoriasis is complex and possibly different from the one from normal skin.

The study by Gao et al. (2008) suggests that the overall bacterial diversity of the microbiota in the psoriatic lesions was greater than in normal skin samples. However, as the authors suggested, this may reflect the heterogeneity of the lesions sampled, whereas the normal skin samples are relatively homogeneous (Gao et al., 2008). The authors also observed significant differences in the distribution of the three major bacterial phyla in the human skin biota, Actinobacteria, Firmicutes, and Proteobacteria. Firmicutes are overrepresented in the lesion whereas the other two are underrepresented. The underrepresentation of Actinobacteria in lesions is most significant in the genus *Propionibacterium* and its species *P. acnes*. The authors hypothesized that the decrease of the species may reflect a disordered ecological niche that these organisms have adapted to and this could lead to the pathogenesis of the disease. Another explanation could be that *P. acnes* plays a protective role functioning as a probiotic (Gao et al., 2008).

Other Skin Diseases and the Skin Microbiome

Many other skin diseases have been linked to microbes as well, such as rosacea, dandruff, seborrheic dermatitis, and folliculitis, but very few have been studied at the microbiome level using molecular and metagenomic approaches. Another actively investigated area is the microbiome associated with wounds. With the rapid development of high-throughput sequencing technologies and the field of metagenomics and bioinformatics, our understanding of the microbiome associated with skin infection and other skin diseases will shed light on the full spectrum of the skin microbes and their roles in skin health and disease and will provide insights on new preventive and therapeutic approaches of infection and other skin diseases.

References

Akdis CA, Akdis M, Bieber T, Bindslev-Jensen C, Boguniewicz M, Eigenmann P, Hamid Q, Kapp A, Leung DY, Lipozencic J et al (2006) Diagnosis and treatment of atopic dermatitis in children and adults: European Academy of Allergology and Clinical Immunology/American Academy of Allergy Asthma and Immunology/PRACTALL Consensus Report. J Allergy Clin Immunol 118:152–169

Baker BS (2006) The role of microorganisms in atopic dermatitis. Clin Exp Immunol 144:1–9

Bataille V, Snieder H, MacGregor AJ, Sasieni P, Spector TD (2002) The influence of genetics and environmental factors in the pathogenesis of acne: a twin study of acne in women. J Invest Dermatol 119:1317–1322

Bieber T (2008) Atopic dermatitis. N Engl J Med 358:1483–1494

Bojar RA, Holland KT (2004) Acne and *Propionibacterium acnes*. Clin Dermatol 22:375–379

Bruggemann H, Henne A, Hoster F, Liesegang H, Wiezer A, Strittmatter A, Hujer S, Durre P, Gottschalk G (2004) The complete genome sequence of *Propionibacterium acnes*, a commensal of human skin. Science 305:671–673

Capon F, Bijlmakers MJ, Wolf N, Quaranta M, Huffmeier U, Allen M, Timms K, Abkevich V, Gutin A, Smith R et al (2008) Identification of ZNF313/RNF114 as a novel psoriasis susceptibility gene. Hum Mol Genet 17:1938–1945

Capon F, Di Meglio P, Szaub J, Prescott NJ, Dunster C, Baumber L, Timms K, Gutin A, Abkevic V, Burden AD et al (2007) Sequence variants in the genes for the interleukin-23 receptor (IL23R) and its ligand (IL12B) confer protection against psoriasis. Hum Genet 122:201–206

Cargill M, Schrodi SJ, Chang M, Garcia VE, Brandon R, Callis KP, Matsunami N, Ardlie KG, Civello D, Catanese JJ et al (2007) A large-scale genetic association study confirms IL12B and leads to the identification of IL23R as psoriasis-risk genes. Am J Hum Genet 80: 273–290

Choi J, Koo JY (2003) Quality of life issues in psoriasis. J Am Acad Dermatol 49:S57–61

Cibula D, Hill M, Vohradnikova O, Kuzel D, Fanta M, Zivny J (2000) The role of androgens in determining acne severity in adult women. Br J Dermatol 143:399–404

Cogen AL, Nizet V, Gallo RL (2008) Skin microbiota: a source of disease or defence? Br J Dermatol 158:442–455

Cookson W (2004) The immunogenetics of asthma and eczema: a new focus on the epithelium. Nat Rev Immunol 4;978–988

Costello EK, Lauber CL, Hamady M, Fierer N, Gordon JI, Knight R (2009) Bacterial community variation in human body habitats across space and time. Science 326:1694–1697

Crespo-Erchiga V, Florencio VD (2006) Malassezia yeasts and pityriasis versicolor. Curr Opin Infect Dis 19;139–147

Cummins CS (1976) Identification of *Propionibacterium acnes* and related organisms by precipitin tests with trichloroacetic acid extracts. J Clin Microbiol 2:104–110

Cunliffe WJ (2002) Looking back to the future – acne. Dermatology 204:167–172

Cunliffe WJ, Holland DB, Clark SM, Stables GI (2000) Comedogenesis: some new aetiological, clinical and therapeutic strategies. Br J Dermatol 142:1084–1091

Dahl MG, McGibbon DH (1979) Complement C3 and immunoglobulin in inflammatory acne vulgaris. Br J Dermatol 101:633–640

Darley CR, Moore JW, Besser GM, Munro DD, Edwards CR, Rees LH, Kirby JD (1984) Androgen status in women with late onset or persistent acne vulgaris. Clin Exp Dermatol 9:28–35

Dawson TL Jr (2007) *Malassezia globosa* and *restricta*: breakthrough understanding of the etiology and treatment of dandruff and seborrheic dermatitis through whole-genome analysis. J Invest Dermatol Symp Proc 12:15–19

Dekio I, Sakamoto M, Hayashi H, Amagai M, Suematsu M, Benno Y (2007) Characterization of skin microbiota in patients with atopic dermatitis and in normal subjects using 16S rRNA gene-based comprehensive analysis. J Med Microbiol 56:1675–1683

Eady EA, Cove JH, Holland KT, Cunliffe WJ (1989) Erythromycin resistant propionibacteria in antibiotic treated acne patients: association with therapeutic failure. Br J Dermatol 121:51–57

Elder JT, Bruce AT, Gudjonsson JE, Johnston A, Stuart PE, Tejasvi T, Voorhees JJ, Abecasis GR, Nair RP (2010) Molecular Dissection of Psoriasis: Integrating Genetics and Biology. J Invest Dermatol 130(5):1213–1226

Farrar MD, Howson KM, Bojar RA, West D, Towler JC, Parry J, Pelton K, Holland KT (2007) Genome sequence and analysis of a *Propionibacterium acnes* bacteriophage. J Bacteriol 189:4161–4167

Fierer N, Hamady M, Lauber CL, Knight R (2008) The influence of sex handedness and washing on the diversity of hand surface bacteria. Proc Natl Acad Sci USA 105:17994–17999

Gao Z, Tseng CH, Pei Z, Blaser MJ (2007) Molecular analysis of human forearm superficial skin bacterial biota. Proc Natl Acad Sci USA 104:927–2932

Gao Z, Tseng CH, Strober BE, Pei Z, Blaser MJ (2008) Substantial alterations of the cutaneous bacterial biota in psoriatic lesions. PLoS One 3:e2719

Gloor M, Franke M (1978) On the propionibacteria in the pilosebaceous ducts of uninvolved skin of acne patients. Arch Dermatol Res 262:125–129

Gloor M, Peters G, Stoika D (1982) On the resident aerobic bacterial skin flora in unaffected skin of patients with atopic dermatitis and in healthy controls. Dermatologica 164.258–265

Goodacre R, Howell SA, Noble WC, Neal MJ (1996) Sub-species discrimination, using pyrolysis mass spectrometry and self-organising neural networks, of *Propionibacterium acnes* isolated from normal human skin. Zentralbl Bakteriol 284:501–515

Goodman H (1943) The hydrogen ion concentration of the skin. Urol Cutan Rev 47:470–477

Gowland G, Ward RM, Holland KT, Cunliffe WJ (1978) Cellular immunity to *P. acnes* in the normal population and patients with acne vulgaris. Br J Dermatol 99:43–47

Grice EA, Kong HH, Conlan S, Deming CB, Davis J, Young AC, Bouffard GG, Blakesley RW, Murray PR, Green ED et al (2009) Topographical and temporal diversity of the human skin microbiome. Science 324:1190–1192

Grice EA, Kong HH, Renaud G, Young AC, Bouffard GG, Blakesley RW, Wolfsberg TG, Turner ML, Segre JA (2008) A diversity profile of the human skin microbiota. Genome Res 18(7):1043–1050

Gupta MA, Schork NJ, Gupta AK, Kirkby S, Ellis CN (1993) Suicidal ideation in psoriasis. Int J Dermatol 32:188–190

Harris WH, Wade HW (1915) The wide-spread distribution of diphtheroids and their occurrence in various lesions of human tissues. J Exp Med 21:493–508

Higaki S, Kitagawa T, Kagoura M, Morohashi M, Yamagishi T (2000) Correlation between *Propionibacterium acnes* biotypes, lipase activity and rash degree in acne patients. J Dermatol 27:519–522

Holland KT, Bojar RA (2002) Cosmetics: what is their influence on the skin microflora? Am J Clin Dermatol 3:445–449

Holmes RL, Williams M, Cunliffe WJ (1972) Pilo-sebaceous duct obstruction and acne. Br J Dermatol 87:327–332

Holt RJ (1971) Aerobic bacterial counts on human skin after bathing. J Med Microbiol 4:319–327

Jong EC, Ko HL, Pulverer G (1975) Studies on bacteriophages of *Propionibacterium acnes*. Med Microbiol Immunol 161:263–271

Katsambas A, Dessinioti C (2008) New and emerging treatments in dermatology: acne. Dermatol Ther 21:86–95

Kersey P, Sussman M, Dahl M (1980) Delayed skin test reactivity to *Propionibacterium acnes* correlates with severity of inflammation in acne vulgaris. Br J Dermatol 103:651–655

Kluytmans J, van Belkum A, Verbrugh H (1997) Nasal carriage of *Staphylococcus aureus*: epidemiology, underlying mechanisms, and associated risks. Clin Microbiol Rev 10:505–520

Krueger G, Ellis CN (2005) Psoriasis – recent advances in understanding its pathogenesis and treatment. J Am Acad Dermatol 53:S94–100

Lee DJ, Van Dyke GS, Kim J (2003) Update on pathogenesis and treatment of acne. Curr Opin Pediatr 15:405–410

Leeming JP, Holland KT, Cuncliffe WJ (1988) The microbial colonization of inflamed acne vulgaris lesions. Br J Dermatol 118:203–208
Leung DY (2003) Infection in atopic dermatitis. Curr Opin Pediatr 15:399–404
Leung DY, Bieber T (2003) Atopic dermatitis. Lancet 361:151–160
Leung DY, Boguniewicz M, Howell MD, Nomura I, Hamid QA (2004) New insights into atopic dermatitis. J Clin Invest 113:651–657
Leyden JJ, McGinley KJ, Vowels B (1998) *Propionibacterium acnes* colonization in acne and nonacne. Dermatology 196:55–58
Lina G, Boutite F, Tristan A, Bes M, Etienne J, Vandenesch F (2003) Bacterial competition for human nasal cavity colonization: role of Staphylococcal agr alleles. Appl Environ Microbiol 69:18–23
Liu PT, Phan J, Tang D, Kanchanapoomi M, Hall B, Krutzik SR, Kim J (2008a) CD209(+) macrophages mediate host defense against *Propionibacterium acnes*. J Immunol 180: 4919–4923
Liu PT, Stenger S, Li H, Wenzel L, Tan BH, Krutzik SR, Ochoa MT, Schauber J, Wu K, Meinken C et al (2006) Toll-like receptor triggering of a vitamin D-mediated human antimicrobial response. Science 311:1770–1773
Liu Y, Helms C, Liao W, Zaba LC, Duan S, Gardner J, Wise C, Miner A, Malloy MJ, Pullinger CR et al (2008b) A genome-wide association study of psoriasis and psoriatic arthritis identifies new disease loci. PLoS Genet 4:e1000041
Marples MJ (1969) The normal flora of the human skin. Br J Dermatol 81(Suppl 1):2–13
Marples RR, McGinley KJ, Mills OH (1973) Microbiology of comedones in acne vulgaris. J Invest Dermatol 60:80–83
Matousek JL, Campbell KL (2002) A comparative review of cutaneous pH Vet Dermatol 13: 293–300
Nair RP, Ruether A, Stuart PE, Jenisch S, Tejasvi T, Hiremagalore R, Schreiber S, Kabelitz D, Lim HW, Voorhees JJ et al (2008) Polymorphisms of the IL12B and IL23R genes are associated with psoriasis. J Invest Dermatol 128:1653–1661
Noble WC (1969) Distribution of the micrococcaceae. Br J Dermatol 81(Suppl):27–32
Norris JF, Cunliffe WJ (1988) A histological and immunocytochemical study of early acne lesions. Br J Dermatol 118:651–659
O'Regan GM, Sandilands A, McLean WH, Irvine AD (2009) Filaggrin in atopic dermatitis. J Allergy Clin Immunol 124:R2–6
Olsen TG (1982) Therapy of acne. Med Clin North Am 66:851–871
Paulino LC, Tseng CH, Blaser MJ (2008) Analysis of *Malassezia* microbiota in healthy superficial human skin and in psoriatic lesions by multiplex real-time PCR. FEMS Yeast Res 8:460–471
Paulino LC, Tseng CH, Strober BE, Blaser MJ (2006) Molecular analysis of fungal microbiota in samples from healthy human skin and psoriatic lesions. J Clin Microbiol 44:2933–2941
Plewig G, Fulton JE, Kligman AM (1971) Cellular dynamics of comedo formation in acne vulgaris. Arch Dermatol Forsch 242:12–29
Puhvel SM, Amirian D, Weintraub J, Reisner RM (1977) Lymphocyte transformation in subjects with nodulo cystic acne. Br J Dermatol 97:205–211
Rieg S, Steffen H, Seeber S, Humeny A, Kalbacher H, Dietz K, Garbe C, Schittek B (2005) Deficiency of dermcidin-derived antimicrobial peptides in sweat of patients with atopic dermatitis correlates with an impaired innate defense of human skin in vivo. J Immunol 174:8003–8010
Ro BI, Dawson TL (2005) The role of sebaceous gland activity and scalp microfloral metabolism in the etiology of seborrheic dermatitis and dandruff. J Invest Dermatol Symp Proc 10:94–197
Rosenberg EW (1969) Bacteriology of acne. Annu Rev Med 20:201–206
Ruocco E, Donnarumma G, Baroni A, Tufano MA (2007) Bacterial and viral skin diseases. Dermatol Clin 25:663–676, xi
Sander HM, Morris LF, Phillips CM, Harrison PE, Menter A (1993) The annual cost of psoriasis. J Am Acad Dermatol 28:422–425

Shaw CM, Smith JA, McBride ME, Duncan WC (1970) An evaluation of techniques for sampling skin flora. J Invest Dermatol 54:160–163
Steiner B, Romero-Steiner S, Cruce D, George R (1997) Cloning and sequencing of the hyaluronate lyase gene from *Propionibacterium acnes*. Can J Microbiol 43:315–321
Strachan DP (1989) Hay fever hygiene, and household size. BMJ 299:1259–1260
Strauss JS, Kligman AM (1960) The pathologic dynamics of acne vulgaris. Arch Dermatol 82: 779–790
Strauss JS, Pochi PE, Downing DT (1976) The role of skin lipids in acne. Cutis 17:485–487
Strauss JS, Thiboutot DM (1999) Diseases of the sebaceous glands. In Fitzpatrick's Dermatology In: Freedberg IM, Eisen AZ, Wolff K, Austen KF, Goldsmith LA, Katz SI, Fitzpatrick TB (eds) General Medicine. McGraw-Hill, New York, NY, pp 769–784
Terui T (2009) Analysis of the mechanism for the development of allergic skin inflammation and the application for its treatment: overview of the pathophysiology of atopic dermatitis. J Pharmacol Sci 110:232–236
Tsunemi Y, Saeki H, Nakamura K, Sekiya T, Hirai K, Fujita H, Asano N, Kishimoto M, Tanida Y, Kakinuma T et al (2002) Interleukin-12 p40 gene (IL12B) 3′-untranslated region polymorphism is associated with susceptibility to atopic dermatitis and psoriasis vulgaris. J Dermatol Sci 30:161–166
Watson W, Cann HM, Farber EM, Nall ML (1972) The genetics of psoriasis. Arch Dermatol 105:197–207
Webster GF (1995) Inflammation in acne vulgaris. J Am Acad Dermatol 33:247–253
Webster GF, Cummins CS (1978) Use of bacteriophage typing to distinguish *Propionibacterium acnes* types I and II. J Clin Microbiol 7:84–90
Wollenberg A, Zoch C, Wetzel S, Plewig G, Przybilla B (2003) Predisposing factors and clinical features of eczema herpeticum: a retrospective analysis of 100 cases. J Am Acad Dermatol 49:198–205
Zhang XJ, Huang W, Yang S, Sun LD, Zhang FY, Zhu QX, Zhang FR, Zhang C, Du WH, Pu XM et al (2009) Psoriasis genome-wide association study identifies susceptibility variants within LCE gene cluster at 1q21. Nat Genet 41:205–210
Zhang YQ, Ren SX, Li HL, Wang YX, Fu G, Yang J, Qin ZQ, Miao YG, Wang WY, Chen RS et al (2003) Genome-based analysis of virulence genes in a non-biofilm-forming *Staphylococcus epidermidis* strain (ATCC 12228). Mol Microbiol 49:1577–1593
Zierdt CH, Webster C, Rude WS (1968) Study of the anaerobic corynebacteria. Int J Syst Bacteriol 18:33–47

Chapter 9
The Human Oral Metagenome

Peter Mullany, Philip Warburton, and Elaine Allan

Introduction

The human oral cavity is estimated to contain more than 750 bacterial species (Jenkinson and Lamont, 2005; Paster et al., 2006). Although this figure is controversial, the fact remains that up to half of the species in the oral microbiota cannot yet be cultivated in the laboratory. Therefore, metagenomics is a powerful way of accessing these unculturable bacteria in order to understand the role of the oral microbiota in health and disease and to mine for useful products such as enzymes, energy sources and antimicrobial agents. The Human Oral Microbiome Database (HOMD http://www.homd.org/) provides comprehensive information on what is known about the composition of the oral microbiota using information derived from cultivation and metagenomic data based on 16S rRNA gene sequencing.

Microbial Ecology of the Human Oral Cavity and Its Role in Health and Disease

There have been a number of detailed reviews of what is known about the microbial ecology of the oral cavity (Avila et al., 2009; Jenkinson and Lamont, 2005; Wilson, 2005), so here we will provide a brief overview. The most abundant cultivable members of the microbiota in the healthy oral cavity are species belonging to the following genera: *Streptococcus, Veillonella, Fusobacterium, Porphyromonas, Prevotella, Treponema, Neisseria, Haemophilus, Eubacteria, Lactobacterium, Capnocytophaga, Eikenella, Leptotrichia, Peptostreptococcus* and *Propionibacterium*.

Within the oral cavity, most bacteria are contained within biofilms that adhere to a wide variety of surfaces, including epithelial cells, saliva-coated enamel and dental devices such as implants and dentures. Biofilm formation is usually initiated

P. Mullany (✉)
UCL Eastman Dental Institute, 256 Gray's Inn Road, London WC1X 8LD, UK
e-mail: p.mullany@eastman.ucl.ac.uk

K.E. Nelson (ed.), *Metagenomics of the Human Body*,
DOI 10.1007/978-1-4419-7089-3_9,

by the attachment of primary colonisers to these surfaces, which are followed by the early colonisers resulting in the formation of multi-layered micro-colonies allowing the formation of a multi-cellular matrix.

The heterogeneity of microenvironments in the oral cavity influences the composition of the local biofilms. Variable factors include temperature, pH, redox potential, salinity and saliva flow. Saliva delivers nutrients and removes waste products and also contains immunoglobulin, enzymes (e.g. amylase) and host-derived antimicrobial peptides. In addition, saliva will contain any drugs the host may be taking. Finally, the composition of the oral environment will be influenced by hygiene measures implemented by the host such as brushing and use of antibacterial mouthwashes. All of the parameters discussed above will influence the composition of the oral microbiota by providing selective pressures for its adaptation to a challenging dynamic environment.

The oral biofilms present in a healthy mouth are thought to protect the host from disease by oral pathogens. This could occur by competition with pathogens for particular ecological niches within the mouth or by more subtle interactions (Jenkinson and Lamont, 2005; Marsh, 2006). Oral bacteria are responsible for two of the most common infectious diseases known to affect humans: dental caries and periodontitis. In addition to oral disease, there is some evidence that oral bacteria can cause diseases beyond the oral cavity, including endocarditis and abscesses in the brain, liver and lungs (Raghavendran et al., 2007; Lockhart et al., 2009; Bahrani-Mougeot et al., 2008; Lockhart et al., 2008).

Interactions Between Oral Bacteria

As can be appreciated from the above discussion, the oral cavity is a diverse environment and the resident bacteria require a range of adaptations to survive and thrive there. For example, in order to be maintained in the oral environment, bacteria need to adhere to host surfaces, to each other and sometimes to inserted dental devices. They also produce bacteriocins to give them a competitive advantage over other bacteria. Like all bacteria, oral organisms need to sense and respond to changes in their environment and to produce signalling molecules to which both the host and other bacteria respond.

Physical Interactions

The physical interactions are mostly mediated by specialised surface structures called adhesins. For example, *Fusobacterium nucleatum* can adhere to a range of bacteria (promoting their co-aggregation), to host tissues and to IgA (Edwards et al., 2007). This makes *F. nucleatum* an important player in the colonisation of the oral cavity.

Adherence and co-aggregation allow the organisms to resist physical forces such as saliva flow and is also important in the formation and maintenance of biofilms.

For an excellent recent review of the role of adhesins in oral bacteria, particularly the streptococci, see Nobbs et al. (2009). For a comprehensive review of interspecies co-aggregation, see Kolenbrander et al. (2002).

Chemical Interactions

There is a huge range of chemical interactions between bacteria and between bacteria and the host. Broadly speaking, these interactions can be divided into synergistic and antagonistic.

An example of synergistic interactions in biofilms is quorum sensing (QS). This is a form of bacterial communication whereby bacteria secrete a signalling molecule that coordinates behaviour in response to increasing cell density. Most bacteria utilise some form of quorum sensing system either through *N*-acyl homoserine lactones, peptides (common in Gram-positive organisms) and/or auto inducer-2 (AI-2). Examples of peptide- and AI-2-mediated QS systems have been described in oral bacteria (see Hojo et al., 2009). One example is the *luxS*-mediated AI-2 signalling system in *Streptococcus gordonii* that controls biofilm formation in the periodontal pathogen, *Porphyromonas gingivalis* (McNab et al., 2003; Irie and Parsek, 2008).

Another example of synergistic interaction is the excretion of a metabolite by one organism that can be used as an energy source by another. A symbiotic association between *Streptococcus* and *Veillonella* species via lactic acid production by the former has been documented (Kumar et al., 2005; Chalmers et al., 2008).

An example of antagonistic interaction in the oral cavity is the production of acid by *S. mutans,* one of the main contributors to dental caries (Gong et al., 2009). The resultant lowered pH is inhibitory to a large number of bacterial species (Beighton, 2005). Another example is bacteriocin production. Bacteriocins are proteinaceous substances produced by bacteria that inhibit the growth of closely related species or strains. *Streptococcus mutans*, a principal causative agent of dental caries, produces several types of bacteriocins known as mutacins, some of which have a relatively broad spectrum of activity. For a summary of mutacin production and regulation, see Hojo et al. (2009).

The final composition of the bacterial microbiota in the oral cavity depends on a complex interplay of synergistic and antagonistic interactions between bacteria and between bacteria and the host. Perturbations of these interactions can lead to disease.

Gene Transfer

The close physical proximity of a large number of bacteria in the oral cavity presents ideal conditions for gene transfer between different bacteria. Horizontal gene transfer has a number of important biological consequences, the most infamous of which is the spread of antibiotic resistance (Walsh, 2000; Alekshun and Levy, 2007). In addition to this, genomic islands that allow organisms to exploit particular ecological niches are frequently capable of transfer between organisms (reviewed in Juhas et al., 2009). As oral organisms come into contact with other organisms from the human microbiome (e.g. the gut) and from beyond (e.g. food), this is a rich area for gene transfer.

There are three major mechanisms for transfer of DNA between bacteria: transformation, transduction and conjugation. The oral cavity contains a number of species that have natural transformation systems, for example, most of the large numbers of streptococcal species are naturally competent (Martin et al., 2006).

Phages have been identified from oral organisms (Mitchell et al., 2010; van der Ploeg, 2008; Hitch et al., 2004) but to date, there have been no reports of phage-mediated transduction.

Conjugation requires specialised genetic elements that encode all the genetic information required for their own transfer. These elements include conjugative plasmids and conjugative transposons (also called Integrative Conjugative Elements, ICE). Small elements that do not encode all the functions required for transfer, but nonetheless can be mobilised by the larger elements, are also common in bacteria; these are called mobilizable elements (usually plasmids or transposons). For a recent review of these elements, see Roberts and Mullany (2009).

Metagenomic Analysis of the Oral Cavity

The previous sections have provided an overview of the diversity of bacteria present in the oral cavity and the interactions between them. Given that at least half of these organisms cannot be cultivated in the laboratory, it can be appreciated that we are almost certainly missing organisms and interesting genes from this environment. In the following section we describe how metagenomic approaches are being used to identify not only bacteria from the oral cavity, but also how functional metagenomics is allowing the isolation of novel biochemical pathways and mobile genetic elements.

Identifying Oral Organisms

The oral cavity contains a very species-rich microbiota, some of which are present at relatively low numbers. In addition, the diet, health status and genetics of the host have an influence on the composition of the microbiota. This makes obtaining a detailed catalogue of all the oral organisms a challenging task. Identification of the non-cultivable organisms has been predominantly obtained through the cloning of 16S rDNA sequences, with a number of databases being constructed that catalogue the oral microbiota, most notably the HOMP (www.homd.org). The recent advances in high-throughput sequencing technologies have allowed greater depth of coverage for the study of diversity in the human oral microbiota (Lazarevic et al., 2009).

The identification of bacteria using methods like 16S sequencing and even whole genome sequencing does not give us the whole picture of the metabolic and pathogenic capabilities of a particular ecological niche. Therefore, in order to gain a comprehensive understanding of a particular environment, such as the oral cavity, it is also important to undertake a functional metagenomic analysis; our progress in this area is outlined below.

Antibiotic Resistance in the Oral Metagenome

The first functional metagenomic studies of the oral cavity were aimed at identifying and determining the relative distribution of antibiotic resistance genes (Diaz-Torres et al., 2003, 2006; Warburton et al., 2009; Seville et al., 2009; Sommer et al., 2009). This approach allowed a novel tetracycline resistance gene, *tet*(37), which has not yet been found in cultivable flora, to be identified (Diaz-Torres et al., 2003). These studies show that the non-cultivable microbiota contains a significant reservoir of diverse antibiotic resistance genes.

The "Mobilome" of the Oral Cavity

The mobilome is defined as the totality of mobile genetic elements in a genome. The meta-mobilome can therefore be defined as the totality of mobile genetic elements in a particular ecological niche. In this part of the chapter we will discuss what is known about the meta-mobilome of the oral cavity.

Mobile genetic elements (MGEs) are a heterogeneous group of nucleic acid-based entities that can move from one part of a genome to another. They can be further sub-divided into elements that are capable of transfer between cells and elements that are capable of only moving from one part of a single genome to another (see Fig. 9.1). Phages are also capable of mediating the transfer of DNA via transduction between cells.

In addition to the conjugal elements, MGEs in the oral cavity also include the mobilizable plasmids, non-conjugative transposons and self-splicing introns (for recent reviews on these elements, see Roberts and Mullany (2009)). The behaviour and possible interactions of these elements is shown in Fig. 9.1. Finally, genomic islands that allow the host organism to exploit different ecological niches may in some cases be mobile (see below).

Discovery of New ICE in Oral Metagenomes

Integrative conjugative elements (ICEs) that include conjugative transposons are a heterogeneous group of mobile genetic elements. These elements are loosely defined as being able to transfer from the genome of one host to the genome of

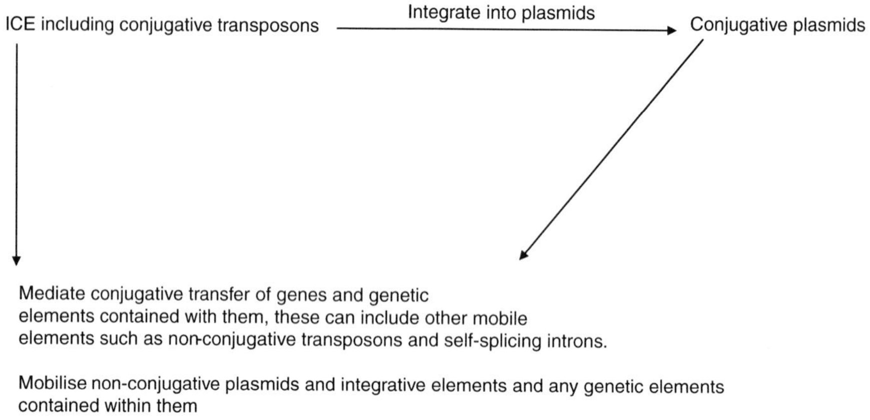
ICE including conjugative transposons
Integrate into plasmids
Conjugative plasmids
Mediate conjugative transfer of genes and genetic elements contained with them, these can include other mobile elements such as non-conjugative transposons and self-splicing introns.
Mobilise non-conjugative plasmids and integrative elements and any genetic elements contained within them

Fig. 9.1 Interactions between mobile genetic elements. Diagram summarising the association between different mobile genetic elements, the arrowheads point to a function carried out by a particular mobile element. Also note that, although not shown on this diagram, phage can also mediate the transfer of genes via transduction

another (Salyers et al., 1995; Roberts and Mullany, 2009). They differ from plasmids in that they do not replicate autonomously but have to integrate into a host replicon. A number of these types of elements have been isolated from the oral cavity (Bentorcha et al., 1992; Hartley et al., 1984; McKay et al., 1995; Warburton et al., 2007).

Discovery of Genomic Islands in Oral Metagenomes

Genomic islands are defined as regions of the genome that differ significantly in the GC content from the host and often have a particular biological role. For example, islands involved in virulence, drug resistance, drug production and ability to survive in particular ecological niches have been identified (Juhas et al., 2009). These islands also frequently have integrase genes and genes that are related to those involved in mobility, although sometimes these contain numerous stop codons indicating that the island is no longer mobile (Juhas et al., 2009). However, some genomic islands retain their mobility. Likely genomic islands have been found in the oral organisms *S. mutans* (Waterhouse and Russell, 2006; Chattoraj et al., 2010) and *Actinobacillus actinomycetemcomitans* (Chen et al., 2005).

Discovery of New Plasmids in Oral Metagenomes

We have recently undertaken a metagenomic analysis of the plasmid content of the oral cavity using a technique called TRACA (Jones and Marchesi, 2007). The system makes use of in vitro transposition of a Tn5, containing an *Escherichia coli* origin of replication and selectable marker, into metagenomic DNA that has been enriched for plasmids. This system has the advantage over previously used methods for the isolation of plasmids from metagenomic DNA in that it does not require any particular plasmid encoded properties for selection. However, the method requires that both the origins of replication present on the Tn5 derivative and the captured plasmid from the metagenomic DNA are compatible, and that there are no additional factors that contribute to plasmid instability in *E. coli* on the target plasmid. Despite these limitations, we have discovered novel plasmids from oral metagenomic DNA (Warburton et al., unpublished).

Functional Analysis of the Oral Metagenome

The oral cavity contains a huge number of different organisms, which in turn have a vast genetic capability. Metagenomics, and in particular functional metagenomics, is an excellent tool for probing this capacity in order to achieve both a complete understanding of oral microbial ecology and to discover new activities useful for biotechnological, industrial and medical applications (Rudney et al., 2010; Warburton et al., 2009).

References

Alekshun MN, Levy SB (2007) Molecular mechanisms of antibacterial multidrug resistance. Cell 23:1037–1050

Avila M, Ojcius DM, Yilmaz O (2009) The oral microbiota: living with a permanent guest. DNA Cell Biol 28:405–411

Bahrani-Mougeot FK, Paster BJ, Coleman S, Ashar J, Barbuto S, Lockhart PB (2008) Diverse and novel oral bacterial species in blood following dental procedures. J Clin Microbiol 46: 2129–2132

Beighton D (2005) The complex oral microflora of high-risk individuals and groups and its role in the caries process. Community Dent Oral Epidemiol 33:248–255

Bentorcha F, Clermont D, de Cespedes G, Horaud T (1992) Natural occurrence of structures in oral streptococci and enterococci with DNA homology to Tn916. Antimicrob Agents Chemother 36:59–63

Chalmers NI, Palmer RJ, Cisar JO, Kolenbrander PE (2008) Characterization of a *Streptococcus* sp.-*Veillonella* sp. Community micromanipulated from dental plaque. J Bacteriol 190: 8145–8154

Chattoraj P, Banerjee A, Biswas S, Biswas I (March 2010) ClpP of *Streptococcus mutans* differentially regulates expression of genomic islands, mutacin production, and antibiotic tolerance. J Bacteriol 192(5):1312–1323

Chen W, Wang Y, Chen C (2005) Identification of a genomic island of *Actinobacillus actinomycetemcomitans*. J Periodontol 76(11 Suppl):2052–2060

Diaz-Torres ML, McNab R, Spratt DA, Villedieu A, Hunt N, Wilson M, Mullany P (2003) Novel tetracycline resistance determinant from the oral metagenome. Antimicrob Agents Chemother 47:1430–1432

Diaz-Torres ML, Villedieu A, Hunt N, McNab R, Spratt DA, Allan E, Mullany P, Wilson M (2006) Determining the antibiotic resistance potential of the indigenous oral microbiota of humans using a metagenomic approach. FEMS Microbiol Lett 258:257–262

Edwards AM, Grossman TJ, Rudney JD (2007) Association of a high molecular weight arginine binding protein of *Fusobacterium nucleatum* ATCC10953 with adhesion of secretory immunoglobulin A and coaggregation with *Streptococcus cristatus*. Oral Microbiol Immunol 22:217–224

Gong Y, Tian XL, Sutherland T, Sisson G, Mai J, Ling J, Li YH (2009) Global transcriptional analysis of acid-inducible genes in *Streptococcus mutans*: multiple two-component systems involved in acid adaptation. Microbiology 155:3322–3332

Hartley DL, Jones KR, Tobian JA, LeBlanc DJ, Macrina FL (1984) Disseminated tetracycline resistance in oral streptococci: implication of a conjugative transposon. Infect Immun 45:13–17

Hitch G, Pratten J, Taylor PW (2004). Isolation of bacteriophages from the oral cavity. Lett Appl Microbiol 39:215–219

Hojo K, Nagaoka S, Ohshima T, Maeda N (2009). Bacterial interactions in dental biofilm development. J Dent Res 88:982

Irie Y, Parsek MR (2008) Quorum sensing and microbial biofilms. In: Romero T (ed) Bacterial biofilms. Springer, Heidelberg, Chapter 4 pp 67–84

Jenkinson HF, Lamont RJ (2005) Oral microbial communities in sickness and in health. Trends Microbiol1 3(12):589–595

Jones BV, Marchesi JR (2007) Transposon-aided capture (TRACA) of plasmids resident in the human gut mobile metagenome. Nat Methods 1:55–61

Juhas M, van der Meer JR, Gaillard M, Harding RM, Hood DW, Crook DW (2009) Genomic islands: tools of bacterial horizontal gene transfer and evolution. FEMS Microbiol Rev 33: 376–393

Kolenbrander PE, Andersen RN, Blehert DS, Egland PG, Foster JS, Palmer RJ Jr (2002) Communication among oral bacteria. Microbiol Mol Biol Rev 66(3):486–505

Kumar PS, Griffen AL, Moeschberger ML, Leys EJ (2005) Identification of candidate periodontal pathogens and beneficial species by quantitative 16S clonal analysis. J Clin Microbiol 43: 3944–3955

Lazarevic L, Whiteson K, Huse S, Hernandez D, Farinelli L, Osteras M, Schrenzel J, Francois P (2009) Metagenomic study of the oral microbiota by illumina high-throughput sequencing. J Microbiol Methods 79:266–271

Lockhart PB, Brennan MT, Sasser HC, Fox PC, Paster BJ, Bahrani-Mougeot FK (2008) Bacteremia associated with tooth brushing and dental extraction. Circulation 117:3118–3125

Lockhart PB, Brennan MT, Thornhill M, Michalowicz BS, Noll J, Bahrani-Mougeot FK, Sasser HC (2009) Poor oral hygiene as a risk factor for infective endocarditis-related bacteremia. J Am Dent Assoc 140:1238–1244

Marsh PD (2006) Dental plaque as a biofilm and a microbial community–implications for health and disease. BMC Oral Health 6(Suppl 1): S14

Martin B, Quentin Y, Fichant G, Claverys JP (2006) Independent evolution of competence regulatory cascades in streptococci? Trends Microbiol 14:339–345

McKay TL, Ko J, Bilalis Y, DiRienzo JM (1995) Mobile genetic elements of *Fusobacterium nucleatum*. Plasmid 33:15–25

McNab R, Ford SK, El-Sabaeny A, Barbieri B, Cook GS, Lamont RJ (2003) Lux S based signaling in *Streptococcus gordonii*: autoinducer 2 controls carbohydrate metabolism and biofilm formation with *Porphyromonas gingivalis*. J Bacteriol 185:274–284

Mitchell HL, Dashper SG, Catmull DV, Paolini RA, Cleal SM, Slakeski N, Tan KH, Reynolds EC (2010) *Treponema denticola* biofilm-induced expression of a bacteriophage, toxin–antitoxin systems and transposases. Microbiology 156:774–788

Nobbs AH, Lamont RJ, Jenkinson HF (2009) Streptococcus adherence and colonization. Microbiol Mol Biol Rev 73:407–450

Paster BJ, Olsen I, Aas JA, Dewhirst FE (2006) The breadth of bacterial diversity in the human periodontal pocket and other oral sites. Periodontol 2000(42):80–87

Raghavendran K, Mylotte JM, Scannapieco FA (2007) Nursing home-associated hospital-acquired pneumonia and ventilator-associated pneumonia: the contribution of dental biofilms and periodontal inflammation. Periodontol 2000(2):1599–1607

Roberts AP, Mullany P (2009) A modular master on the move: the Tn916 family of mobile genetic elements. Trends Microbiol 17:251–258

Rudney JD, Xie H, Rhodus NL, Ondrey FG, Griffin TJ (2010) A metaproteomic analysis of the human salivary microbiota by three-dimensional peptide fractionation and tandem mass spectrometry. Mol Oral Microbiol 25:38–49

Salyers AA, Shoemaker NB, Stevens AM, Li LY (1995) Conjugative transposons: an unusual and diverse set of integrated gene transfer elements. Microbiol Rev 59:579–590

Seville LA, Patterson AJ, Scott KP, Mullany P, Quail MA, Parkhill J, Ready D, Wilson M, Spratt D, Roberts AP (2009) Distribution of tetracycline and erythromycin resistance genes among human oral and fecal metagenomic DNA. Microb Drug Resist 15:159–166

Sommer MGA, Dantas G, Church GM (2009) Functional characterization of the antibiotic resistance reservoir in the human microflora. Science 325:1128–1131

van der Ploeg JR (2008) Characterization of *Streptococcus gordonii* prophage PH15: complete genome sequence and functional analysis of phage-encoded integrase and endolysin. Microbiology 154:2970–2978

Walsh C (2000) Molecular mechanisms that confer antibacterial drug resistance. Nature 406: 775–781

Warburton P, Roberts AP, Allan E, Seville L, Lancaster H, Mullany P (2009) Characterization of tet(32) genes from the oral metagenome. Antimicrob Agents Chemother 53:273–276

Warburton PJ, Palmer RM, Munson MA, Wade WG (2007) Demonstration of in vivo transfer of doxycycline resistance mediated by a novel transposon. J Antimicrob Chemother 60:973–980

Waterhouse JC, Russell RR (2006) Dispensable genes and foreign DNA in *Streptococcus mutans*. Microbiology 152:1777–1788

Wilson (2005) Microbial inhabitants of humans: their ecology and role in health and disease. Cambridge University Press, Cambridge

Chapter 10
The Human and His Microbiome Risk Factors for Infections

Marie-France de La Cochetière and Emmanuel Montassier

Abbreviations

IM	Intestinal Microbiota
NEC	Necrotizing Enterocolitis
VLBW	Very low-birth-weight infant
MRSA	Methicillin-resistant *Staphylococcus aureus*
VRE	Vancomycin-resistant *Enterococci*
VISA	Vancomycin-intermediate *S. aureus*
ESBL	Extended-spectrum beta-lactamase
PBPs	Penicillin-binding proteins
QSARs	Quantitative structure activity relationships
MDRPA	Multidrug-resistant *Pseudomonas aeruginosa*

Introduction

If we take humans as "supra-organisms" (as written by Turnbaugh et al.), they could be considered as composed of human and microbial cells. If we apply ecological and evolutionary principles, we may improve our current understanding of both health and disease (Turnbaugh et al., 2007). Humans are essentially interlopers in a microbial world. Their evolution has been completely dependent on microbes. Thus, even if humans are the dominant species on our planet, they are not the most important from a purely biological point of view. The omnipresent microbial population is not only the oldest form of life on Earth, having emerged more than 3 billion years ago, but also the most numerous and biologically diverse. They are also the most

M.-F. de La Cochetière (✉)
INSERM, UFR Médecine, Université Nantes, Thérapeutiques Cliniques et Expérimentales des Infections, EA 3826, 1 Rue Gaston Veil, Nantes, France
e-mail: mfdlc@nantes.inserm.fr; marie-france.de.la-cochetiere@inserm.fr

The two authors have had equal contribution to this work. E Montassier as medical doctor, MF de La Cochetiere as senior scientist

K.E. Nelson (ed.), *Metagenomics of the Human Body*,
DOI 10.1007/978-1-4419-7089-3_10,

able to adapt to environmental changes. According to many pundits, microbes will be the sole survivors of any global catastrophe short of the complete destruction of all life on Earth. Because of this, humans should be characterized by an aggregate of the genes from the human genome, and the microbiota, and human metabolic features and microbiome. To understand this "supra-organism," and how they can be affected, the constituent microorganisms must be characterized from the first colonization onward.

Infectious diseases kill more than 11 million people each year and diminish the lives of countless others. The WHO definition of infectious diseases is infections caused by pathogenic microorganisms, such as bacteria, viruses, parasites, or fungi; the diseases can be spread directly or indirectly from one person to another. The causes of death (millions) in low- and middle-income countries are lower respiratory infections 3.4, HIV/AIDS 2.6, diarrheal diseases 1.8, tuberculosis 1.6, and malaria 1.2 (Disease control project – US National Institutes of Health – The World Bank – World Health organisation – Population reference Bureau – Bill & Melinda Gates Foundation – 2006). The most efficient way to reduce global pathogen prevalence is not to eradicate diseases, but rather to control them. In recent years, our understanding of infectious-disease epidemiology and control has been greatly increased through mathematical modeling. This plays a key role in study design, analysis (including parameter estimation), and interpretation.

The original definition of an antibiotic was a substance produced by one microorganism that selectively inhibits the growth of another microorganism. However, wholly synthetic antibiotics (usually chemically related to natural antibiotics) have since been produced that accomplish comparable tasks. Antibiotics have been shown to be efficacious in the treatment of infectious diseases caused by a variety of bacterial pathogens. The continued emergence of significant antimicrobial resistance among microorganisms responsible for infectious diseases, however, makes the judicious use of antibiotics, together with the understanding of resistance mechanisms, a critical part of research (Livermore and Pearson, 2007). But the insidious effects of antibiotics go much further than gut dysbiosis. Because humans rely on their native microbiota for nutrition and resistance to colonization by pathogens, symbiotic microbes make essential contributions to the development, metabolism, and immune response of the host (Dethlefsen et al., 2008). In particular, the use of broad-spectrum antibiotics in pediatric practices alters the gut colonization and consequently may impair the barrier function (Schumann et al., 2005). In order to understand the effect of antibiotics, the range of human physiologic diversity within the microbiome and the factors that influence the distribution and evolution of the constituent microbiota must be characterized.

The Host–Microbiota Ecosystem

The microbial communities of humans (known as microbiota) are complex mixtures of microorganisms that have co-evolved with their human hosts (Dethlefsen et al., 2007). This microbiota and its collective genome (microbiome) provides us with genetic and metabolic attributes that we have not evolved on our own, including the ability to harvest otherwise inaccessible nutrients (Backhed et al., 2005). To date, the human microbiota has not been fully described. Microorganisms are present in site-specific communities on mucosal surfaces, in the gut, on the skin, and on specific sites, including the human vagina and oral cavity. Thus it is of utmost importance to understand how these communities help maintain human health and how disturbances of the community structure and function could increase susceptibility to infectious disease.

The microbiota is estimated to outnumber human somatic and germ cells by a factor of ten (Turnbaugh et al., 2007). In this chapter, we will focus on the human gut because it harbors the largest collection of microorganisms, and its role in health is very important. The human gut is populated with as many as 100 trillion (10^{14}) cells, whose collective genome, the microbiome, is a reflection of evolutionary selection pressures acting at the level of the host and at the level of the microbial cell. The vast majority reside in our colon where densities approach 10^{11}–10^{12} cells/ml. There are 6.5 billion humans on Earth, equating to a gut reservoir of 10^{23}–10^{24} microbial cells. This number is just five orders of magnitude less than the world's oceans that hold an estimated 10^{29} cells (Ley et al., 2006). Therefore, the human gut constitutes a substantial microbial habitat in our biosphere. The shared evolutionary fate of humans and their microbiota has selected for mutualistic interactions that are essential for human health (Dethlefsen et al., 2007). The species that make up these communities vary between hosts as a result of restricted migration of microorganisms between hosts and strong ecological interactions within hosts, as well as host variability in terms of diet, genotype, and colonization history (McCracken and Lorenz, 2001). The ecological rules that govern the shape of microbial diversity in the gut apply to mutualists and pathogens alike (Ley et al., 2006).

Homeostasis

Homeostasis is the ability or tendency of an organism or cell to maintain internal equilibrium by adjusting its physiological processes. It describes the relatively constant conditions within the organism (or the physiological processes) by which, in the face of external variation, stability is maintained while adjusting to changing conditions. Systems in dynamic equilibrium reach a balance in which internal changes continuously compensate for external changes in a feedback control process, to keep conditions relatively uniform. As for the human host and ones microbiome, homeostasis can be considered as being the condition for health. Any shock brings a disorder. As long as the disequilibrium stands under a specific threshold, the resilience of the ecosystem will allow homeostasis to be reached again. Thus the host will stay healthy. But when the shock overcomes the specific individual threshold, dysbiosis develops: a disturbed balance of beneficial bacteria that could promote intestinal inflammation (diseases). A very common cause of bacterial or fungal dysbiosis is often the repeated or long-term use of antibiotics. Antibiotics kill both the pathogenic species you want killed and the "friendly" bacteria in the intestine (and other areas as the vagina), leaving these areas open to colonization by yeast, unfriendly bacteria, and parasites.

Resilience

Resilience is the capacity of each fecal microbiota to return to a community structure made of many of the same dominant species (De La Cochetiere et al., 2005).

Functional redundancy in the human gut microbiota: The rapid return to the pretreatment (preshock) community composition is indicative of factors promoting community resilience (Fig. 10.1).

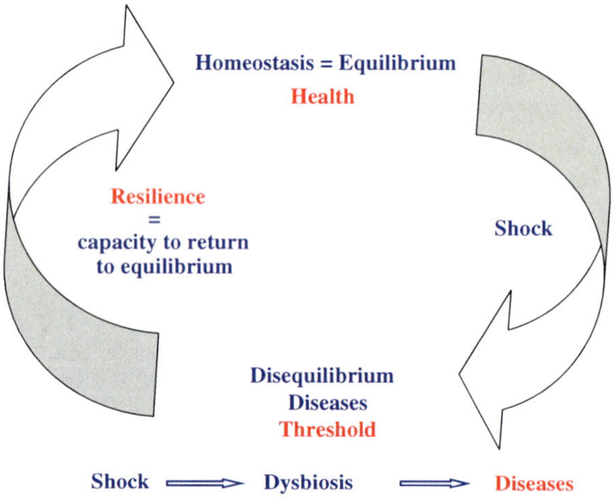

Fig. 10.1 The Human host and ones microbiome are at equilibrium: A condition in which all acting influences are canceled by others, resulting in a stable, balanced, or unchanging system. Any shock (e.g., antibiotherapy) brings disequilibrium and if over a specific threshold, induces dysbiosis, and disease develops (e.g., inflammation)

Threshold

In the quest to understand and combat infectious diseases, this threshold has to be understood. The reason why each individual has his own threshold, and capacity of resilience, and why the pervasive effect of antibiotics is not equal between individuals are long-standing questions. The observation that similar functional repertoires can be attained in the gastrointestinal tract microbiota in the face of substantial differences in phylogenetic composition supports the notion that diseased states of this community may be best identified by atypical distributions of functional gene categories (Vaishampayan et al., 2010), as recently shown for obesity (Turnbaugh et al., 2009). Humans have been associated with complex microbial communities throughout evolution (Mshvildadze and Neu, 2010). To understand the range of human genetic and physiologic diversity, to prevent or treat disease, and improve overall health, the microbiome and the factors that influence the distribution and evolution of the constituent microorganism must be considered. Thus the characterization of this immensely diverse ecosystem is the first step in elucidating its role in health and disease.

To uncover the basis of nonpathogenic microbial colonization, microbes have been found to produce a multitude of factors that either confer virulence or promote colonization by other means. The actions of these factors are countered by the equally diverse responses of the host immune system. This is a dynamic interplay between host and microbes, which start at birth.

The Human Gut Microbiota

The endogenous intestinal microbiota has a fundamentally important role in health and disease (Eckburg et al., 2005). Each adult's gut appears to have a unique microbial community, with a structure that remains stable on the timescale of months (Palmer et al., 2007). The gastrointestinal ecosystem is a precious alliance among epithelium, immunity, and microbiota. This collection of microbes known as intestinal microbiota (IM) is dominated numerically by obligatory anaerobic species. The human IM is essential to the health of the host and plays a role in nutrition, development, metabolism, pathogen resistance, and regulation of immune responses. Co-evolved, beneficial, human–microbe interactions can be altered by many aspects of a modern lifestyle, including urbanization, global travel, and dietary changes (Dethlefsen et al., 2007). The single-cell epithelial layer of the intestinal mucosa confronts the largest antigenic microbial challenge of any other mucosal surface in the human body.

Intestinal Microbiota in Neonates and Preterm Infants

In contrast to adults, the neonate's IM is more variable in its composition and less stable over time. In the first year of life, their intestinal area progresses from sterility to extremely dense colonization. Currently little data exist about microbial DNA in the first stools of term and premature babies using non-culture-based methods, although the extreme importance of this first core microbiota formation is likely to be critical for normal development (Mshvildadze and Neu, 2010; Mshvildadze et al., 2009).

Thus, the neonatal period is crucial for intestinal colonization. The processes involved in the establishment of microbial populations are complex and involve both microbial succession as well as interactions between the newborn, its mother, and the immediate surroundings (Wall et al., 2009). The molecular mechanisms that are involved in shaping and selecting a stable microbiota for a precise individual, together with long-term consequences on the adult health of this intestinal colonization are areas of considerable research interest (Mshvildadze and Neu, 2010; Palmer et al., 2007; Sharma et al., 2010). The immature immune system and frequent exposure to antibiotics render neonates particularly susceptible to dysbiosis. As a result of which, in this nonexhaustive paragraph, data that contribute directly to the understanding of the establishment and development of IM are presented (De La Cochetiere et al., 2007).

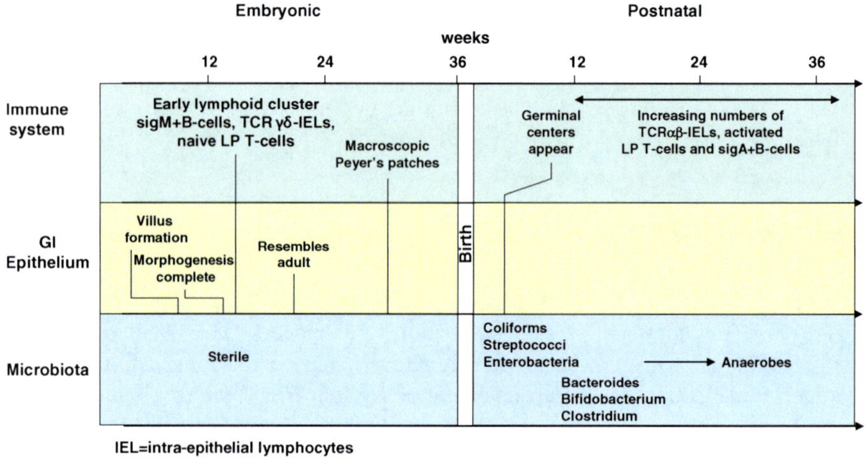

Fig. 10.2 Schematic representation of the development of the human gastrointestinal ecosystems. Humans develop fully activated and functional mucosa immunity until after exposure to normal intestinal microbiota (McCracken and Lorenz, 2001). sIgM, surface Immunoglobulin M; TCR, T-cell receptor; IELs, intraepithelial lymphocytes; LP, lamina propria; sIgA, surface immunoglobulin A

The establishment of the gut microbial population is not strictly a succession in the ecological sense. Rather, this colonization is a complex process influenced by microbial and host interactions as well as by internal and external factors (Fig. 10.2). The climax intestinal microbiota is attained in successive stages (Fanaro et al., 2003). The enteric microbiota contributes to health by facilitating carbohydrate assimilation and interaction with the developing immune system and also contributes to disease (Palmer et al., 2007; Schwiertz et al., 2003). The environmental conditions under which babies are born and nurtured may affect their exposure to microbes and may subsequently influence the composition of their gut microbiota. The environment of the intestine is influenced by three main factors: dietary intake, bacterial ecology, and factors such as peristalsis and glandular secretions that are intrinsic to the intestine (De La Cochetiere et al., 2007).

Humans maintain a diverse, dynamic, and complex IM that performs a multitude of vital functions, such as growth and development of the epithelial barrier, stimulation of intestinal angiogenesis, regulation of nutrition, metabolic functions, and education of naïve neonatal innate immunity. After birth, host environment and attributes lend and opt for a stable IM for long-term symbiosis. Dysregulation (dysbiosis) of this process during the early postnatal period can form the basis for later inflammatory, immune, and allergic disorders. As a result, the initial immunologic exposure of neonates has long-term consequences.

Physiological Colonization

The human fetus receives nutrients, growth factors, and immunoglobulins via active or passive placental transport. Swallowing of amniotic fluid nourishes the fetal intestine and prepares this organ for birth. Preterm delivery interrupts the transfer of these factors that are critical to prepare and protect the newborn infant from bacteria that will colonize the intestinal tract postnatally.

The Process of Colonization

The process of colonization is greatly influenced by the successive shift from formula feeding to weaning. Culture studies have indicated that, in general, infants are initially colonized by *Enterobacteria* and Gram-positive cocci, which are thought to create a reduced environment that is favorable for the establishment of *Bacteroides*, *Bifidobacterium*, and *Clostridium* by day 7. A full-term breast-fed infant has an intestine microbiota in which *Bifidobacteria* and *Lactobacillus* predominate over potentially harmful bacteria, whereas in formula-fed infants, coliforms, enterococci, and *Bacteroides* predominate. In full-term infants, a diet of breast milk induces the development of a microbiota rich in *Bifidobacterium* sp. Other obligate anaerobes such as *Clostridium* sp. and *Bacteroides* sp. are isolated less frequently, and enterobacteria and enterococci are also rare. *Clostridia* have consistently been found at lower levels in breast-fed babies; thus the presence of this group of bacteria may indicate that the babies have been fed formula. The intestinal microbiota in breast-fed infants can be followed by different biochemical parameters. Acetic acid is found at higher concentrations in breast-fed infants than in formula-fed infants. Degradation of mucin begins later in breast-fed infants than in formula-fed infants. The conversion of cholesterol to coprostanol is also delayed by breastfeeding.

Park et al. employed a molecular approach to study the feces of one infant on the first, third, and sixth days after birth and showed that microbiotic diversity changes very rapidly in the days following birth. In addition, the acquisition of unculturable bacteria expanded rapidly after the third day (Park et al., 2005). Of the 325 isolated clones, 220 were characterized as known species, whereas the other 105 clones were characterized as unknown species. On the first day of the life, *Enterobacter*, *Lactococcus lactis*, *Leuconostoc citreum*, and *Streptococcus mitis* were present in the infant's feces with the largest taxonomic group in number of clones isolated being *L. lactis*. On the third day of life, *Enterobacter*, *Enterococcus faecalis*, *Escherichia coli*, *S. mitis*, and *Streptococcus salivarius* were present. On the sixth day, *Citrobacter*, *Clostridium difficile*, *Enterobacter* sp., *Enterobacter cloacae*, and *E. coli* were present. At this point the largest taxonomic group was *E. coli*. Geographical differences in the composition of the intestinal microbiota in infants have also been reported. For example, *Enterobacteria*, *Enterococci*, *Bifidobacteria*, *Lactobacilli*, and *Bacteroides* show a different distribution in developed and developing countries.

Peristalsis

Peristalsis is developmentally regulated and controls microbiota changes along the length of the intestine. In term infants, as in adults, migrating motor complexes pass as waves along the gastrointestinal tract. Feeding results in further complexes superseding the background wave pattern. In preterm neonates, however, migrating complexes are not present until about 34 weeks' gestation. In these infants, the mechanisms necessary for maintaining a stable temporospatial relation in the intestine are not fully developed. In the fetus, the environment of the intestine is mostly controlled by the amniotic fluid, and thus, the role of peristalsis in regulating lumenal homeostasis is correspondingly less important. In preterm infants, however, the intestinal environment is affected by the outside world. Thus, the possibility of build-up of substances within the intestine exists because the propulsive action of the intestine is not yet fully developed. Characteristics of upper esophageal sphincter and primary peristalsis are present as early as 33 weeks, undergo further maturation during the postnatal period, and are significantly different from those in adults. Fetal swallowing contributes greatly to amniotic fluid homeostasis and fetal somatic development. Fetal gastric emptying cycles normalize during the early third trimester. The near-term evidence of delayed emptying may contribute to newborn infant feeding satiation.

The Environment of the Epithelial Surface

In preterm neonates, the degree of mixing of lumenal contents may be small due to immature peristaltic activity. This may result in an unstirred layer of greater thickness than is present in neonates born at term. Changes in the gastrointestinal tract longitudinally are determined to a large extent by peristalsis. Molecules passing from the contents of the intestine are propelled by peristalsis of the intestine to the epithelial cell apex. These molecules encounter the unstirred mucus layer and the deep mucus layer, both of which are present in the neonatal intestine. The effect of each layer on the absorption of the molecules depends on the chemical nature of the molecules. The unstirred layer is a significant barrier to lipid-soluble molecules, whereas the acidic microclimate has a large effect on weak electrolyte uptake. The unstirred layer may not be a distinct layer on the mucosal surface but may serve as a barrier in which molecules diffuse at a rate different from that predicted by the diffusion coefficient of water. Increased agitation of the content in this layer enhances the diffusion barrier. In preterm neonates, the degree of mixing of lumenal contents may be small because of immature peristaltic activity and therefore might result in an unstirred layer of greater thickness than is present in term neonates. In premature neonates, pancreatic and biliary functions are not as well developed as in adults. As a result, the unstirred layer is a significant barrier to lipid absorption.

Mucus Secretion

Mucus secretion is well developed in neonates although its composition changes during development. Although the role of mucus secretion is thought to be important, quantification in neonates is difficult. Therefore, studies have been conducted using rat models. The deep mucus layer is significantly more acidic than the lumen, and changes at the surface of the epithelium are less variable than those in the bulk phase in different parts of the gastrointestinal tract. The acid microclimate has a direct effect on transport of dipeptides, which, unlike amino acids, are transported into the cell in association with hydrogen ions. Human neonates produce a microclimate sufficient for these absorptive functions, but little is known about the microclimate in preterm neonates.

Intestinal Permeability

In preterm infants (26–36 weeks' gestation), intestinal permeability is greater during the first 2 days of life than during days 5 through 8. Permeability is greater in preterm infants than in term infants only when measured within 2 days of birth. These results suggest rapid postnatal adaptation of the small intestine in preterm infants. The barrier function of the intestinal epithelium transiently decreases during the first week after birth in preterm neonates that are enterally fed. Both a diminished barrier function and a low absorptive capacity during the early postnatal period, particularly in neonates born at less than 28 weeks' gestation, may underline the high vulnerability of these patients to intestinal complications. Given the finding that epithelial integrity was restored on initiation of enteral feeding, early administration of enteral nutrition may offer an effective strategy to support intestinal adaptation to extra-uterine life in preterm neonates.

Marker of Intestinal Permeability

Immaturity of the intestinal epithelial barrier function and absorptive capacity may play a role in the pathophysiology of intestinal complications in preterm neonates during the early postnatal period. Thus, identification of noninvasive markers would be of utmost clinical relevance. Lactulose and mannitol have been used to test the passive intestinal permeability as neither of these molecules is metabolized, and both are wholly and solely excreted by the kidney. Urinary recovery is a measure of the intestinal uptake. Lactulose is thought to pass across the gut wall by a paracellular pathway, whereas mannitol passes across the gut wall by a transcellular pathway. The studies by Weather et al. indicated that the immaturity of the gut in preterm infants was responsible, as opposed to a process for adaptation to enteral nutrition. In addition, Mills used 2,3 butanediol, which was detected in urine samples of premature infants by capillary gas chromatography. The presence of this biochemical marker indicated bacterial fermentation of pyruvate in the gut by abnormal gut colonization with acetone-producing microorganisms, an abundant supply of nutrient lactose in the colon, and an increase in intestinal permeability. The lactulose-to-rhamnose ratio was determined as a marker of intestinal permeability. The urinary excretion percentages of D-xylose and 3-*O*-methyl-D-glucose were determined as markers of passive and active carrier-mediated monosaccharide absorption, respectively (Rouwet et al., 2002; Weaver et al., 1984; Mills and Walker, 1989).

Gradient from Stomach to Colon

The establishment and succession of bacterial communities in preterm infants produce an increasing gradient from the stomach to the colon and provides spatial distribution within each gut compartment. Basically, the intestine is comprised of four microhabitats: the intestinal lumen, the unstirred mucus layer, the deep mucus layer, and the surface of mucosal epithelial cells. Blakey et al. studied the developing microflora in the throat, stomach, and feces of 28 preterm babies during their first 3 weeks of life using classical culture methods. The flora at all levels of the gastrointestinal tract differed from that of healthy breast-fed and artificially fed full-term babies. Colonization of the throat and stomach was delayed beyond the first 4 days of life in 74% of the preterm babies studied. Flora of the stomach was sparse and resembled fecal flora. The fecal flora was established more rapidly although only in 70% of the babies during the first 4 days of life. Initially, *Bacteroides* sp. was predominant, but *E. coli* and other aerobic Gram-negative bacilli gradually increased in frequency. Lactic acid-producing bacteria usually appeared late in the third week of life (Blakey et al., 1982).

Translocation

Translocation describes the transmucosal passage of viable and nonviable microbes and their by-products (endotoxins) across the intact intestinal barrier. Predisposing factors in the pathogenesis of systemic infections, such as prematurity, promote impaired mucosal barrier function and consequently foster gut permeability. Under these conditions, indigenous bacteria, viruses, and toxins, which are normally confined within the gastrointestinal tract, may reach systemic organs and tissues. Human infants who received nutrition solely by the parenteral route have an increased risk of Gram-negative bacterial translocation from the gastrointestinal tract into the systemic circulation and other organs. Moy et al. (1999) gives confirmation of translocated gastrointestinal bacteria in a neonatal model by demonstrating that transformed *E. coli* K1 fed to healthy rabbit pups spontaneously translocated from the intestinal lumen and subsequently disseminated to the mesenteric lymph nodes, spleen, and liver. Bacterial translocation is one important cause of nosocomial infections following major abdominal surgery. Seehofer et al. (2004) showed that synchronous liver resection and colon anastomosis led to increased bacterial translocation compared to the single operations in a rat model. Oral administration of probiotics was shown to minimize this translocation. From these studies, the authors proposed that bacterial overgrowth in the cecum and impaired hepatic regeneration, but not histological changes or alterations of paracellular permeability, are the potential pathogenic mechanisms for translocation following the surgeries. A high proportion of bacterial translocation in neonates results not only from immaturity of host defense functions, but also from the dominant colonization of aerobic bacteria in the intestine. Bacteria colonization develops differently in breast-fed, formula-fed, premature, and full-term infants. In a model of neonate rats, Yajima et al. (2001) showed that the frequency of isolation of bacteria from mesenteric lymph nodes and other peripheral sites did not mirror the composition of the intestinal flora. Among the translocated bacteria, *Staphylococcus* may be especially hard to recognize and difficult for the host defense system to destroy. Furthermore, breast-feeding inhibited systemic bacterial translocation in the suckling period of the rat. Additionally, Katalaya demonstrated that (1) the adherence of bacteria to the intestinal mucosal surface is an important factor in bacterial translocation, (2) the intestinal mucus modulates bacterial adherence, and (3) increased levels of mucosa-associated bacteria are associated with a loss of intestinal barrier function to bacteria. The mechanisms by which probiotic agents, such as enteral *Lactobacillus*, enhance the intestinal defenses against potential luminal pathogens have been examined in a neonatal animal model. Enterally administered *Lacto* GG decreases the frequency of *E. coli* K1A translocation in a neonatal rabbit model (Moy et al., 1999; Seehofer et al., 2004; Yajima et al., 2001; Katayama et al., 1997).

The Intestinal Mucosal Immune System

The gastrointestinal tract of the premature infant has a large but fragile surface area covered by a thin monolayer of epithelial cells that overlies a highly immunoreactive submucosa (Mshvildadze and Neu, 2010). Interactions in the lumen between microbiota, nutrients, and the intestinal mucosa can range from the healthy homeostasis to an uncontrolled dysbiosis (systemic inflammatory response syndrome) that leads to multiple organ system failure. The preterm infant is particularly sensitive to colonization patterns as inherent intestinal defense mechanisms are immature, and immature intestinal epithelial cells are known to have exaggerated inflammatory responses to microbiota (Nanthakumar et al., 2000). The immune system is composed of the innate, the specific immunity, and with regard to newborns, the immunity passively acquired from the mother by means of IgG antibodies and human milk. The innate immunity involves humoral elements such as complement system proteins, acute phase proteins, cytokines, and cellular elements such as monocytes, macrophages, granulocytes, dentritic cells, and natural killer lymphocytes. The innate immunity has a limited capacity to distinguish between microorganisms, and often has a similar response to different microorganisms. The components of specific immunity are the lymphocytes and their products (e.g., antibodies). It responds specifically to each microorganism and has a memory. All newborns have an increased risk of microbial infections as compared with older children and young adults. Extremely premature newborns (<28 weeks' gestation) have a 5- to 10-fold higher incidence of microbial infection than even term newborns. Whether near-term newborns have an intermediate risk of acquiring sepsis immediately after birth or within the first few months of life is unknown. There is an intriguing possibility that the well-known immaturity of the fecal immune system has a biological protective purpose. It helps to prevent "premature rejection" by the host – the mother. This immaturity may therefore represent an adaptive response to preventing premature birth. Yet, the ontogeny and sequences of maturation of the immune system in the late preterm infant have not been well studied (Clapp, 2006). The protective function of the gut requires the microbial stimulation of initial bacterial colonization. Breast milk contains prebiotic oligosaccharides, including insulin-type fructans, which are not digested in the small intestine but enter the colon as intact large carbohydrates that are then fermented by the resident bacteria to produce short-chain fatty acids. The nature of this fermentation and the resulting pH of the intestinal contents dictate proliferation of specific resident bacteria. For example, infants fed breast milk containing prebiotics support increased proliferation of *Bifidobacteria* and *Lactobacilli* (probiotic), whereas formula-fed infants produce more *Enterococci* and *Enterobacteria*. Probiotics, stimulated by prebiotic fermentation, are important to the development and sustainment of intestinal defenses. Probiotics, for instance, can stimulate the synthesis and secretion of polymeric IgA, the antibody that coats and protects mucosal surfaces against harmful bacterial invasion. In addition, appropriate colonization with probiotics helps to produce a balanced T-helper-cell response (Th1=Th2=Th3/Tr1) and to prevent a T-cell imbalance (Th1>Th2 or Th2>Th1) that may contribute to clinical disease (a Th2 imbalance contributes to atopic dermatitis while a Th1 imbalance contributes to Crohn's disease and *Helicobacter pylori*-induced gastritis). Furthermore, toll-like receptors on gut lymphoid and epithelial cells recognize bacterial molecular patterns (e.g., endotoxin, lipopolysaccharide, and flagellin) and modulate the intestinal innate immunity and an appropriate adaptive immune response. Both animal and clinical studies have shown that insulin-type fructans will stimulate an increase in commensal bacteria, which have been shown to modulate the development and persistence of an appropriate mucosal immune response. These results are compelling. Thus, prebiotics could potentially be used as a dietary supplement to stimulate a balanced and effective mucosal immune system in newborns and infants (Forchielli and Walker, 2005a, b).

Through a process of "cross talk" with the mucosal immune system, the microbiota negotiates mutual growth, survival, and inflammatory control of the intestinal ecosystem. The IM regulates the intestinal innate immune system by modulating expression of TLRs and NOD/CARD (caspase recruitment domain)-mediated activation of immunosensory cells through MAMPs. Sharma et al. present molecular microbial–mucosal interactions by providing a summary of the different strata of the intestinal mucosa, beginning with the microbiota as the outermost layer, and describe how different layers of the intestinal mucosa form a physical and immune barrier influenced by the microbiota (Sharma et al., 2010).

Context Influences and Pathology

Intestinal Microbiota and Allergy Development

The infant's immature intestinal immune system develops as it comes into contact with dietary and microbial antigens in the gut. The evolving indigenous IM has a significant impact on the developing immune system. Disturbance in the mucosal immune system is reflected in the composition of the gut microbiota and vice versa. Distinctive alterations in the composition of the gut microbiota appear to precede the manifestation of atopic disease, which suggests a role for the interaction between the intestinal immune system and specific strains of the microbiota in the pathogenesis of allergic disorders. Furthermore, dietary lipids as immunomodulators may prevent allergic sensitization by downregulating inflammatory response while protecting the epithelial barrier, and probiotic bacteria have been shown to reinforce the different lines of gut defense (immune exclusion, immune elimination, and immune regulation) (Rautava et al., 2005, 2009). A Swedish team has related IM in early infancy, notably *Bifidobacteria* and *Lactobacilli,* at species level, to allergy development during the first 5 years of life and study if environmental factors influence the early IM of infants: They found that children who developed allergy are significantly less often colonized with *Lactobacilli* group I (*Lactobacillus* (L.) *rhamnosus, L. casei, L. paracasei*), *Bifidobacterium adolescentis*, and *C. difficile* during their first 2 months. Furthermore, they found that infants colonized with several *Bifidobacterium* species are exposed to higher amounts of endotoxin and grew up in larger families than those infants harboring fewer species. Thus, a more diverse IM early in life might prevent allergy development and may be related to the previously suggested inverse relationship between allergy, family size, and endotoxin exposure (Sjogren et al., 2009a, b).

Breast Milk

Breast milk is associated with a lower risk of necrotizing enterocolitis (NEC) and slower growth in the early postnatal period. Exclusive breastfeeding protects against asthma. In breast-fed infants, *Bifidobacterium* predominates with *Lactobacillus* and *Streptococcus* as minor components, whereas in formula-fed infants, Gram-negative organisms such as *E. coli* and *Klebsiella* are more likely to colonize the gut. In fact, additional comprehensive non-culture-based analyses are needed to provide the critical data required to evaluate differences in IM in formula- versus breast-fed premature infants.

Type of Birth

The type of delivery of the neonate has a significant effect on the development of the IM (Gronlund et al., 1999). The primary IM in infants born by cesarean delivery may be disrupted for up to 6 months after birth. A longer vaginal delivery increases the likelihood that viable microbes can be isolated from the stomach and mouth of the infant. Although infants delivered by cesarean section are also exposed to their mother's microbiota, their initial exposure is most likely to be environmental isolates from equipment, air, and other infants, with the nursing staff serving as a vector for transfer. In industrial countries, obstetric and hygienic procedures aimed at reducing the spread of pathogenic bacteria in maternity and neonatal facilities may result in delayed development of the gut microbiota or even to the absence of certain groups of intestinal bacteria during succession (Huurre et al., 2008).

Premature Neonatal Gut

The pattern of bacterial colonization in the premature neonatal gut is different from that in the healthy, full-term infant gut. Infants requiring intensive care acquire intestinal organisms slowly. The establishment of *Bifidobacterium* sp. is retarded, and delayed bacterial colonization with a limited number of bacterial species tends to be common. Schwiertz et al. (2003) showed an increase in similarity of the bacterial communities in hospitalized preterm infants in contrast to breast-fed, full-term infants. A strikingly high similarity was observed between bacterial communities from different preterm infants regardless of birth weight, feeding regime, and antibiotic therapy. This work underscored the fact that the initial colonization of the newborn gastrointestinal tract is highly dependent on the environment and that cross-transmission of bacteria is a serious problem in the hospital (Schwiertz et al., 2003).

The Problem of Low-Birth-Weight (LBW) Infants

Very low birth weight (VLBW) contributes substantially to infant mortality and to childhood disabilities. The principal determinant of LBW in the US is preterm delivery. Poverty is strongly and consistently associated with LBW. As in full-term infants, the intestine of the VLBW infant is first colonized by *Enterobacteria* and *Streptococci*; however, both microorganisms predominate for a longer period of time, and the establishment of *Bifidobacterium* sp. is delayed in the VLBW infants. Emergence of *Bacteroides*, *Clostridium*, and *Lactobacillus* is also delayed. The observed decreased milk intake of the VLBW infant can contribute to this delay. One study on duodenal microbiota in VLBW neonates showed a high incidence of duodenal colonization with *Enterobacteriaceae*. This Gram-negative predominance may be due to the immaturity of the gastrointestinal tract, as it occurred beyond 4 days of age in infants that had been fed enterally, increased with age, and was associated with a longer stay in pediatric units. The state of health at a given gestational age probably depends on the balance between the many developing structures and the achievement of their functions. The diversity of possible influences explains the difficulties in clarifying the etiology of disease states such as NEC. The use of animal models and relatively noninvasive clinical methods will dramatically improve our knowledge of the development of bacto-intestinal function, and thus, the medical care of preterm infants, especially nutritional support, will continue to improve (Sakata et al., 1985; Hoy et al., 2000).

Necrotizing Enterocolitis (NEC)

NEC is a devastating condition with high morbidity and mortality that specifically affects preterm and VLBW infants. NEC may be the consequence of synergy among three of the major risk factors: prematurity, enteral feeding, and bacterial colonization. Together these factors result in an exaggerated inflammatory response that often leads to ischemic bowel necrosis. Human milk may reduce the incidence of NEC by decreasing pathogenic bacterial colonization and promoting growth of nonpathogenic microbiota as well as by maturation of the intestinal barrier and amelioration of the pro-inflammatory response. The pathogenesis of NEC is unclear with many possible factors including stress that provokes a mesenteric ischemia with digestive stasis and exogenous bacterial pathogens. Since the bacteria multiply as a result of both digestive stasis and immature immune function of the intestinal barrier, a weakened intestinal wall could easily be overwhelmed. Food ingestion also aggravates the problem. Both vascular and infectious factors often co-exist in premature infants, whereas in full-term newborns, infection is the main factor (epidemic trend). Parenteral antibiotic therapy should commence after appropriate cultures (blood, cerebrospinal, urine, and stool) are obtained. The initial therapy should include an extended spectrum cephalosporin and vancomycin. Inclusion of clindamycin in the management of NEC has been questioned. The original observation that bacterial proliferation was a factor for NEC prompted suppression of the IM by administration of topical antibiotics in order to prevent the condition. Evidence suggests that oral antibiotics reduce the incidence of NEC in LBW infants; however, concerns about possible adverse outcomes, specifically the development of resistant bacteria, persist. Feeding these infants with breast milk has been suggested to reduce the colonization by pathogenic organisms and induce colonization by commensal organisms by modulating inflammatory reactions and decreasing intestinal injury. Surprisingly enough, using noncultured approaches, Mshvildadze et al. found that the overall microbiota profiles in cases with NEC were not distinguishable from that in control subjects (Mshvildadze et al., 2009), whereas de La Cochetiere et al. found a *Clostridium* sp. only in the cases of NEC (De La Cochetiere et al., 2004). Both found *Klebsiella* sp. more frequently in control subjects. These data provide important foundation for further studies including the effects of antibiotherapies on the early intestinal microbial ecology, and correlations with clinical effects and subsequent development of disease.

Antibiotics

Antibiotics have been called the single most important therapeutic discovery in the history of medicine. An interesting feature of their historic discoveries is that they occurred within the lifetime of many of the population living today. But the seemingly endless miracles attributed to these drugs have led to their mis use and overuse. Antibiotics are critical in the treatment of bacterial infections. The discovery of penicillin was followed by an extraordinary progress in research related to antibiotics and their extensive use. Drastic improvement in mortality and morbidity due to infectious diseases during the 1980s led to great euphoria and complacence among medical fraternity. The result of this was misuse or inappropriate use of antibiotics with emphasis of curative medicine at the cost of disease-preventive measures. Excessive use of antibiotics resulted in the emergence of bacterial resistance. The resistant strains had a survival advantage, and under the selective pressure of antibiotics propagated and spread throughout the world.

Antibiotic resistance is no longer a problem of the developing countries alone. Today even after all the advances in therapeutics and the availability of a large number of antibiotics, a person can also die in a developed country from infection with resistant bacteria. So antibiotic resistance has become a serious public health concern with economic and social implications throughout the world, and the community-acquired infections like *Streptococcal* infections, pneumonia, typhoid fever, etc., or hospital-acquired infections due to methicillin-resistant *S. aureus* (MRSA), vancomycin-resistant *Enterococci* (VRE), vancomycin intermediate *S. aureus* (VISA), or extended spectrum beta-lactamase (ESBL) enzyme producing Gram-negative bacteria. These infections lead to higher rates of hospitalization, longer hospital stay, and increase in the cost of treatment and thus increased economic burden on the community. The problem is much larger in developing countries. The economic consequences have greater implications on the already overburdened economy of these countries. There are many factors that could be responsible for the increase in antibiotics resistance in developing countries. The combination of a heavy disease burden, huge populations, rapid spread through crowding, poor sanitation, and inappropriate use of the available drugs all contribute to this problem. This is further complicated by the availability of antibiotics in open markets without proper prescriptions in majority of these countries. Many a time, the amount of antibiotics given is inadequate to treat serious infections due to poverty or lack of education. Infectious diseases, and therefore antibiotic resistance, also thrive in conditions of civil unrest, mass migration, and unhygienic environmental conditions where large numbers of people are exposed to infectious diseases with little health-care infrastructure.

For sometime, progress in the development of newer and effective antibiotics kept pace with the emerging antibiotic resistance. But the present scenario is such that the available antibiotics have become ineffective in diseases of proven bacterial etiology especially in a hospital setting. Thus, in-depth analysis of molecular and metabolic signaling through organization and structure of new molecules is needed.

Mechanism of Antibiotic Action

Most antimicrobial agents used for the treatment of bacterial infections can be categorized according to their main mechanism of action. There are different modes of action for different antibiotics, including inhibition of protein synthesis, interference with cell wall synthesis, interference with nucleic acid synthesis, and inhibition of metabolic pathways. Thus, antibiotics can be classified based on the cellular component or system they affect, in addition to whether they induce cell death (bactericidal drugs) or merely inhibit cell growth (bacteriostatic drugs). A very nice review from Kohanski et al., (2010) focuses on most current bactericidal antimicrobials that inhibit DNA, RNA, cell wall, or protein synthesis (Kohanski et al., 2010). The current knowledge of the drug–target interactions and the associated

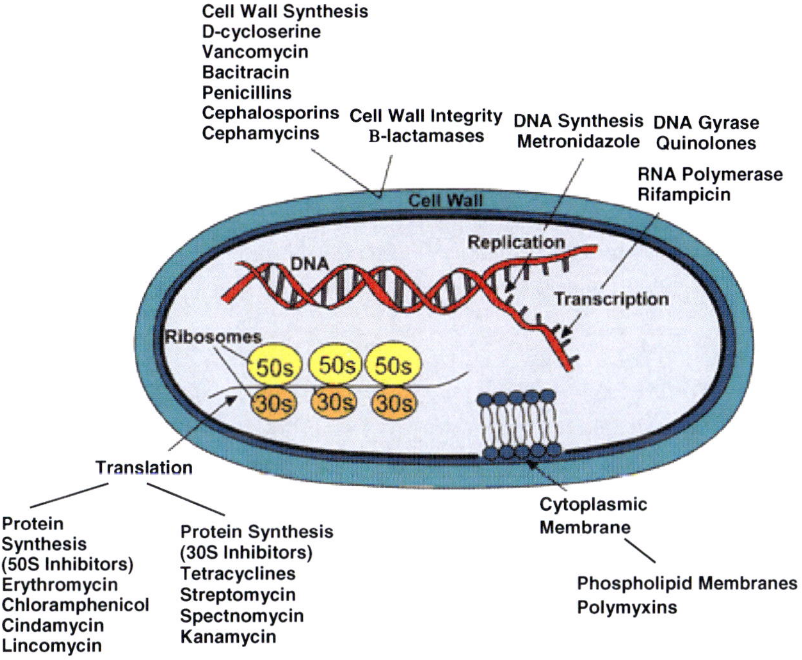

Fig. 10.3 Schematic diagram of the different modes of antibiotic action (Epelboin and Macey, 2009)

mechanisms by which the main classes of bactericidal antibiotics kill bacteria are described (Fig. 10.3).

DNA Synthesis Inhibitors

Quinolones interfere with the maintenance of chromosomal topology by targeting topoisomerase II and topoisomerase IV, trapping these enzymes at the DNA cleavage stage and preventing strand rejoining. The resulting formation of stable complexes with DNA have substantial negative consequences for the cell in terms of its ability to deal with drug-induced DNA damage (Kohanski et al., 2010).

Trimethoprim–sulfamethoxazole blocks cell metabolism by inhibiting enzymes, which are needed in the biosynthesis of folic acid, which is a necessary cell compound.

RNA Synthesis Inhibitors

Rifamycins inhibit one of the enzymes (DNA-dependent RNA polymerase) needed in this process. They stably bind with high affinity to the β-subunit of a DNA-bound and actively transcribing RNA polymerase (Kohanski et al., 2010). This has a catastrophic effect on nucleic acid metabolism.

Protein Synthesis Inhibitors

Tetracyclines bind to the subunit of the bacterial ribosome (30S subunit).
Aminoglycosides bind to a portion of the bacterial ribosome (30S subunit).
Macrolides (e.g., erythromycin) and *Lincosamides* (e.g., clindamycin) bind to the subunit of the bacterial ribosome (50S subunit).
Chloramphenicol binds to a subunit of bacterial ribosome (50S subunit).
Streptogramins binds to a subunit of bacterial ribosome (50S subunit).
Lincosamides binds to a subunit of bacterial ribosome (50S subunit).
Oxazolidinones binds to a subunit of bacterial ribosome (50S subunit).

Cell Wall Synthesis Inhibitors

β-lactams (including penicillins, carbapenems, and cephalosporins) block the cross-linking of peptidoglycan units by inhibiting the peptide bond formation reaction that is catalyzed by penicillin-binding proteins.
Glycopeptides inhibit peptidoglycan synthesis by binding peptidoglycan units (at the D-alanyl-D-alanine dipeptide) and by blocking transglycosylase and penicillin-binding proteins.

Antibiotic-mediated cell death is a complex process that only begins with the drug–target interaction and the primary effects of these respective interactions. So a better understanding of the specific sequences of events beginning with the binding of a bactericidal drug to its target and ending in bacterial cell death is urgently needed. Bioinformatics approaches, which use high-throughput genetic screening or gene expression profiling, are valuable tools to explore the accurate response of bacteria to different antibiotic molecules. Bacterial genomic sequencing is now being used to find previously unidentified genes and their corresponding proteins. Genes and proteins essential for the survival of bacteria can be considered as potential drug targets. Finally, antibiotics with new mechanisms of action are urgently required to combat the growing health threat posed by resistant pathogenic microorganisms, as peptidomimetic antibiotics based on the antimicrobial peptide protegrin I, which is effective against multi-drug-resistant Gram-negative bacteria (Srinivas et al., 2010; Shanahan, 2010).

Contribution of Structural Molecular Biology

The 2D (or 3D) molecular representation of antibiotic molecules is an approach to understanding detailed models of the molecule, its efficiency, simplicity, and usability in order to compare, correlate, and predict their biological properties. It gives a molecular vision of life. The binding affinity of the drug molecules toward a common receptor, the affinities of a particular ligand toward a given receptor, are based on structural features of both the ligand and the receptor (Klebe and Mietzner, 1994; Sotriffer and Klebe, 2002; Guan et al., 2003).

The Example of β-Lactam Antibiotics

The group of antibiotics known as the β-lactams includes penicillins, cephaplosporins, monobactams, and the carbapenems gives a simple example**:**

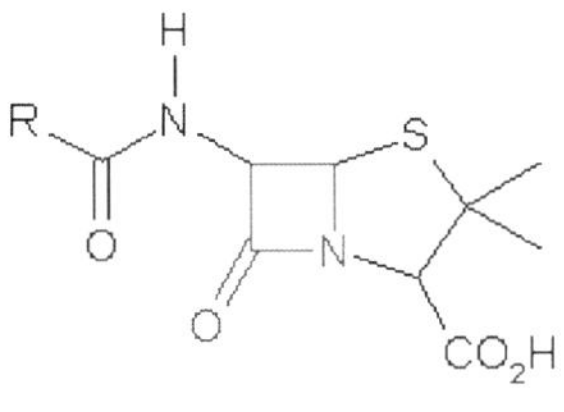

Penicillin nucleus

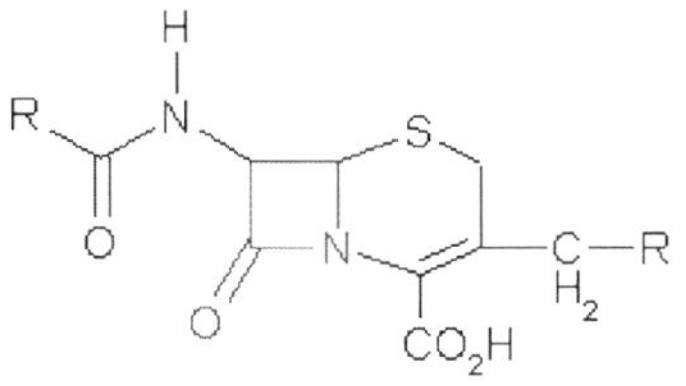

Cephalosporin nucleus

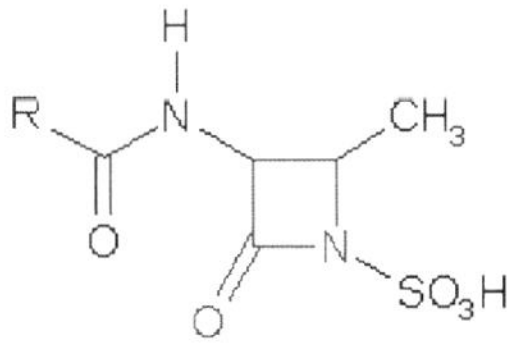

Monobactam nucleus

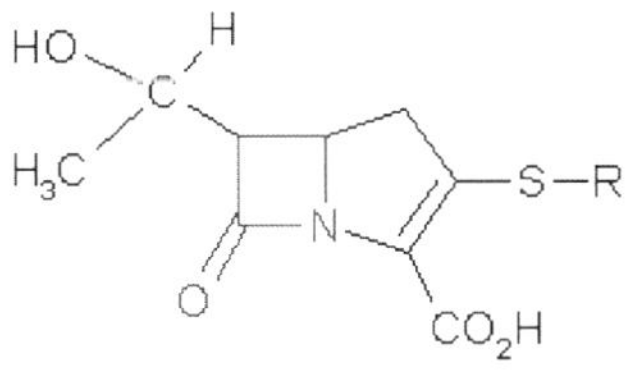

Carbapenem nucleus

They all share a common β-lactam ring (in *blue*). The *ring* is very strained and the bond between the carbonyl and the nitrogen in the β-lactam ring is very labile, hence making the molecule reactive. The R-group can be changed to give the molecule different antibacterial properties.

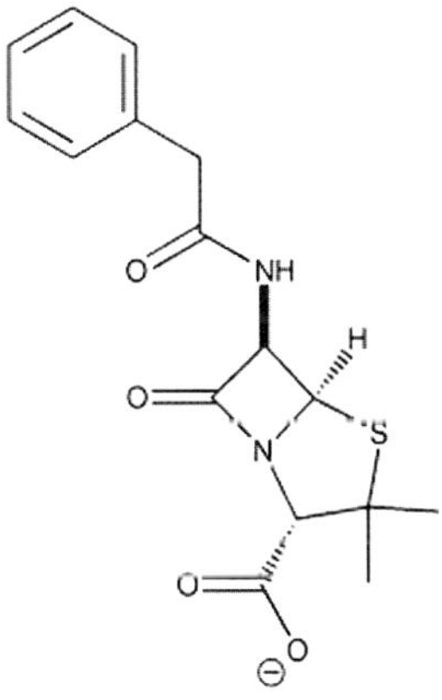

Penicillin (benzylpenicillin) structure C16H17N2O4S-

The development of antimicrobials has advanced tremendously over the past century. However, as our production capacity increases, the threat of resistance is ever present. To combat this resistance, two main avenues of drug discovery are being pursued: identifying new microbial proteins for which to direct drug discovery efforts, and designing innovative drugs that target existing proteins. The advent of structural genomics research has advanced to the point of rapidly discovering novel microbial protein targets. In addition, modern tools of computational biology greatly enhance the speed and reliability of antimicrobial discovery.

The Synchrotron Radiation

Structural molecular biology uses the synchrotron radiation facilities for research on the structure and dynamics of biomolecules (Davidovich et al., 2008). Knowledge of the structure of biomolecules is an essential prerequisite for understanding biological processes on a molecular and cellular level, with applications in biotechnology, medicine, and pharmacology. Because of the high-intensity and variable wavelength of the synchrotron light it is possible to carry out X-ray structure investigations of biomolecules rapidly and accurately.

Community Dynamics and Ecology

More knowledge is needed to understand the composition, structure, and function of the large number of organisms present in natural systems. Advanced molecular and biochemical methods should be applied to identify populations, evaluate evolution in response to exposure to antibiotics, and understand the influence of a variety of environmental factors on biotransformation rates and processes. In addition, better understanding of the factors influencing the survival and effectiveness of introduced organisms needs to be developed.

Biomolecular Science and Engineering

The potential of molecular manipulation to enhance bioremediation remains untapped. To realize this potential, more information is needed to analyze genes, proteins, and regulatory elements of critical molecules for bioremediation. Knowledge of structure and function relationships is also needed to understand the enzymatic mechanisms for detoxification. Building on this foundation, organisms could be engineered with superior degradative capabilities. However, ethical, legal, and social issues associated with the development and utilization of engineered organisms must be addressed to the satisfaction of the public and regulatory agencies before they can be used.

Ada Yonath was awarded the 2009 Nobel Prize in Chemistry for her studies of the structure and function of the ribosome, a part of the cell that synthesizes protein and translates genetic code in the production of protein. The contribution of synchrotron radiation was decisive in this discovery because it enabled the researchers to obtain immediately usable diffraction data (http://www.synchrotron-soleil.fr). Her work has revolutionized the field of structural biology and has given it wide-ranging applications, particularly in the development and design of new antibiotics. In the late 1970s, Prof. Yonath began her unique pioneering work on protein biosynthesis by ribosomal crystallography, long before others thought it possible to crystallize such large complex structures. Her studies made it possible for scientists to determine the detailed structure of the ribosome, which has led to the discovery of the unified mechanism for the production of proteins. The experiments she conceived, including the development of novel techniques, were based on the introduction of the state-of-the-art, ultra-cold bio-crystallography, an innovative method whereby temperatures of $-185°C$ are used to minimize damage to the extremely sensitive ribosomal crystal during structural studies involving X-rays. Her achievements include a singular pioneering work on the structure of over a dozen antibiotics and their interaction with ribosomes on the molecular level. This has paved the way to the understanding of antibiotic selectivity and the mechanism of drug action, synergy, and resistance, thus inspiring the development and design of new antibiotic drugs (Auerbach et al., 2010; Yonath, 2005).

Resistance

The emergence of resistance is profoundly unsurprising. But what is remarkable is how long it has taken for the problem to become a source of public, as well as scientific, concern. The amount of antibiotics use in the world is continuously increasing. The fraction devoted to human care is only about half the total amount. There are multiple other fields of usage, in agriculture, breeding, and veterinary medicine (Andremont, 2000). Antibacterial use disrupts the microbial ecology of the individual. Natural selection favors mechanisms that confer resistance determinants that prevent their own counter selection and resistant strains with enhanced survival ability or virulence. Resistance can result from modification of an antibacterial target or from functional bypassing of that target, or it can be contingent on impermeability, efflux, or enzymatic inactivation. All members of a species may be resistant. Alternatively, resistance may arise in hitherto susceptible organisms via mutation or DNA transfer (Livermore, 2007).

Mechanism of Resistance

The use of antibiotics should have created a catastrophic situation for microbial populations but genetic flexibility has allowed bacteria to survive and multiply under antibiotic pressure. Bacteria can resist antibiotics as a result of chromosomal mutation or by exchange of genetic material, which carries resistance genes, through transformation, transduction, or conjugation by plasmids. The mechanism of resistance to antimicrobial agents can be due to:

(i) impermeability of the drug: this is the most frequent cause of intrinsic resistance. Resistance in *Enterococcus* sp. and *Pseudomonas aeruginosa* is a good example of such mechanisms;
(ii) alteration in target molecules: this is one of the most important mechanisms of resistance to clinically used antibacterial drugs, for example, methicillin-resistant *S. aureus* with altered penicillin-binding proteins;
(iii) Enzymatic drug modifications: β-lactamase enzymes currently account for most of the resistance to penicillins and cephalosporins. β-Lactamases affect a

common drug site, i.e., β-lactam ring. Penicillins, cephalosporins, monobactams, and carbapenems can all be hydrolyzed by multiple members of the beta lactamase family of enzymes, resulting in a microbiologically ineffective compound. The other important class of antibiotics, which are destroyed by enzymes, are aminoglycosides due to the action of aminoglycoside-modifying enzymes produced by the bacteria;

(iv) Efflux: the role of efflux of drug from the bacterial cell as a resistance mechanism is comparatively less common in clinical practice. Although both chromosomal mutations and genetic transfer can be responsible for the resistance acquisition, it is the transferable resistance that poses a great threat as it can achieve much larger dimensions due to wide and rapid dissemination. This transferable resistance is carried on R plasmids. A single plasmid can carry a number of genes coding for multiple drug resistance. It has been suggested that evolution of multi-drug-resistant plasmids in pathogens is a comparatively recent phenomenon, which came into existence after the introduction of antibiotics after the 1940s. This further supports the observation that the use of antibiotics itself has been responsible for emergence of resistance in the pathogenic bacteria in clinical practice. While plasmids act as vectors of resistance genes, the genes themselves are most often located on discrete movable DNA elements called transposons. The important process in the gene pick up is done by transposons carrying multiple antibiotic-resistance genes. Integrons are key structural constituents of transposons. The integron is a mobile DNA element with a specific structure consisting of two conserved segments flanking a central region – "cassette." Genes encoding functions like resistance can be inserted in this region. Transposases carrying (R) genes have the ability to enter a conjugative plasmid or a chromosome. The composition of integrons is rather stable over time, although their importance should not be overtaken relative to that of other vehicles of resistance (Livermore, 2007). Many of the resistance determinants now found on plasmids are believed to have originated in the chromosomes of other species, although only a few of their source organisms have been identified definitively. The plasmid-mediated SHV β-lactamases are derived from the chromosomal β-lactamases of *Klebsiella pneumoniae*; the plasmid-borne AmpC enzymes emerging in *Klebsiella* spp. and *E. coli* are chromosomal escapes from *Citrobacter freundii*, *Hafnia alvei*, *Morganella morganii*, and *Enterobacter cloacae* and several CTX-M cefotaximases are chromosomal escapes from *Kluyvera* spp. (Table 10.1).

Impact of Resistance

The consequences of resistance are difficult to define. Increased morbidity and mortality are the most dramatic consequences of resistance. Other effects are more insidious. Physicians and surgeons are forced to use previously reserved agents as first-line therapy that could be less effective or more toxic. Finally, resistance adds cost: longer hospital stay because of treatment failures, repeated physician visits,

Table 10.1 Sources of β-lactamase resistance genes now found on transferable DNA

Gene(s) or product(s)	Sources
SHV β-lactamases	*Klebsiella pneumoniae* chromosome
CTX-M2,4,5,6,7 and Toho-1 β-lactamases	*Kluyvera* spp. chromosome
CMY-2, 3, 4, 5, 6, 7, LAT-1, -2, -3, -4, BIL-1 AmpC β-lactamases	*Citrobacter freundii* chromosome
ACC-1 AmpC β-lactamases	*Hafnia alvei* chromosome
DHA-1 and -2 AmpC β-lactamases	*Morganella morganii* chromosome
TEM, OXA,[a] PSE, staphylococcal penicillinase	Unknown

[a]Although source organisms for plasmid-mediated OXA enzymes have not been identified, some *Aeromonas* spp. have chromosomal enzymes belonging to this family

Source: From Dr. David M. Livermore, Antibiotic Resistance Monitoring and Reference Laboratory, Central Public Health Laboratory, 61 Colindale Ave., London NW9 5HT, United Kingdom (DLivermore@phls.nhs.uk) (Livermore and Pearson, 2007)

blocked hospital beds. Antimicrobial combination therapy must often be used in severe infections to provide broad-spectrum coverage, to decrease the emergence of resistant strains, and to increase the antibacterial activity of an antimicrobial agent used alone (Jacqueline et al., 2006).

Evolution of Resistance

The initial evolution of resistance is a fluid process, forever generating random combinations of genes and strains. The subsequent accumulation of resistance reflects the degree of selection pressure and the fitness of the strains that have acquired resistance. It also reflects the new opportunities that arise for pathogens through social and demographic changes and as a result of advances elsewhere in medicine, which expand the pool of vulnerable patients (Livermore, 2007).

The origin of resistance genes could be due to a natural process whereby the resistant genes are maintained in nature because of the presence of antibiotics producing bacteria in soil. These antibiotics act on bacterial species other than the producer bacteria. There has to be a mechanism of protection in the host bacteria against the antibiotics that it produces, which could be the source of genes encoding resistance. Resistance to penicillin in *S. aureus* was observed just a few years after penicillin came into use. As the next generations of antibiotics were developed to overcome the problems of resistance against available antibiotics, bac teria developed mechanisms to resist the newer antimicrobial also. For example, the resistance to penicillin initially was due to production of an enzyme penicillinase by bacteria. Antibiotics like cloxallin were developed, which was resistant to penicillinase enzyme. To resist these antibiotics, bacteria altered the target site for binding of β-lactam antibiotics, i.e., PBPs, and this led to the development of MRSA. Simultaneously attempts were being made to develop newer generations of cephalosporins with a wide range of activity and resistance to existing known β-lactamase enzymes. The third and fourth generations of cephalosporins were introduced, which were not destroyed by the β-lactamases produced by the Gram-negative bacteria.

However, after the wide use, the bacteria responded by developing mechanisms producing ESBL to destroy these drugs. To overcome this problem carbapenems were introduced, which were resistant to ESBL enzymes. These agents were the best in the β-lactam group of antibiotics but no sooner had they come into use, the bacterial populations started producing carbapenemases which hydrolyze carbapenems. Thus, it is evident that although numerous β-lactam antibiotics have been developed during the past 40 years in an attempt to circumvent the activity of β-lactamases, the prime result of this has been the selection of more diverse and potentially more deleterious enzymes hydrolyzing all β-lactams. The situation is alarming because β-lactams are otherwise the best bactericidal agents, which, when combined with aminoglycosides another bactericidal agent, make a synergistic combination and are an ideal choice for all types of critical infections. This is also the best choice of empirical therapy in infections in an immunocompromised host and in patients in high-risk units.

Another classical example of emergence of resistance due to abuse of antibiotics is the extensive use of vancomycin. As the infections due to MRSA in hospitals all over the world increased, vancomycin became the drug of choice to treat these infections. This led to the selection of VRE present as part of normal microbiota of the gut of the patient, and possibly contributed to the emergence of VRE besides other factors.

Various contributing factors responsible for emergence of antibiotic resistance are:

(i) Lack of education – The combination of poverty and ignorance makes the ground perfect for resistance development. An important reason for irrational therapy is inability to buy adequate quantity of antibiotics or to reach to qualified doctors due to poverty or ignorance for rational prescriptions of antibiotics.
(ii) Hospital-acquired infections – Hospitals are the places where the selective pressure of antibiotics is the highest as the hospital bacteria are mostly multi-drug resistant. The main reason for this is the increase in hospital-associated infections because of the disregard to standard isolation precautions in most of the busy hospitals with limited resources.
(iii) Use of antibiotics in agriculture or aquaculture – Antibiotics are used widely in agriculture and aquaculture for therapeutic, prophylactic, and growth-promoting purposes. The presence of residual antibiotics in the flesh of animals may result in direct exposure of the consumers to these drugs. In addition, the presence of low levels of antibiotics may select for resistant bacteria in the intestines of animals intended for human consumption. The animals can be contaminated with fecal bacteria during the slaughter process and therefore contaminate the meat reaching the consumer. Emergence of VRE is one particular example of appearance of resistant bacteria in animals that have affected susceptible human populations. Antibiotic-resistant bacteria can also be found

on fruits and vegetables due to spreading of sewage sludge on farmland or use of antibiotics directly on fruit and vegetable crops.

(iv) Environmental factors – The presence of antibiotic-resistant bacteria in fresh water sources has been documented from different parts of the world. Selection of resistant organisms in nature may result from the natural production of antibiotics by soil organisms, or contamination from animal feed or crops or waste products from treated animals or humans. Resistant organisms from farming practices may be transferred into rivers and other water sources through waste disposal systems or by drainage or rainwater from farm land. All these factors contribute to the natural reservoirs of resistance genes, which may provide a source of transferable genes.

(v) Use in household products – There is an increase in the use of surface antibacterial agents over the years into healthy households. The antibacterial substances added to diverse household cleaning products are similar to antibiotics in many ways, and these products can also select for resistant strains.

Three interesting studies were done on specific remote populations of Guyana, Nepal, and Bolivia:

(i) Guyana–Andremont's work on a remote community of French Guyana shows that antibacterial-resistant bacteria can spread in persons not taking antibacterial agents: Resistance rates in the predominant flora were from 95% to tetracycline to 0% to ceftazidime and nalidixic acid, with no significant differences between adults and children, men and women, or villages. Because data confirmed the lack of direct antibacterial drug exposure in their study, the results demonstrate that, once resistance elements are introduced into a population, moderate use of antibacterial drugs in the environment is enough to maintain them in intestinal bacteria when sanitary conditions are poor (Grenet et al., 2004).

(ii) Nepal–Within Nepal, geographic, social, and economic barriers greatly limit access to allopathic health care. The country therefore offered the opportunity to evaluate the effect of antibiotic accessibility (as measured by allopathic medicine consumption) on antibiotic resistance in the normal intestinal microbiota. The aerobic Gram-negative fecal microbiota of healthy adults from each of three villages with different access to health-care facilities in Kathmandu was examined for antibiotic susceptibility. The frequency of antibiotic resistance decreased significantly with increasing distance from Kathmandu and decreasing population density but did not reflect contact with health-care providers or individual medicine consumption. The findings suggest that an individual's overall exposure to antibiotics and antibiotic-resistant bacteria (resulting from close proximity to other community members and to sources of accessible allopathic health care, such as in the vicinity of Kathmandu), has an equal or greater impact on an individual's carriage of antibiotic-resistant bacteria than does direct consumption of antibiotics (Walson et al., 2001).

(iii) Bolivia – The incidence of antimicrobial-resistant, nonpathogenic *E. coli* among healthy children aged 6–72 months in Camiri town and a rural village, Javillo, in southeastern Bolivia showed a high prevalence of carriage of nonpathogenic *E. coli* resistant to antimicrobials. The prevalence of resistance to ampicillin and TMP/SMX was higher than that previously reported in developing countries (Bartoloni et al., 1998).

Conclusion: The selective pressure generated by the use of antibiotics in clinical, veterinary, husbandry, and agricultural practices is considered the major factor responsible for the emergence and spread of antibiotic-resistant bacteria since the beginning of the antibiotic era. However, those recent studies have consistently demonstrated that acquired resistance traits can also be found in bacteria isolated from humans and wild animals not subjected to significant antibiotic exposure and living in remote areas of the planet. The existence of a large reservoir of resistance genes in healthy individuals in developing countries represents a threat to the success of antimicrobial therapy throughout the world. Programs to improve rational and effective drug use in developing countries are urgently needed.

Control of Resistance

If an infection is addressed in a comprehensive and timely manner, resistance can be contained. Appropriate use of antibiotics will delay and in many cases prevent the emergence of resistance. Historically, several approaches to antibiotic prescribing have been employed to address antimicrobial resistance. One approach is to use a newer, more potent antimicrobial in settings where resistance to an older agent has emerged. However, if newer agents are overused or used inappropriately, resistance will invariably develop to the newer drug. Another approach to combating resistance is to continue using older agents as first-line choices, in preference to newer, more potent drugs, in an effort to preserve the activity of the new drugs. Thus the newer agents are reserved for infections caused by mutated multi-resistant strains. However, as resistance continues to increase to the first-line agents, poor outcomes and secondary costs associated with clinical failures increase. Efforts to overcome bacterial resistance range from judicious and rational use of antibiotics, an effective hospital infection control program, and research in the field related to development of newer antibiotics. The use of antibiotics in the community can be restricted by implementing laws to stop over-the-counter sale of antibiotics.

Another approach is to use combination therapy. The most recent interest has been in rotating use of antibiotics. Some studies have suggested the usefulness of this strategy, but this has to be done with close microbiological monitoring. Appropriate antimicrobial use is an integral component of any program to slow the emergence and spread of antimicrobial-resistant microorganisms in the health-care setting. Although many possible interventions have been proposed, deciding which one is the most effective in a particular setting can be difficult. Despite guidelines from governmental and professional groups, many hospitals have yet to institute any antimicrobial use policies or programs to improve antimicrobial agent prescribing.

Limiting the spread of antibiotic-resistant bacteria in hospitals now envisages restrained and limited use of antimicrobial agents and improved hospital hygiene – both easier to state than to implement. Hospital infection control will prevent the spread of infection and colonization of patients especially in high-risk areas with multi-drug-resistant strains of bacteria. This will minimize the use of antibiotics in hospitals. More importantly, rational policies are urgently needed to promote rational use of antibiotics in poultry, animal husbandry, and agriculture.

Adequate knowledge of the types of microorganisms and events that influence the timing of colonization may provide opportunities to modulate the microbiota when needed to prevent or treat disease and improve overall health. The turnover of genera, species, and strains suggest that the phylogenetic composition of this community is malleable and should be responsive to probiotic intervention.

MRSA control in these ICUs characterized by a high prevalence of MRSA at admission was achieved via multiple factors, including screening, contact precautions, and use of alcoholic hand-rub solution. Our results after adjustment of risk factors for MRSA acquisition and the steady improvement in MRSA over several years strengthen these findings. MRSA spreading can be successfully controlled in ICUs with high colonization pressure (Lucet et al., 2005).

Another example is given by the work of Aubry-Damon et al. They assessed the quantitative contribution of pig farming to antimicrobial resistance in the commensal microbiota of pig farmers by comparing healthy pig farmers from the major French porcine production areas to nonfarmers, each matched for sex, age, and county of residence. All reported that they had not taken antimicrobial agents within the previous month. Throat, nasal, and fecal swabs were screened for resistant microorganisms on agar containing selected antimicrobial agents. Nasopharyngeal carriage of *S. aureus* was significantly more frequent in pig farmers, as was macrolide resistance of *S. aureus* from carriers. Nongroupable *Streptococci* from the throat were more resistant to the penicillin in pig farmers. The intestinal isolation of *Enterococci* resistant to erythromycin or vancomycin was not significantly higher in pig farmers in contrast to that of enterobacteria resistant to nalidixic acid, chloramphenicol, tetracycline, and streptomycin. Prevalence of resistance in predominant fecal *Enterobacteria* was also significantly higher in pig farmers for cotrimoxazole, tetracycline, streptomycin, and nalidixic acid. They could determine a significant association between pig farming and isolation of resistant commensal bacteria (Aubry-Damon et al., 2004).

Example of Resistant Organisms

Salmonella typhi and *S. paratyphi* A. Enteric fever continues to be a major public health problem in our country. Chloramphenicol remained the drug of choice for the treatment of this infection till plasmid-mediated chloramphenicol resistance was encountered. Following this, ciprofloxacin became the mainstay of treatment. Being a safer and more effective drug, it was used even when the bacteria were sensitive to chloramphenicol. The isolates of *S. typhi* and *S. paratyphi* A showed higher MIC to ciprofloxacin, and there is clinical resistance to treatment with ciprofloxacin in the patients suffering from enteric fever. The choice left now is an expensive drug such as ceftriaxone or cefexime. Infections with *S. paratyphi* A, which were always considered to be mild, have also shown similar trends and complications occurring if the treatment is delayed.

Shigella sp. The emergence of multi-drug-resistant *Shigella* has remained a cause of concern in endemic regions. The nalidixic acid resistance has increased for *Shigella sonnei* in some of the Indian isolates. This is noteworthy, since it has been recommended for the empirical treatment of patients suspected to have shigellosis. Multi-drug-resistant *Shigella dysenteriae* serotype 1 strains have re-emerged in patients hospitalized with diarrhea, which were multi-drug resistant (resistant to norfloxacin and ciprofloxacin).

Methicillin-resistant Staplylococcus aureus (MRSA). MRSA is an important cause of nosocomial infections worldwide. These are also resistant to most of the other antibiotics, and in many cases the only choice left is vancomycin. There is evidence to show that MRSA is not only limited to hospital environment but can also cause infections in a community, a fact which is alarming. This could be due to the carriers in the hospitals being discharged or health-care workers carrying it with them. Similarly, methicillin-resistant coagulase negative staphylococci (CoNS) are also increasing in numbers in such an environment, and the multi-drug-resistant strains are higher among CoNS. Yet no vancomycin or teichoplanin resistance has been reported from India. But continuous monitoring is required as the presence of VISA strains has been reported as causing infections in some parts of the world.

Entercoccus sp. The enterococci are inherently resistant to many antibiotics but the combination of penicillin and gentamicin being synergistic remained the treatment of choice for infections related to this bacteria. Now strains have emerged which do not respond to treatment with this combination as they have a high level of resistance of aminoglycosides and are called as HLAR. Vancomycin is the only alternative left for the treatment of infections caused by HLAR strains. But a major problem is that vancomycin use is a risk factor for colonization and infection with VRE. Vancomycin resistance, emerging amidst the increasing incidence of high-level resistance to penicillins and aminoglycosides, has limited treatment options for bacteremia due to *E. faecalis* or *E. faecium* in the hospitals, which is increasingly being encountered. Although at present there is a limited number of reports of VRE from India, it needs continuous monitoring. All these strains are presently sensitive to linezolid. Emergence of VRE may also increase the possibility of the emergence of vancomycin-resistant *S. aureus*. Conjugative transfer of high-level vancomycin resistance from *E. faecalis* to *S. aureus* in the laboratory has been possible, and there is a possibility that this resistance may be transferred to wild-type *S. aureus*.

Streptococcus pyogenes. S. pyogenes, the group A streptococcus, has remained sensitive to penicillin until now, but it continues to cause invasive infections and toxic shock syndrome. Indian isolates have remained sensitive to penicillin, but the resistance to macrolide is being encountered. The penicillin susceptibility of this organism now needs continuous monitoring.

Streptococcus pneumoniae. There is a concern of the spread of the penicillin-resistant *S. pneumoniae* strain throughout the world. Though the prevalence of this resistance is at present not a major problem in our country, intermediate resistance to penicillin has been reported. Resistance to cotrimoxazole and chloramphenicol is seen more frequently. The isolates showing total resistance to penicillin have been reported from India, which were also multi-drug resistant, being resistant to cefotaxime, erythromycin, chloramphenicol, and trimethoprim-sulfamethoxazole.

Haemophilus influenzae. In India, β-lactamase-mediated resistance in *H. influenzae* is increasingly being encountered. Recently, a multicentric study showed an increasing resistance to ampicillin, chloramphenicol, erythromycin, and trimethoprim-sulfamethoxazole in *H. influenzae* strains isolated from different parts of India. All the strains, however, were sensitive to cefotaxime.

Gram-negative Bacilli The most important cause of hospital-acquired infections is Gram-negative bacteria. These bacteria have acquired resistance to multiple antibiotics. The Gram-negative bacteria pose a therapeutic problem not only in the hospital settings but also in the community-acquired infections. *E. coli* is an important cause of community-acquired urinary tract infections, but resistance is seen in nearly 70–80% of the strains to the commonly used antibiotics. In patients suffering from cystic fibrosis, colonization with *P. aeruginosa* in the community setup is very common. These patients can harbor multi-drug-resistant strains over a period of time as they are on long-term antibiotic prophylaxis and need frequent antibiotic treatment. Among the nosocomial pathogens, multi-drug-resistant Gram-negative bacteria are the important cause of hospital-associated infections. These bacteria can survive for a long period of time in an adverse environment, and once having entered the host, can lead to long-term colonization. Nosocomial infections due to extended-spectrum β-lactamases (ESBL) producing *K. pneumoniae*, *P. aeruginosa*, *Acinetobacter baumanii*, *Serratia marsecens*, *E. coli*, etc. are the main threat in the present scenario. The therapeutic options are limited in infections caused by these agents.

The Pervasive Effects of Antibiotics

Antibiotics save lives and reduce morbidity, but can cause significant unwanted problems such as allergic reactions, thrush, and diarrhea. Important interactions with other medicines or foods can also occur.

Common Side Effects

Antibiotics are one of the most common groups of drugs to which patients report allergies, and some of these are clinically important. The most common example is penicillin, to which 1–10% of patients report allergies. About 8% of these show cross-sensitivity to cephalosporins, and in some cases also to other β-lactam antibiotics, such as imipenem and meropenem, that may be administered intravenously in severe infections.

There are many important interactions between antibiotics and medicines or food. Here are examples of some common, but important, antibiotic interactions. This list is not exhaustive:

- Broad-spectrum antibiotics may reduce the efficacy of combined oral contraceptives and additional contraceptive precautions should be used while taking the antibiotic and for 7 days afterwards.
- The oral absorption of tetracycline and quinolone (for example, ciprofloxacin) is reduced when taken with milk or zinc, iron, or calcium salts. These should be taken 2 h apart.
- Ciprofloxacin, rifampicin, erythromycin, and metronidazole can affect drugs with a narrow therapeutic index (such as warfarin, digoxin, theophylline, and phenytoin).
- Interactions occur with enzyme-inducing antibiotics (e.g., rifabutin and rifampicin) and enzyme-inhibiting antibiotics (e.g., co-trimoxazole, erythromycin, isoniazid, metronidazole, and quinolones).
- Alcohol must not be consumed during treatment with metronidazole and for 48 h after, because this may lead to a disulfiram-like reaction in some individuals. Alcohol may be taken in moderation with other antibiotics.
- Children: Tetracyclines are contraindicated, because of deposition of the drug in growing bone and teeth, and are unlicensed in children. Quinolones must be used with caution due to the risk of arthropathy.
- Elderly are often prescribed numerous medications with which antibiotics may interact. Care should also be taken that the appropriate dosage form is supplied; some older patients may find liquid medication easier to take. Renal function is often reduced, which may result in slow excretion and accumulation of the drug if the dosage is not reduced appropriately. This applies to antibiotics excreted mainly by the kidneys, such as penicillins, cephalosporins, and nitrofurantoin. Antibiotics that can impair hepatic function, such as flucloxacillin and coamoxiclav, should be avoided if possible in those with severe liver impairment. Elderly patients are also more prone to side effects, including *C. difficile*-associated diarrhea, which can be caused by any antibiotic, but particularly cephalosporins and clindamycin. Prolonged courses of antibiotics should therefore be avoided wherever possible in older patients.

Antibiotherapy and Microbiota

Because many chemical transformations in the gut are mediated by specific microbial populations, with implications for cancer, obesity, allergies, diarrhea, among other conditions, changes in the composition of the gut microbiota would have important health effects. The human intestinal microbiota is essential to the health of the host and plays a role in nutrition, development, metabolism, pathogen resistance, and regulation of immune responses (Dethlefsen et al., 2008).

- First of all, it is important to examine the microbiota at different sites within the intestinal tract in order to appreciate possible local antibiotic effects that are not revealed in fecal samples. Antimicrobial agents not only affect the pathogens to which they are directed but also impact other members of the intestinal microbiota. Thus, it is important to characterize the nature and amplitude of the antibiotic-related modification.
- Then, the resilience has to be taken into account. After an antibiotic treatment, an approximate return to pretreatment conditions can occur within days or weeks after cessation of antibiotic treatment, as assessed by subjective judgments of bowel function and characterizations of microbial diversity. After a 5-day course of amoxicillin, one of the most prescribed antibiotics in Europe, it has been found that the dominant fecal microbiota tended to return to its initial state within 60 days, but not in all cases (De La Cochetiere et al., 2005). Relative abundance levels are affected, although these changes may be the direct effect of varying sensitivity to the antibiotic among the taxa comprising the gut microbiota. The taxa that increased in relative abundance following treatment may represent taxa with intrinsic resistance to the antibiotic – strains that are typically sensitive but had already acquired resistance prior to the study, or strains that developed resistance due to the current exposure. After clindamycin administration, long-term ecological stability, with a focus on the *Bacteroides* group, has been studied by Jernberg et al. (2007). *Bacteroidetes* is one of the three main phyla of the intestinal microbiota. *Bacteroides* species ferment a variety of carbohydrates, and some, such as *B. thetaiotaomicron*, can ferment plant polysaccharides. *Bacteroides* can be an opportunistic pathogen and are resistant to a wide range of antibiotics, including animoglycosides, β-lactam, and tetracycline. Many of the changes in the community are likely to be explained by indirect effects, mediated by ecological interactions among taxa such as resource competition, cross-feeding, or the cooperative lysis of polymeric substrates (Flint et al., 2007; Belenguer et al., 2006; De La Cochetiere et al., 2008; De La Cochetiere et al., 2010).

The microbial community provides *functional redundancy* because of apparent continuity of gut functions. All communities supporting similar functions are equivalent. Dethlefsen et al. present a study from three healthy adults before, during, and after a short course of the antibiotic ciprofloxacin (Cp), which is reported to have a lower rate of common gastrointestinal side effects than some other broad-spectrum antibiotics (Dethlefsen et al., 2008): (1) The relative abundance levels of about 30% of the taxa in the gut were affected; (2) Despite a persuasive disturbance of the gut microbiota, gut function remained normal as assessed subjectively by the participants, and the community composition in samples taken 4 weeks after treatment were within the range of pretreatment samples. But continuity in the predominant metabolic activity of the community (i.e., hydrolysis and fermentation of polysaccharides) does not necessary imply the continuity of more specialized activities (i.e., bile transformation, immune modulation, pathogen resistance due to specific inhibition).

The Case of *Pseudomonas aeruginosa*

Antibiotic resistance is a major concern of contemporary medicine. The ongoing emergence of resistant strains that cause nosocomial infections contributes substantially to the morbidity and mortality of hospitalized patients. *P. aeruginosa* is one of the main organisms responsible for drug-resistant nosocomial infections and is a leading cause of bacteremia and nosocomial pneumonia. In addition to being intrinsically resistant to several antimicrobial agents, *P. aeruginosa* often acquires mechanisms of resistance to other antibiotics. Previous treatment with antibiotics that are characterized by high antipseudomonal activity and prolonged antibiotic treatment are both recognized risk factors for the emergence of drug-resistant *P. aeruginosa*. Acquisition of strains resistant to ceftazidime, imipenem, piperacillin, or ciprofloxacin is associated with significantly longer hospital stays and an increased rate of secondary bacteremia in patients with *P. aeruginosa* infection (Coenye, 2010; Abbo et al., 2010). A matched case-control study was performed to identify risk factors for acquiring multi-drug-resistant *P. aeruginosa* (MDRPA) in ICU patients during a 2-year period. MDRPA was defined as *P. aeruginosa* with combined decreased susceptibility to piperacillin, ceftazidime, imipenem, and ciprofloxacin. Thirty-seven patients who were colonized or infected with MDRPA were identified, 34 of whom were matched with 34 control patients who had cultures that showed no growth of *P. aeruginosa*. Matching criteria were severity of illness and length of ICU stay, with each control patient staying in the ICU for at least as long as the time period between the corresponding case patient's admission to the ICU and the acquisition of MDRPA. Baseline demographic and clinical characteristics and the use of invasive procedures were similar for case patients and control patients. Multivariate analysis identified duration of ciprofloxacin treatment as an independent risk factor for MDRPA acquisition, whereas the duration of treatment with imipenem was of borderline significance. These data support a major role for the use of antibiotics with high antipseudomonal activity, particularly ciprofloxacin, in the emergence of MDRPA (Paramythiotou et al., 2004).

Neonates

The use of broad-spectrum antibiotics in pediatric practices alters the gut colonization and consequently may impair the barrier function (Mshvildadze and Neu, 2010). But little is known about how antibiotics alter the establishment of gastrointestinal microbiota in premature infants. Our research team has done some descriptive work on the first implantation of the microbiota, in the gut of newborns, and is getting data on the impact of antibiotherapy (Caillaux et al., 2010).

Antibiotic administration early in life impairs specific humoral responses to an oral antigen and increases intestinal mast cell numbers and mediator concentrations (Nutten et al., 2007). It affects the gut colonization in three important ways: (1) antimicrobial agents can have specific effects on individual components of the microbiota rather than a general nonspecific suppression of all microbes, (2) the resultant microbial profile influences the populations that emerge after treatment has stopped, and (3) the effect of antibiotic therapy can persist beyond treatment.

The postnatal maturation of the gut, which is partially modulated by bacterial colonization, results in the establishment of an efficient barrier to luminal antigens and bacteria. The use of broad-spectrum antibiotics in pediatrics alters the gut bacterial colonization and, consequently, may impair the maturation of the gut barrier function. Animal model studies have shown that clamoxyl treatment altered the normal colonization pattern of the gut microbiota and the normal maturation profile of 10–30% of the genes in the different intestinal segments (Schumann et al., 2005).

Influence of the Antibiotic Therapy on Intestinal Microbiota

Therapy with broad-spectrum antibiotics is frequently observed in pediatric practices, children within their first year of life being particularly affected. One major consequence of such early antibiotherapy is the alteration of the normal colonization process by the gut microbiota. Neonatal antibiotic treatment has been shown to reduce the biodiversity of the fecal microbiota, to delay the colonization by beneficial species such as *Bifidobacterium* or *Lactobacillus*, and to induce colonization by antibiotic-resistant opportunistic strains. Penders et al. demonstrate on fecal samples from 1032 infants that oral use of antibiotics during the first month of life resulted in a decreased number of *Bifidobacteria* and *Bacteroïdes fragilis* species and may have a major effect on the composition of the gut microbiota, particularly on obligate anaerobes (Penders et al., 2006). An experimental study in the rat clearly demonstrates that amoxicillin deeply affects the maturation of 10–30% of the genes involved in the intestinal barrier function at the suckling–weaning interface, a period during which the gut is challenged by several novel food-borne antigens (Schumann et al., 2005). Changed patterns of early-life gut colonization are reported to be associated with allergic sensibilization, metabolic priming, and development of regulatory lymphocyte population (Penders et al., 2007). There is rapidly increasing evidence from experimental studies that the initial colonization of the intestine is a moment of pivotal importance in long-term health. The potential for long-term persistence of early colonizing bacteria suggests that much more thought should be given to the late consequences of perinatal antibiotherapy (Mshvildadze and Neu, 2010).

Microbiota and Microbiome Analysis Methods

Our aim is not to give an exhaustive description of all used methodologies, but to give important points that should help readers.

Analytical Methodologies

Until recently, microbial ecologists relied largely on techniques that required cultivation of organisms on selective media. The use of selective media specific for different types of bacteria has the disadvantage of being time and labor intensive and also imposes an a priori bias on the types of bacteria that can be enumerated. Additionally, only 20–40% of bacterial species from mammalian gastrointestinal tracts can be cultured and identified using known cultivation techniques and, therefore, the majority of intestinal bacterial species will not be represented at all using cultivation-based techniques (Vaughan et al., 2000).

Bacterial genome sequencing and its use in infectious diseases. The availability of genome sequences is revolutionizing the fields of microbiology and infectious diseases (Fournier et al., 2007). Further, high-throughput sequencing is revolutionizing microbial ecology studies (Qin et al., 2010).

Genome sequences provide researchers with a unique source of information. 16S rRNA divergence provides useful information about average conservation of gene content, even within bacterial species or genera (Zaneveld et al., 2010). Recently, molecular ecology techniques that are based on the 16S rRNA gene (rDNA) have become increasingly popular and useful. These methods have proved to be reliable for the detection and identification of bacteria. The rDNA analysis is based on physical and chemical proprieties of DNA molecules. Utilization of these molecular techniques bypass the cultivation step and enable characterization and quantification of the microbiota, while providing a classification scheme to predict phylogenetic relationships. Microbial community structure and evolution can be analyzed via fingerprinting techniques, whereas dot blot hybridization or fluorescent in situ hybridization can be used to measure the abundance of particular taxa. Emerging approaches, such as those based on functional genes and their expression, and the combined use of stable isotopes and biomarkers, are being developed and optimized as a means to study the metabolic activity of groups or individual organisms in situ. Fingerprinting techniques have great potential for microbial ecology as these methods allow the application of statistical analysis. These culture-independent molecular methods have proven most appropriate in dynamic studies of dominant species diversity within complex ecosystems like the colon (Lepage et al., 2005; Seksik et al., 2005; Huse et al., 2008). In our laboratory, we utilize the denaturing high performance liquid chromatography (dHPLC) because it has two applications, a descriptive and a comparative analysis thus constitutes a high-potential molecular tool to study microbiota dynamics (Rougè et al., 2010)

Metagenomics. Much of the extant microbial genetic diversity, referred to as the metagenome, remains unexploited. Metagenomics utilizes a whole-genome shotgun sequencing approach to study the genomes of all microbes, regardless of their ability to be cultured. This method, which is both cost-effective and culture independent, is used to identify microbes and analyze microbial genomes. By treating the microbial community as a single dynamic entity, metagenomics explores the genome content of the whole community and provides analysis of changes in content and expression as a function of location, time, and various states of perturbation, e.g., progression toward and regression from disease following treatment. From this type of analysis, a better understanding of the biology of the organisms found in the gastrointestinal tract and of the process of adaptation to co-exist and interact with their host will undoubtedly arise. These new types of data will further our understanding of the impact the microbiota has on preterm human physiology. The completed genomes of *Bifidobacterium longum* NCC2705, *Lactobacillus acidophilus* NCFM, *Bacteroides thetaiotaomicron* VPI-5482, *Lactobacillus johnsonii* NCC533, and *Lactobacillus plantarum* WCFS1 have already been sequenced (Altermann et al., 2005).

MetaHIT is a project financed by the European Commission under the 7th FP program. The consortium gathers 13 partners from academia and industry, from a total of eight countries. Its total cost has been evaluated at more than 21.2 million €. The central objective of the MetaHIT project is to establish associations between the genes of the human intestinal microbiota and our health and disease. It focuses on two disorders of increasing importance in Europe: inflammatory bowel disease (IBD) and obesity. The incidence of IBD has been growing steadily during the past 5 decades in Western Europe and is now increasing dramatically in Eastern Europe. IBD is becoming an important burden on the young populations. The global epidemic of obesity is well recognized and imposes a huge and rapidly growing challenge for the public health services.

Mathematical Network

Mathematical network theory is ideal for studying the ecological networks of interacting species that comprises the human microbiome. Microbial communities can be viewed as mathematical networks with structural features that reflect how the networks developed and predict their responses to perturbations because, theoretically, networks require little consortia-specific data to provide insight into both normal and disturbed microbial community functions (Foster et al., 2008). However, these approaches are exploratory, and thus introduced a close work with clinicians for validation.

Analytical Methodologies

Mathematical networks provide a system-level approach to characterize microbiota and microbial interactions. Networks may capture specific biological information, such as how the antibiotic molecule acts on the bacteria cell. Ecological principles applied to such microbiome-treatment networks are likely to constrain how the microbiome will respond to specific antibiotic treatment, how it will then respond to invasive species, and give an idea of possible and long-term consequences. Besides, network structure may allow us to identify keystone species, even when specific interactions have not been quantified. This is particularly useful when detailed data on the immune constituents and species interactions in a consortium are unavailable.

Many microbial consortia associated with the human body constitute rich ecosystems with hundreds of species finely tuned to functions relevant to human health. The collective effects of the intestinal microbiota are dictated by a complex network of interactions that span the cellular, immunological, and environmental levels. An important question is whether a simple system can be applied 1) to predict and 2) to control the effects of this heterogeneous population composed of different subgroups.

Microbial communities can be viewed as mathematical networks with structural features that reflect how the networks developed and predict their responses to perturbations (Foster et al., 2008). This approach to understanding the dynamic of microbial communities is admittedly speculative, because of (1) lack of knowledge about community composition and (2) species interactions, in the human microbiome. An investigation into the factors responsible for community resilience is warranted. A mixture of selective forces intrinsic to the community (e.g., a competitive hierarchy based on relative growth rates and substrate affinity or interference mechanisms) and imposed by the environment (e.g., composition of the diet and of host-derived substrates) is likely to be involved, but nonselective forces such as recolonization of the gut lumen from protected environments (perhaps the mucosa) must also be considered.

Theories

Ecological network analysis (ENA) is a systems-oriented methodology to analyze within-system interactions used to identify holistic (relating to or concerned with wholes or with complete systems rather than with the analysis of, treatment of, or dissection into parts) properties that are otherwise not evident from direct observations. As such, it is necessary that the network model be a partition of the environment being studied, i.e., be mutually exclusive and exhaustive. For such approaches it is rather difficult to quantify the wholeness and consequent indirectness in the system. Ecosystems comprise a rich web of many interactions and it would be remiss to exclude, a priori, any of them. Therefore, network models aim to include all ecological compartments and interactions and the analysis determines the overall relationships and significance of each. The difficulty of course lies in obtaining the data necessary to quantify all the ecological compartments and interactions (Fath et al., 2001; Borrett et al., 2007).

To further elaborate, regarding an ecological network as a communication channel, the information contained in the structure of an ecological network may be theoretically defined by the concept of mutual information.

Chemometric methods derive mathematical models (quantitative structure activity relationships: QSARs) describing a biological activity of interest (or a profile of such activities) in terms of chemical parameters (measured, from tabulations, or computed). Here is a variety of such methods: multiple regression analysis, principal component and factor analysis, principal component regression, PLS, cluster analysis, discriminant analysis, SIMCA, support vector machines, neuronal nets, or topology-based logical approaches. QSAR models are allowed to make estimates of biological properties of new compounds and can thus aid in decision making for further syntheses and/or biological testing. They can also provide information on the mechanism of action (Richard et al., 2006).

Examples

Phylogenetic diversity in the bacterial domain. In 1987, Woese described the bacterial domain as comprised of about 12 natural relatedness groups, based mainly on analyses of familiar cultivated organisms such as cyanobacteria, spirochetes, and Gram-positive bacteria. Despite considerable interpersonal variation in species content, there is surprising functional convergence in the microbiome of different humans (Zaneveld et al., 2010). Zaneveld et al. characterized the relationship between 16S rRNA gene sequence similarity and overall levels of gene conservation in four groups of species: gut specialists and cosmopolitans, each of which can be divided into pathogens and nonpathogens. At short phylogenetic distances, specialist or cosmopolitan bacteria found in the gut share fewer genes than is typical for genomes that come from non-gut environments, but at longer phylogenetic distances gut bacteria are more similar to each other than are genomes at equivalent evolutionary distances from non-gut environments, suggesting a pattern of short-term specialization but long-term convergence. They have developed the first atlas of bacterial diversity across the human body, charting wide variations in microbe populations that live in different regions of the human body and which aid us in physiological functions that contribute to our health. Eckburg et al. examined 13,355 prokaryotic rRNA gene sequences from multiple colonic mucosal sites and feces of healthy subjects to improve our understanding of gut microbial diversity as the human endogenous intestinal microbiota is an essential "organ" in providing nourishment, regulating epithelial development, and instructing innate immunity (Eckburg et al., 2005); yet, surprisingly, basic features remain poorly described. A majority of the bacterial sequences corresponded to uncultivated species and novel microorganisms. They discovered significant inter-subject variability and differences between stool and mucosa community composition. Our research team pointed out those complex statistical analyses that provided further information and thus could highlight risk factors for development of *C. difficile* and antibiotic-associated diarrhea (de La Cochetiere et al., 2010; de La Cochetiere et al., 2008).

Conclusion

Because of the remarkable influence of microbiota on health and disease, learning about characteristics of microbiota and the factors that modulate it has become the focus of ongoing research. The reported observations further support that a given gut microbial ecosystem contributes to the development of intestinal dysfunction. Further knowledge in this field should assist in developing strategies to promote health and prevent diseases.

The microbiota directs myriad biotransformations, ranging from synthesis of essential vitamins to the metabolism of the xenobiotics that we ingest and the lipids that we produce.

The microbiota modulates the maturation and activity of the innate and adaptive immune system: an immune system educated to allow the host to tolerate a great degree of microbial diversity provides a selective advantage since this diversity ensures the stable functioning of a microbiota in the face of environmental stresses.

The intestinal ecosystem is characterized by dynamic and reciprocal interactions between the host and its microbiota. Although the importance of the gut microbiota for human health has been increasingly recognized, the early bacterial colonization in the neonatal gut is not yet completely understood. The mechanisms underlying these interactions are complex and influenced by many factors. The relative importance of these factors is difficult to organize into a hierarchy. A better knowledge of the microbiota and the impact of antibiotics will provide an essential step toward understanding the development of this important bacterial community. Recent research in the area of probiotics and prebiotic oligosaccharides is leading to a more targeted development of functional food ingredients. Improved molecular techniques for analysis of the gut microbiota and its development, increased understanding of metabolisms and interaction between host and environment, and new manufacturing biotechnologies are facilitating the production of such food supplements. Thus, our increased understanding is fostering our ability to modulate the gastrointestinal microbiota for therapeutic outcomes.

Finally, evaluation of the intestinal microbiota, together with metabolic functions, should be a challenge for clinicians in order to get early patient data and, thus, to work with personalized therapy.

Acknowledgments The authors would like to thank Professor Gilles Potel without whom this work could not have been done; Françoise Le Vacon, Thomas Carton, and Christophe Dufour, Silliker company, for their work with dHPLC and outstanding assistance; Professor Christian Bréchot and Christine Mrini, Institut Mérieux, for helpful discussions; Jocelyne Caillon and all EA 3826 research team for helpful advice, Christele Gras-Leguen and Professor Jean-Christophe Rozé for their guideline as pediatricians, Stephane Bonacorsi for support.

References

Abbo O et al (2010) Necrotizing fasciitis due to *Pseudomonas aeruginosa* in immunocompromised children. Pediatr Blood Cancer 55:213–214. doi:10.1002/pbc.22502

Altermann E et al (2005) Complete genome sequence of the probiotic lactic acid bacterium *Lactobacillus acidophilus* NCFM. Proc Natl Acad Sci USA 102:3906–3912. doi:0409188102 [pii]10.1073/pnas.0409188102

Andremont A (2000) Consequences of antibiotic therapy to the intestinal ecosystem. Ann Fr Anesth Reanim 19:395–402

Aubry-Damon H et al (2004) Antimicrobial resistance in commensal flora of pig farmers. Emerg Infect Dis 10:873–879

Auerbach T et al (2010) The structure of ribosome–lankacidin complex reveals ribosomal sites for synergistic antibiotics. Proc Natl Acad Sci USA 107:1983–1988. doi:0914100107 [pii]10.1073/pnas.0914100107

Backhed F, Ley RE, Sonnenburg JL, Peterson DA, Gordon JI (2005) Host–bacterial mutualism in the human intestine. Science 307:1915–1920. doi:307/5717/1915 [pii]10.1126/science.1104816

Bartoloni A et al (1998) Patterns of antimicrobial use and antimicrobial resistance among healthy children in Bolivia. Trop Med Int Health 3:116–123

Belenguer A et al (2006) Two routes of metabolic cross-feeding between *Bifidobacterium adolescentis* and butyrate-producing anaerobes from the human gut. Appl Environ Microbiol 72:3593–3599. doi:72/5/3593 [pii]10.1128/AEM.72.5.3593–3599

Blakey JL et al (1982) Development of gut colonisation in pre-term neonates. J Med Microbiol 15:519–529

Borrett SR, Fath BD, Patten BC (2007) Functional integration of ecological networks through pathway proliferation. J Theor Biol 245:98–111. doi:S0022-5193(06)00434-6 [pii]10.1016/j.jtbi.2006.09.024

Caillaux G et al (2010) Application of denaturing high-performance liquid chromatography for intestinal microbiota analysis of newborns. J Perinat Med 38:339–341. doi:10.1515/JPM.2010.035

Clapp DW (2006) Developmental regulation of the immune system. Semin Perinatol 30:69–72. doi:S0146-0005(06)00030-9 [pii]10.1053/j.semperi.2006.02.004

Coenye T (2010) Response of sessile cells to stress: from changes in gene expression to phenotypic adaptation. FEMS Immunol Med Microbiol 59:239–252. doi:FIM682 [pii]10.1111/j.1574-695X.2010.00682.x

Davidovich C, Bashan A,Yonath A (2008) Structural basis for cross-resistance to ribosomal PTC antibiotics. Proc Natl Acad Sci USA 105:20665–20670. doi:0810826105 [pii]10.1073/pnas.0810826105

De La Cochetiere MF, Rouge C, Darmaun D, Rozé JC, Potel G, Gras-Leguen C. (2007) Intestinal microbiota in neonates and preterm infants: a review. Curr Pediatr Rev 3:21–34

De La Cochetiere MF et al (2004) Early intestinal bacterial colonization and necrotizing enterocolitis in premature infants: the putative role of Clostridium. Pediatr Res 56:366–370

De La Cochetiere MF et al (2008) Effect of antibiotic therapy on human fecal microbiota and the relation to the development of *Clostridium difficile*. Microb Ecol 56:395–402. doi:10.1007/s00248-007-9356-5

De La Cochetiere MF et al (2010) Human intestinal microbiota gene risk factors for antibiotic-associated diarrhea: perspectives for prevention: risk factors for antibiotic-associated diarrhea. Microb Ecol 59:830–837. doi:10.1007/s00248-010-9637-2

De La Cochetiere MF et al (2005) Resilience of the dominant human fecal microbiota upon short-course antibiotic challenge. J Clin Microbiol 43:5588–5592

Dethlefsen L, Huse S, Sogin ML, Relman DA (2008) The pervasive effects of an antibiotic on the human gut microbiota, as revealed by deep 16S rRNA sequencing. PLOS Biol 6:e280

Dethlefsen L, McFall-Ngai M, Relman DA (2007) An ecological and evolutionary perspective on human–microbe mutualism and disease. Nature 449:811–818. doi:nature06245 [pii]10.1038/nature06245

Eckburg PB et al (2005) Diversity of the human intestinal microbial flora. Science 308:1635–1638

Epelboin L , Macey J (2009) Maladies infectieuses et transmissibles. Cahier des ECN. Masson Ed

Fanaro S, Chierici R, Guerrini P, Vigi V (2003) Intestinal microflora in early infancy: composition and development. Acta Paediatr Suppl 91:48–55

Fath BD, Patten BC, Choi JS (2001) Complementarity of ecological goal functions. J Theor Biol 208:493–506. doi:10.1006/jtbi.2000.2234S0022-5193(00)92234-3 [pii]

Flint HJ, Duncan SH, Scott KP, Louis P (2007) Interactions and competition within the microbial community of the human colon: links between diet and health. Environ Microbiol 9:1101–1111. doi:EMI1281 [pii]10.1111/j.1462-2920.2007.01281.x

Forchielli ML, Walker WA (2005a) The role of gut-associated lymphoid tissues and mucosal defence. Br J Nutr 93(Suppl 1):S41–48. doi:S0007114505000796 [pii]

Forchielli ML, Walker WA (2005b) The effect of protective nutrients on mucosal defense in the immature intestine. Acta Paediatr Suppl 94:74–83. doi:Q032784285111516 [pii]10.1080/08035320510043592

Foster JA, Krone SM, Forney LJ (2008) Application of ecological network theory to the human microbiome. Interdiscip Perspect Infect Dis 2008:839501. doi:10.1155/2008/839501

Fournier PE, Drancourt M, Raoult D (2007) Bacterial genome sequencing and its use in infectious diseases. Lancet Infect Dis 7:711–723. doi:S1473-3099(07)70260-8 [pii]10.1016/S1473-3099(07)70260-8

Grenet K et al (2004) Antibacterial resistance, *Wayampis amerindians, French Guyana*. Emerg Infect Dis 10:1150–1153

Gronlund MM, Lehtonen OP, Eerola E, Kero P (1999) Fecal microflora in healthy infants born by different methods of delivery: permanent changes in intestinal flora after cesarean delivery. J Pediatr Gastroenterol Nutr 28:19–25

Guan P, Doytchinova IA, Flower DR (2003) A comparative molecular similarity indices (CoMSIA) study of peptide binding to the HLA-A3 superfamily. Bioorg Med Chem 11:2307–2311. doi:S0968089603001093 [pii]

Hoy CM, Wood CM, Hawkey PM, Puntis, JW (2000) Duodenal microflora in very-low-birth-weight neonates and relation to necrotizing enterocolitis. J Clin Microbiol 38:4539–4547

Huse S M et al (2008) Exploring microbial diversity and taxonomy using SSU rRNA hypervariable tag sequencing. PLOS Genet 4:e1000255. doi:10.1371/journal.pgen.1000255

Huurre A et al (2008) Mode of delivery – effects on gut microbiota and humoral immunity. Neonatology 93:236–240. doi:000111102 [pii]10.1159/111102

Jacqueline C et al (2006) In vitro and in vivo assessment of linezolid combined with ertapenem: a highly synergistic combination against methicillin-resistant *Staphylococcus aureus*. Antimicrob Agents Chemother 50:2547–2549. doi:50/7/2547 [pii]10.1128/AAC.01501-05

Jernberg C, Lofmark S, Edlund C, Jansson JK (2007) Long-term ecological impacts of antibiotic administration on the human intestinal microbiota. ISME J 1:56–66

Katayama M, Xu D, Specian RD, Deitch EA (1997) Role of bacterial adherence and the mucus barrier on bacterial translocation: effects of protein malnutrition and endotoxin in rats. Ann Surg 225:317–326

Klebe G, Mietzner T (1994) A fast and efficient method to generate biologically relevant conformations. J Comput Aided Mol Des 8:583–606

Kohanski MA, Dwyer DJ, Collins JJ (2010) How antibiotics kill bacteria: from targets to networks. Nat Rev Microbiol 8:423–435. doi:nrmicro2333 [pii]10.1038/nrmicro2333

Lepage P et al (2005) Biodiversity of the mucosa-associated microbiota is stable along the distal digestive tract in healthy individuals and patients with IBD. Inflamm Bowel Dis 11:473–480

Ley RE, Peterson DA, Gordon JI (2006) Ecological and evolutionary forces shaping microbial diversity in the human intestine. Cell 124:837–848

Livermore DM, Pearson A(2007) Antibiotic resistance: location, location, location. Clin Microbiol Infect 13(Suppl 2):7–16. doi:CLM1724 [pii]10.1111/j.1469-0691.2007.01724.x

Livermore DM (2007) Introduction: the challenge of multi-resistance. Int J Antimicrob Agents 29(Suppl 3):S1–7. doi:S0924-8579(07)00158-6 [pii]10.1016/S0924-8579(07)00158-6

Livermore D (2007) The zeitgeist of resistance. J Antimicrob Chemother 60(Suppl 1):i59–61. doi:60/suppl_1/i59 [pii]10.1093/jac/dkm160

Lucet JC et al (2005) Successful long-term program for controlling methicillin-resistant *Staphylococcus aureus* in intensive care units. Intensive Care Med 31:1051–1057. doi:10.1007/s00134-005-2679-0

McCracken VJ, Lorenz RG (2001) The gastrointestinal ecosystem: a precarious alliance among epithelium, immunity and microbiota. Cell Microbiol 3:1–11. doi:cmi90 [pii]

Mills GA, Walker V (1989) Urinary excretion of 2,3-butanediol and acetoin by babies on a special care unit. Clin Chim Acta 179:51–59

Moy J, Lee DJ, Harmon CM, Drongowski RA, Coran AG (1999) Confirmation of translocated gastrointestinal bacteria in a neonatal model. J Surg Res 87:85–89. doi:10.1006/jsre.1999.5745 S0022-4804(99)95745-1 [pii]

Mshvildadze M, Neu J (2010) The infant intestinal microbiome: friend or foe? Early Hum Dev. doi:S0378-3782(10)00020-4 [pii]10.1016/j.earlhumdev.2010.01.018

Mshvildadze M et al (2009) Intestinal microbial ecology in premature infants assessed with non-culture-based techniques. J Pediatr 156:20–25. doi:S0022-3476(09)00623-4 [pii]10.1016/j.jpeds.2009.06.063

Nanthakumar NN, Fusunyan RD, Sanderson I, Walker WA (2000) Inflammation in the developing human intestine: a possible pathophysiologic contribution to necrotizing enterocolitis. Proc Natl Acad Sci USA 97:6043–6048. doi:97/11/6043 [pii]

Nutten Set al (2007) Antibiotic administration early in life impairs specific humoral responses to an oral antigen and increases intestinal mast cell numbers and mediator concentrations. Clin Vaccine Immunol 14:190–197. doi:CVI.00055-06 [pii]10.1128/CVI.00055-06

Palmer C, Bik EM, DiGiulio DB, Relman DA, Brown PO (2007) Development of the human infant intestinal microbiota. PLOS Biol 5:e177. doi:07-PLBI-RA-0129 [pii]10.1371/journal.pbio.0050177

Paramythiotou E et al (2004) Acquisition of multidrug-resistant *Pseudomonas aeruginosa* in patients in intensive care units: role of antibiotics with antipseudomonal activity. Clin Infect Dis 38:670–677. doi:10.1086/381550CID32086 [pii]

Park HK et al (2005) Molecular analysis of colonized bacteria in a human newborn infant gut. J Microbiol 43:345–353. doi:2255 [pii]

Penders J et al (2006) Factors influencing the composition of the intestinal microbiota in early infancy. Pediatrics 118:511–521. doi:118/2/511 [pii]10.1542/peds.2005-2824

Penders J et al (2007) Gut microbiota composition and development of atopic manifestations in infancy: the KOALA birth cohort study. Gut 56:661–667. doi:gut.2006.100164 [pii]10.1136/gut.2006.100164

Qin J et al (2010) A human gut microbial gene catalogue established by metagenomic sequencing. Nature 464:59–65. doi:nature08821 [pii]10.1038/nature08821

Rautava S, Kalliomaki M, Isolauri E (2005) New therapeutic strategy for combating the increasing burden of allergic disease: probiotics-A nutrition, allergy, mucosal immunology and intestinal microbiota (NAMI) Research Group report. J Allergy Clin Immunol 116:31–37. doi:S0091674905003635 [pii]10.1016/j.jaci.2005.02.010

Rautava S, Salminen S, Isolauri E (2009) Specific probiotics in reducing the risk of acute infections in infancy – a randomised, double-blind, placebo-controlled study. Br J Nutr 101:1722–1726. doi:S0007114508116282 [pii]10.1017/S0007114508116282

Richard C et al (2006) Evidence on correlation between number of disulfide bridges and toxicity of class IIa bacteriocins. Food Microbiol 23:175–183. doi:S0740-0020(05)00031-6 [pii]10.1016/j.fm.2005.02.001

Rougè C et al (2010) Investigation of the intestinal microbiota in preterm infants using different methods. Anaerobe (Epub ahead of print)

Rouwet EV et al (2002) Intestinal permeability and carrier-mediated monosaccharide absorption in preterm neonates during the early postnatal period. Pediatr Res 51:64–70

Sakata H, Yoshioka H, Fujita K (1985) Development of the intestinal flora in very low birth weight infants compared to normal full-term newborns. Eur J Pediatr 144:186–190

Schumann A et al (2005) Neonatal antibiotic treatment alters gastrointestinal tract developmental gene expression and intestinal barrier transcriptome. Physiol Genomics 23:235–245. doi:00057.2005 [pii]10.1152/physiolgenomics.00057.2005

Schwiertz A et al (2003) Development of the intestinal bacterial composition in hospitalized preterm infants in comparison with breast-fed, full-term infants. Pediatr Res 54:393–399. doi:10.1203/01.PDR.0000078274.74607.7A01.PDR.0000078274.74607.7A [pii]

Seehofer D et al (2004) Probiotics partly reverse increased bacterial translocation after simultaneous liver resection and colonic anastomosis in rats. J Surg Res 117:262–271. doi:10.1016/j.jss.2003.11.021S0022480403007510 [pii]

Seksik P et al (2005) Search for localized dysbiosis in Crohn's disease ulcerations by temporal temperature gradient gel electrophoresis of 16S rRNA. J Clin Microbiol 43:4654–4658

Shanahan F (2010) 99th Dahlem conference on infection, inflammation and chronic inflammatory disorders: host–microbe interactions in the gut: target for drug therapy, opportunity for drug discovery. Clin Exp Immunol 160:92–97. doi:CEI4135 [pii]10.1111/j.1365-2249.2010.04135.x

Sharma R, Young C, Neu J (2010) Molecular modulation of intestinal epithelial barrier: contribution of microbiota. J Biomed Biotechnol 2010:305879. doi:10.1155/2010/305879

Sjogren YM et al (2009a) Influence of early gut microbiota on the maturation of childhood mucosal and systemic immune responses. Clin Exp Allergy 39:1842–1851. doi:CEA3326 [pii]10.1111/j.1365-2222.2009.03326.x

Sjogren YM, Jenmalm MC, Bottcher MF, Bjorksten B, Sverremark-Ekstrom E (2009b) Altered early infant gut microbiota in children developing allergy up to 5 years of age. Clin Exp Allergy 39:518–526. doi:CEA3156 [pii]10.1111/j.1365-2222.2008.03156.x

Sotriffer C, Klebe G (2002) Identification and mapping of small-molecule binding sites in proteins: computational tools for structure-based drug design. Farmaco 57:243–251

Srinivas N, Jetter P, Ueberbacher BJ, Werneburg M, Zerbe K, Steinmann J, Van der Meijden B, Bernardini F, Lederer A, Dias RLA, Misson PE, Henze H, Zumbrunn J, Gombert FO, Obrecht D, Hunziker P, Schauer S, Ziegler U, Käch A, Eberl L, Riedel K, DeMarco SJ, Robinson JA (2010) Peptidomimetic antibiotics target outer-membrane biogenesis in *Pseudomonas aeruginosa*. Science 327:1010–1013. doi:327/5968/1010 [pii]10.1126/science.1182749

Turnbaugh PJ et al (2009) A core gut microbiome in obese and lean twins. Nature 457:480–484. doi:nature07540 [pii]10.1038/nature07540

Turnbaugh PJ et al (2007) The human microbiome project. Nature 449:804–810

Vaishampayan PA et al (2010) Comparative metagenomics and population dynamics of the gut microbiota in mother and infant. Genome Biol Evol 6:53–66. doi:10.1093/gbe/evp057

Vaughan EE et al (2000) A molecular view of the intestinal ecosystem. Current issues in intestinal microbiology 1:1–12

Wall RR, Ross RP, Ryan CA, Hussey S, Murphy B, Fitzgerald GF, Stanton C (2009) Role of gut microbiota in early infant development. Clin Med Pediatr 3:45–54

Walson JL, Marshall B, Pokhrel BM, Kafle KK, Levy SB (2001) Carriage of antibiotic-resistant fecal bacteria in nepal reflects proximity to Kathmandu. J Infect Dis 184:1163–1169. doi:JID990706 [pii]10.1086/323647

Weaver LT, Laker MF, Nelson R (1984) Intestinal permeability in the newborn. Arch Dis Child 59:236–241

Yajima M et al (2001) Bacterial translocation in neonatal rats: the relation between intestinal flora, translocated bacteria, and influence of milk. J Pediatr Gastroenterol Nutr 33:592–601

Yonath A (2005) Ribosomal crystallography: peptide bond formation, chaperone assistance and antibiotics activity. Mol Cells 20:1–16. doi:871 [pii]

Zaneveld JR, Lozupone C, Gordon JI, Knight R (2010) Ribosomal RNA diversity predicts genome diversity in gut bacteria and their relatives. Nucleic Acids Res 38:3869–3879. doi:gkq066 [pii]10.1093/nar/gkq066

Chapter 11
Infectogenomics: Aspect of Host Responses to Microbes in Digestive Tract

Zongxin Ling and Charlie Xiang

Introduction

An enormous number of microorganisms, the vast majority of which are bacterial species, are known to colonize and form complex communities, or microbiota, at various sites within the human body (Dethlefsen et al., 2007; Turnbaugh et al., 2007). Microbial cells that thrive on and within the human body are approximately 10 times more numerous than our own cells and contain, in aggregate, about 100 times more genes, leading to the suggestion that humans and our microbial symbionts be considered "superorganisms" (Gill et al., 2006). A growing body of evidence suggests that the composition and function of microbiota in different human body habitats play a vital role in human development, physiology, immunity, and nutrition (Mazmanian et al., 2005; Cash et al., 2006; Ley et al., 2006a; Turnbaugh et al., 2006; Dethlefsen et al., 2007).

The human gastrointestinal tract is home to a complex consortium of trillions (approximately 1×10^{13}–1×10^{14}) of microbes, thousands of bacterial phylotypes, as well as hydrogen-consuming methanogenic archaea, colonizing the entire length of the gut with a collective genome (Gill et al., 2006). Although largely unexplored and under-appreciated, our gastrointestinal microbiota plays an intricate and sometimes pivotal role for our health and well-being. The taxonomic composition of the microbial community may affect the propensity to develop obesity (Ley et al., 2006; Turnbaugh et al., 2009), inflammatory bowel disease (Frank et al., 2007), Type 1 diabetes (Wen et al., 2008), cardiovascular diseases (Ordovas and Mooser, 2006), and so on. With the advent of molecular techniques, more and more bacterial species, which were uncultivated by traditional microbiological techniques, were found in the gastrointestinal tract. So far, there were more than 1,000 genera found in this microhabitat.

The gastrointestinal tract harbors an ecosystem composed of the gastrointestinal mucosa and the commensal microbiota. The interaction between host and

C. Xiang (✉)
State Key Laboratory for Diagnosis and Treatment of Infectious Diseases, The First Affiliated Hospital, School of Medicine, Zhejiang University, Hangzhou, China 310003
e-mail: cxiang@zju.edu.cn

K.E. Nelson (ed.), *Metagenomics of the Human Body*,
DOI 10.1007/978-1-4419-7089-3_11, © Springer Science+Business Media, LLC 2011

microbiota plays an important role in maintaining the balance of the ecosystem and preventing pathogens or opportunistic pathogens to colonize the microhabitat. The normal gastrointestinal ecosystem can efficiently block intrusion of many pathogenic bacteria. This has been termed "microbial interference" or "colonization resistance." The lack of an intact microbiota (e.g., in axenic or germfree mice raised under sterile conditions and antibiotic-treated mice) dramatically increases susceptibility to enteric infection (e.g., *Salmonella* spp., *Streptococcus mutans*, *Clostridium difficile*, and *Shigella flexneri*) (Stecher and Hardt, 2008). Conversely, selected commensal species, such as *Lactobacillus* spp. or *Bifidobacterium* spp., have therapeutic and/or prophylactic effects against enteric infection (Gill, 2003). Some have been commercialized as probiotics or live microbial food supplements with health-promoting attributes.

In the normal gastrointestinal tract, the relationship between the microbiota and the host is mutually beneficial. The microbiota is provided with steady growth conditions and a (somewhat limited) nutrient supply. In return, the microbiota contributes to the host's nutrition, immune system development, angiogenesis, and fat storage (Stecher and Hardt, 2008). In healthy individuals, the gastrointestinal microbiota exists in a state of eubiosis. However, this dynamic equilibrium can be "disturbed" by various stresses. When eubiosis of the gastrointestinal tract was disturbed, infection of the microhabitat occurred. However, the mechanism or molecular basis of these interactions is still unknown. With the few years of research in the human postgenomic era, we have gained considerable knowledge about host–pathogen interactions through host genomes. With "Omics" research, study of host–pathogen interactions can be defined as "Infectogenomics." Infectogenomics studies the interaction between host genetic factors and the composition of the microbiota (Kellam and Weiss, 2006; Nibali et al., 2009). Infectogenomics approaches might be an effective way to explore the nature of the occurrence of infections in the gastrointestinal tract when the gastrointestinal ecosystem changes from eubiosis to dysbiosis.

Through technological and conceptual innovations in metagenomics, the complex microbial habitat of the human gastrointestinal tract is now amenable to detailed ecological analysis. Large-scale shifts in gut commensal populations, rather than occurrence of particular microorganisms, are associated with several gastroenterological conditions; redress of these imbalances may ameliorate the conditions. In addition to exploring the bacterial diversity of the gastrointestinal tract in different health states, such as is a major part of the Human Microbiome Project (HMP), it is also necessary and important to understand host responses to the bacterium that enable it to evade or protect itself from host immunity.

The Microecological Barrier of the Gastrointestinal Tract

The mucosal surface of the gastrointestinal tract is the largest body surface in contact with the external environment (200–300 m^2). It is a complex ecosystem combining the gastrointestinal epithelium, immune cells, and resident microbiota (McCracken and Lorenz, 2001). The gastrointestinal mucosa is the main interface between the immune system and the external environment and therefore has an important role in the host–microbiota interplay. As we all know, the mucosa of the gastrointestinal tract is exposed to various microbial pathogens. These potentially harmful enteric microorganisms can hijack the cellular molecules and signaling pathways of the host and become pathogenic. The commensal microbiota profoundly influences the development of humoral components of the gastrointestinal mucosal immune system, acting as a crucial factor in the prevention of exogenous pathogen intrusion.

With the development of molecular microecology, researchers found that the microecological barrier of the gastrointestinal tract played a vital role in the human body. The microecological barrier of the gastrointestinal tract is constituted with biological barrier, physical barrier, immune barrier, and chemical barrier (Fig. 11.1). The microbiota (mainly commensal bacteria) is composed of the biological barrier of gastrointestinal tract. The normal microbiota in the gastrointestinal tract is sufficient to suppress the growth of pathogenic bacteria. While the colonization resistance is disturbed and changed from eubiosis to dysbiosis, the infection will occur. The physical barrier is composed of the mucus/glycocalyx layer and four epithelial cell lineages, including the enterocytes, enteroendocrine, goblet, and Paneth cells present in the intestinal villi. In addition, M cells are present in the follicle-associated epithelia of Peyer's patches (PPs) and serve as a transcytotic receptor for mucosal antigens. The mucus layer is like a mesh of networking fibers made primarily of mucins, glycoproteins, and lipids that allows the passage of definite-size molecules (Neutra et al., 1996) and excludes bacteria from contacting

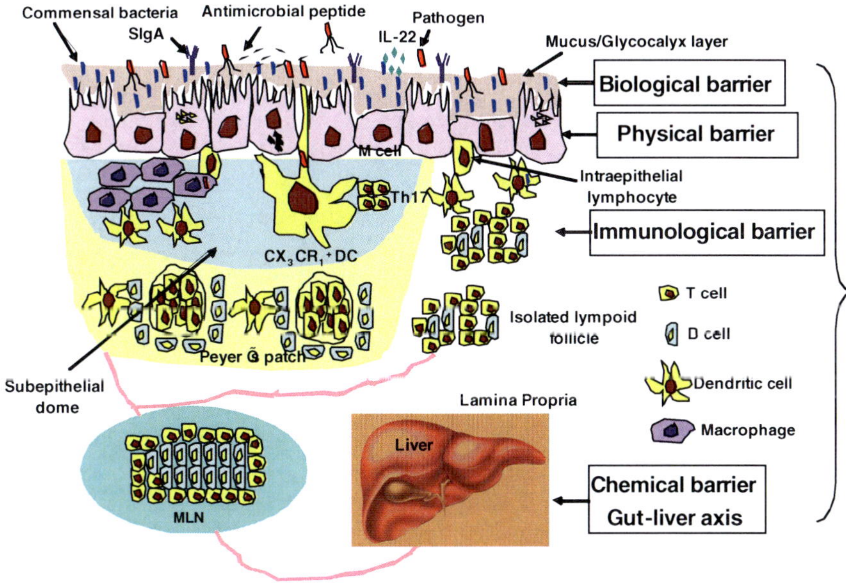

Fig. 11.1 The schematic diagram of the microecological barrier of the gastrointestinal tract

the epithelial barrier. The integrity of the layer of epithelial cells is maintained by intercellular junctional complexes composed of tight junctions (TJs), adherens junctions (AJs), and desmosomes, whereas gap junctions allow intercellular communication to occur. TJs, the most apical components of the junctional complex (Anderson et al., 1993; Schneeberger and Lynch, 2004), create a semipermeable diffusion barrier between individual cells, which can be regulated, and serve as the permeability barrier. Moreover, cells of this villous epithelium produce a variety of functional molecules, such as defensins (Ayabe et al., 2000), trefoil factors (Kindon et al., 1995), and mucins, which help protect the human host. Mucins are the principal components of mucus, which lines the surface epithelium throughout the intestinal tract (Lamont, 1992). Microbes of all types (bacteria, viruses, protozoa) become trapped in the mucus layer and are expelled from the intestine by peristalsis. Other important components of mucus include various proteolytic enzymes, which serve not only to facilitate digestion of polypeptides but also to alter/diminish the immunogenic properties of these peptides because peptides of less than 8–10 amino acids are poor immunogens (Mayer, 2003).

The immune barrier of the gastrointestinal tract comprises an immunological network termed the gut-associated lymphoid tissue (GALT) that consists of unique arrangements of B cells, T cells, and phagocytes, which sample luminal antigens through specialized epithelia termed the follicle-associated epithelia (FAEs) and orchestrate coordinated molecular responses between immune cells and other components of the mucosal barrier. GALT consists partly of organized tissue representing both solitary and multiple lymphatic follicles (PPs, appendix) and freely dispersed lamina propria lymphocytes (LPLs). These components, in cooperation with components of the innate mucosal immune system, accomplish specific (adaptive) immune responses. Organized lymphatic tissue (PPs and lymphatic follicles) represents an inductive site of the mucosal immune system. Germinal centers of lymphatic follicles consist mainly of differentiating B cells. T lymphocytes occupy interfollicular space preferably around venules with a high endothelium. Organized lymphatic tissue is covered with an epithelial layer (FAE) containing a special type of membranous epithelial cells called "M" cells. Specialized M cells are most effective in absorbing particular antigens and transporting these from lumen to follicular environment (PPs), where these antigens activate T lymphocytes and thus induce mucosal immunity. Diffuse lymphocytes represent an efficient effector component of the mucosal immune system. Lymphocytes are present in the epithelium on the basolateral side of enterocytes – these are called intra-epithelial lymphocytes (IELs)-T cells, predominantly $CD8^+$ in nature and differ from cells present in the bloodstream. IELs have several features in common: phenotype $CD8^+$, $CD45RO^+$, adhesive molecules (integrin $\alpha E\beta 7$), and cytoplasmatic granules containing cytolytic proteins (perforin). It was suggested that IELs may be capable of identifying proteins that generally are not present in the epithelium and to react cytolytically to damaged or changed epithelial cells. Diffuse lymphocytes of lamina propria are the most numerous and the most active mucosal effector cells. They are represented by T cells, predominantly CD4, and B cells producing polymeric IgA. Mucosal T cells produce various cytokines; an important and interesting population of regulatory T cells is noted mainly for its suppressive activity. These regulatory T cells express CD25 and were shown to produce inhibitory cytokines IL-10 and TGF-β (Tlaskalova-Hogenova et al., 2004). IgA is one of the most important humoral defense factors on mucosal surfaces. Its polymeric form dominates in secretions, and its monomeric form in circulation. The main source of monomeric serum IgA is bone marrow, of polymeric IgA plasma cells in mucosa and exocrine glands. The GALT is the largest lymphatic organ of the body. Total amount of IgA-producing cells of the intestine (7×10^{11}) and the daily IgA production in intestine (2–5 g) indicate that IgA is the most represented class of the immunoglobulins in the body. Most IgA specifities are directed against mucosal microbiota. The molecular structure of polymeric IgA enables this molecule to penetrate into secretions, resist enzymatic activities, and function as an effector molecule on mucosal surfaces. Secretory IgA, i.e., dimeric IgA molecule, is resistant to proteolysis and its primary task is to prevent both adherence of bacteria to mucosal surfaces and penetration of antigens to the internal environment of the organism. Moreover, IgA is capable of reacting with several nonspecific bactericidal substances present in secretions (e.g., lactoperoxidase and lactoferrin) and transport these to bacterial surfaces (Fujihashi et al., 2001; Tlaskalova-Hogenova et al., 2004). Migration of lymphocytes from inductive parts of mucosal immune system and their homing in effector parts of mucosa and exocrine glands ("common mucosal immune system") is a good example of a multiphasic process of interaction between lymphocytes and their original environment. It was confirmed that this selective homing of lymphocytes to epithelial surfaces concerns mainly recirculating, activated, blastic forms of lymphocytes and small, memory lymphocytes. These lymphocytes (of both man and mouse) have a characteristic α4β7 integrin on their surface and act as receptor of specific mucosal venous adresin Mad-CAM/1 present on endothelial cells of mucosal capillaries. A good example of the effect of migration and selective colonization by cells from intestinal mucosa is offered by the composition of mammary gland secretion–maternal milk. Apart from nutritive components, mother's milk contains a number of immunologically nonspecific and specific factors and a large quantity of immune cells: All these milk components protect the not yet completely mature intestine of the infant. A consequence of a colonization of the mammary gland by cells from the intestine is a presence of IgA antibody and of cells directed against antigens present in maternal gut ("enteromammary axis"), i.e., mainly bacteria belonging to maternal microbiota that colonize the intestine of the infant within the first days of life (Fujihashi et al., 2001).

The gut–liver barrier is the chemical barrier of the microecological barrier of the gastrointestinal tract. The gut and the liver are the key organs in nutrient absorption and metabolism. Bile acids, drugs, and toxins undergo extensive enterohepatic circulation. Bile acids play a major role in several hepatic and intestinal diseases. Endotoxins like LPS deriving from gastrointestinal Gram-negative bacteria are important in the pathogenesis of liver and systemic diseases. Chronic liver diseases can influence gastrointestinal motility, which together with other factors may contribute to bacterial overgrowth and in patients with ascites to an increased risk of spontaneous bacterial peritonitis. Patients with end-stage liver disease frequently develop portal hypertension, leading to varices, gastric vascular ectasia, and portal hypertensive gastroenteropathy. Several liver and biliary abnormalities are observed in patients with inflammatory bowel disease (primary sclerosing cholangitis, autoimmune hepatitis, and cholelithiasis). The primary defect in hemochromatosis is located in the intestine, causing an inappropriate increase in iron absorption, and the liver is the site of earliest and heaviest iron deposition. Elevated transaminases are observed in many patients with celiac disease, and steatohepatitis frequently develops in patients with jejunoileal bypass and short bowel syndrome. Furthermore, the liver is the primary organ for metastasis of intestinal cancer. Many viral, bacterial, fungal, and parasitic diseases affect the intestine as well as the liver and the biliary tract (Zeuzem, 2000).

In a word, these four barriers are connected by a complex network in the gastrointestinal tract, which sense microbiota, and are combat with pathogen or opportunistic pathogens in nature and constitute the integral microecological barrier. It plays a vital role in the host and microbiota interplay.

Host–Bacterium Interactions in the Gastrointestinal Tract

Host defense requires an accurate interpretation of the microenvironment to distinguish commensal organisms from episodic pathogens and a precise regulation of subsequent responses. The epithelium provides the first sensory line of defense, and active sampling of resident bacteria, pathogens, and other antigens is mediated by three main types of immunosensory cell (M cells, dendritic cells (DCs), and intestinal epithelial cells). First, surface enterocytes serve as afferent sensors of danger within the luminal microenvironment by secreting chemokines and cytokines, which alert and direct innate and adaptive immune responses to the infected site (Shanahan, 2005). Second, M cells that overlie lymphoid follicles sample the environment and transport luminal antigens to subadjacent DCs and other antigen-presenting cells. Third, intestinal DCs themselves have a pivotal immunosensory role and can directly sample gut contents by either entering or extending dendrites between surface enterocytes without disrupting TJs (Shanahan, 2005). DCs (especially $CX_3CR_1^+$ DCs) can ingest and retain live commensal bacteria and travel to the mesenteric lymph node where immune responses to commensal bacteria are induced locally (Macpherson and Uhr, 2004). Thus, acting as a gatekeeper, the mesenteric lymph node prevents access of commensal bacteria to the internal milieu.

Innate immune recognition (also known as pattern recognition) is based on the detection of molecular structures that are unique to microorganisms. Pattern recognition is unusual, in that each host receptor (pattern recognition receptor (PRR)) has a broad specificity and can potentially bind to a large number of molecules that have a common structural motif or pattern. The targets of PRRs are sometimes referred to as pathogen-associated molecular patterns (PAMPs), although they are present on both pathogenic and nonpathogenic microorganisms. PAMPs are well suited to innate immune recognition for three main reasons. First, they are invariant among microorganisms of a given class. Second, they are products of pathways that are unique to microorganisms, allowing discrimination between self and nonself molecules. Third, they have essential roles in microbial physiology, limiting the ability of the microorganisms to evade innate immune recognition through adaptive evolution of these molecules. Bacterial PAMPs are often components of the cell wall, such as lipopolysaccharide (LPS), peptidoglycan, lipoteichoic acids, and cell-wall lipoproteins. An important aspect of pattern recognition is that PRRs themselves do not distinguish between pathogenic microorganisms and symbiotic (nonpathogenic) microorganisms, because the ligands of the receptors are not unique to pathogens. Yet, despite humans being colonized by trillions of symbiotic bacteria, homeostasis is somehow maintained under normal conditions (Medzhitov, 2007).

The ability of immunosensory cells to discriminate pathogenic from commensal bacteria is mediated, in part, by two major host PRR systems – the family of Toll-like receptors (TLRs) and the nucleotide-binding oligomerization domain/caspase recruitment domain isoforms (NOD/CARD) (Macpherson and Uhr, 2004). These PRRs have a fundamental role in immune-cell activation in response to specific microbial-associated molecular patterns. For example, TLR2 is activated by peptidoglycan and lipotechoic acids, TLR4 by LPS, TLR5 by flagellin, and NOD1/CARD4 and NOD2/CARD15 function as intracellular receptors of peptidoglycan subunits, and TLR9 by unmethylated CpG-ODN. TLRs and NOD proteins are expressed by surface enterocytes and DCs (Abreu et al., 2005), and in the gut PRRs seem to be crucial for bacterial–host communication. Decreased enterocyte proliferation and levels of cytoprotective factors have been observed in TLR-defective mice, and TLR signals mediated by commensal bacteria or their ligands are essential for intestinal barrier function and repair of the gut (Rakoff-Nahoum et al., 2004; Fukata et al., 2005). Many PRR ligands are expressed by commensal bacteria, yet the healthy gut does not evoke inflammatory responses to these bacteria. Conversely, some commensal bacteria exert protective effects by attenuating pro-inflammatory responses induced by various enteropathogenic bacteria (Kelly et al., 2004; Ma et al., 2004; O′Hara et al., 2006). The host and bacterial mechanisms that underpin these effects are being explored.

TLRs/NODs sense microbial infection and engage multiple mechanisms that control the initiation of adaptive immune responses. DCs are the key cell type that couples TLR/NOD-mediated innate immune recognition to the initiation of T cell and B cell activation. Recent studies suggest that DCs may be specialized to accomplish distinct functions in the course of the immune response, such that multiple DC subtypes are engaged in response to a single infection. This functional specialization may explain the differential TLR expression patterns in different DC subsets. Type I interferons, which are best known for their antiviral activities, have many essential functions in the control of adaptive immunity. Future studies will need to address the biological importance of the known differences in TLR/NOD function, including expression patterns, ligand specificities, and the differentially engaged signaling pathways.

Functional Genomics Approaches in Studying Host Immunoresponses

The host genome ultimately manifests its function via differences in expression levels of genes, or in protein sequence differences. Gene expression profiling, using microarrays, can show how cells respond to pathogen infection. Not only does this indicate potential candidate loci for susceptibility studies, but it also can suggest new ways to treat or manage infections.

DNA microarrays are devices that measure the expression of many thousands of genes in parallel (Schena et al., 1995). They have revolutionized molecular biology, and during the past decade, their use has grown very rapidly throughout academia, medicine, and the pharmaceutical, biotechnology, agrochemical and food industries. Shi et al. (2008) used a whole-genome cDNA microarray (Fig. 11.2) to study DC cell differentiation after LPS induction and bacterial infection. Their study found that the tripartite-motif protein TRIM30α, a RING protein, was induced by TLR agonists and interacted with the TAB2-TAB3-TAK1 adaptor–kinase complex involved in the activation of transcription factor NF-κB. TRIM30a promoted the degradation of TAB2 and TAB3 and inhibited NF-κB activation induced by TLR signaling. In vivo studies showed that transfected or transgenic mice overexpressing TRIM30a were more resistant to endotoxic shock. Consistent with that, in vivo "knockdown" of TRIM30a mRNA by small interfering RNA impaired LPS-induced tolerance. The expression of TRIM30α depended on NF-κB activation. These results collectively indicate that TRIM30α negatively regulates TLR-mediated NF-κB activation by targeting degradation of TAB2 and TAB3 by a "feedback" mechanism (Shi et al., 2008).

In another study, Liu et al. (2009) applied the same kind of microarrays to identify a novel transcription factor Dec2, a member of basic helix–loop–helix (bHLH) superfamily that is progressively induced during the course of Th2 differentiation, especially at the late stage. The upregulated Dec2 can strongly promote Th2 development under Th2-inducing conditions, as evidenced by retrovirus-mediated gene transfer or transgenic manipulation. In addition, an enhancement of Th2 responses

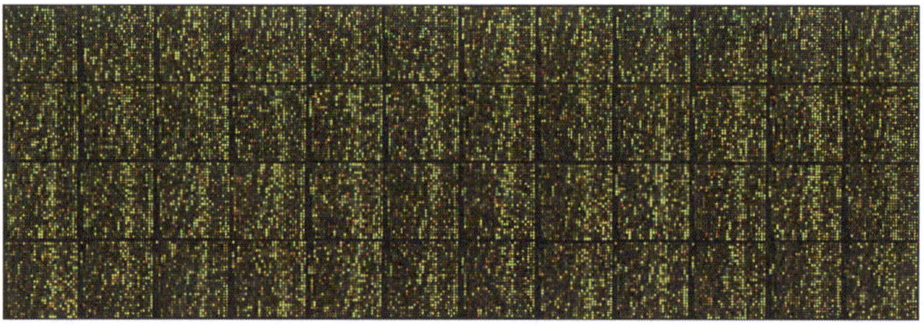

Fig. 11.2 Mouse whole-genome cDNA microarray containing 36,000 elements representing 25,000 unique genes

is also detectable in Dec2 transgenic mice in vivo. RNA interference-mediated suppression of endogenous Dec2 could attenuate Th2 differentiation. Finally, the enhanced Th2 development is at least in part due to substantial upregulation of CD25 expression elicited by Dec2, thereby resulting in hyper-responsiveness to IL-2 stimulation (Liu et al., 2009).

Whole-genome sequencing or next-generation sequencing (mainly 454 pyrosequencing, Illumina, and SOLiD) has been applied in a variety of basic and clinical studies of diseases, especially in cancer research (Mardis, 2009). In a recent study of acute myeloid leukemia (AML), researchers used whole-genome resequencing to study an AML genome and the genome of its matched normal (skin) sample and found eight single-nucleotide nonsynonymous variants in genes that essentially would never have been on a candidate gene list for AML (Ley et al., 2008). This result has made a strong case for whole-genome resequencing as an unbiased approach by which medical research can pursue the genomic basis of cancer.

The same massively whole-genome resequencing technique has also been used in the study of complex mechanisms of human innate immunity. Lazarus et al. (2002) examined 16 genes by resequencing 93 unrelated subjects from three ethnic samples (European American, African-American, and Hispanic American) and a sample of European American asthmatics. They found the innate immune genes sequenced demonstrated substantial interindividual variability predominantly in the form of single nucleotide polymorphisms (SNPs). Genetic variations in these genes are suggested to play a role in determining susceptibility to a range of common, chronic human diseases, which have an inflammatory component (Lazarus et al., 2002). In a recent study, Kuningas et al. (2009) resequenced the IL10 gene region ($\sim$23.5 kb) in 37 individuals of Ghana and identified 58 variants of which 33 were previously identified SNPs (present in dbSNP) and 20 undescribed SNPs. In addition, three insertions and two deletions were observed. The association among genetic variation in the IL10 gene, cytokine production, and mortality, the skewing of allele frequencies over age and environmental conditions, all provide strong arguments for selection of genetic variation inducing pro-inflammatory responses under adverse environmental conditions in a Ghanaian population (Kuningas et al., 2009) http://www.ncbi.nlm.nih.gov/pubmed/17376198.

Taken together from the above studies, the functional genomics approaches are dramatically accelerating the pace of human disease research and are already impacting patient care.

Host Responsiveness to Microbiota – Metabonomic Approaches

Altered gut microbiota has recently been suggested to be critical in the development of obesity, diabetes, and hypertension (Turnbaugh et al., 2006; Cani et al., 2007; Holmes et al., 2008; Wen et al., 2008). Gene expression profiling and proteomic approaches have been applied to elucidate the molecular mechanisms underlying symbiotic host–bacterial relationships (Schell et al., 2002; Xu et al., 2003; Pridmore et al., 2004). However, gene expression and proteomic data might only indicate the potential for physiological changes because many pathway feedback mechanisms are simply not reflected in protein concentration or gene expression. On the other hand, metabolite concentrations and their kinetic variations in tissues or biological matrixes represent real end-points of physiological regulatory processes (Rezzi et al., 2007; Holmes et al., 2008). Recently, complementary metabonomic approaches have been employed for the biochemical characterization of metabolic changes triggered by gut microbiota, dietary variation, and stress interactions (Wang et al., 2005; Marchesi et al., 2007; Martin et al., 2007a, b, 2008). Solid phase microextraction followed by gas chromatography and mass spectrometry represents a novel method for studying metabolic profiles of biological samples. This approach has been used to compare neonates and adult feces and to identify volatile markers of gastrointestinal disease (Garner et al., 2007).

One example of how the gut microbiota influenced the host metabolic profiles was demonstrated by Vitali and colleagues (2010). In this study, they investigated the global impact of a dietary intervention on the gut ecology and metabolism in healthy humans. The participants administered a synbiotic food, containing fructooligosaccharides and the probiotic strains *Lactobacillus helveticus* Bar13 and *Bifidobacterium longum* Bar33 for 1 month, which were selected on the basis of their adhesion and immune-regulation properties, as assessed by both in vitro (Candela et al., 2008) and in vivo studies on animal models (Roselli et al., 2009). Although the synbiotic food did not modify the overall structure of the gut microbiome, they demonstrated that the intake of a synbiotic food leads to a modulation of the gut metabolic activities with a maintenance of the gut biostructure. The extent of short-chain fatty acids (SCFA), ketones, carbon disulfide, and methyl acetate was significantly affected by the synbiotic food consumption. The significant increase of SCFA, ketones, carbon disulfide, and methyl acetate suggests potential health, promoting effects of the synbiotic food.

Recently, Vijay Kumar et al. (2010) demonstrated the mechanism that the shift of gastrointestinal microbiota influenced the metabonomics of host significantly and finally led to metabolic syndrome. In fact, the gastrointestinal microbiota is shaped by both environment and host genetics with the innate immune system in particular, long appreciated for its role in defending against infection by pathogenic microbes, now suggested to play a key role in regulating the gastrointestinal microbiota (Slack et al., 2009). Thus, in addition to its role in infection/inflammation, innate immunity may play a key role in promoting metabolic health. TLR5 is a transmembrane protein that is highly expressed in the intestinal mucosa and that recognizes bacterial flagellin. With mice genetically deficient in TLR5 (T5KO mice), they have shown that loss of TLR5 results in a phenotype reminiscent of human metabolic syndrome. The underlying molecular mechanisms remain to be defined, but they speculate that loss of TLR5 produces alterations in the gut microbiota that induce low-grade inflammatory signaling. This signaling may in turn cross-desensitize insulin receptor signaling, leading to hyperphagia, which then drives other aspects of metabolic syndrome. The results suggest that the specific composition of microbiota to which individuals are first exposed may be an important means by which early environment exerts a lasting influence on metabolic phenotype. They also suggest that the excess caloric consumption driving the current epidemic of metabolic syndrome may be caused, at least in part, by alterations in host–microbiota interactions (Vijay Kumar et al., 2010).

Another example of how the gut microbiota influenced the host metabolic profiles was the development of obesity. The incidence of obesity has increased exponentially over the past 3 decades and cannot exclusively be explained by genetic factors. Could the gut microbial community contribute to the obesity epidemic? Although the majority of mouse gut species are unique, mouse and human microbiota are dominated by the same major groups of bacteria: *Bacteroidetes* and *Firmicutes*. Recent studies have shown that obese mice have dramatically higher levels of *Firmicutes* and lower levels of *Bacteroidetes* compared with their lean counterparts (Ley et al., 2005). Interestingly, a recent twin study showed that the gut microbiota of obese humans is characterized by phylum-level changes (e.g., reduced levels of *Bacteroidetes*), reduced diversity, and an increased capacity to absorb energy from the diet (Turnbaugh et al., 2009). Kalliomaki et al. (2008) showed that decreased levels of *Bifidobacteria* during infancy may predict overweight (Kalliomaki et al., 2008). Taken together, these results suggest that variations in the gut microbiota early in life may confer an increased risk of developing obesity later in life. To understand how the gut microbiota affects host physiology, Backhed et al. (2007) recently investigated how the gut microbiota regulates the metabolome and transcriptome in germfree and conventionally raised mice. Metabolomic analysis revealed that the gut microbiota affects several important metabolic processes including energy metabolism, amino acid, and lipid metabolism. The serum metabolome was associated with increased hepatic transcription of genes involved in proteolysis, energy, and xenometabolism. Surprisingly, they detected increased levels of neurotransmitters in serum of conventionally raised animals, which suggests that the gut microbiota may affect animal behavior. These studies suggest that variations in an individual's gut microbiota may have profound effects on host metabolism and physiology and will be an important factor when considering personalized medicine (Backhed et al., 2007).

In conclusion, infectogenomics, which focus on host–microbe interaction, may ultimately contribute to a more comprehensive understanding of human health, disease susceptibilities, and the pathophysiology of infectious and immune-mediated gastrointestinal diseases.

References

Abreu MT, Fukata M, Arditi M (2005) TLR signaling in the gut in health and disease. J Immunol 174:4453–4460

Anderson JM, Balda MS, Fanning AS (1993) The structure and regulation of tight junctions. Curr Opin Cell Biol 5:772–778

Ayabe T, Satchell DP, Wilson CL et al (2000) Secretion of microbicidal alpha-defensins by intestinal Paneth cells in response to bacteria. Nat Immunol 1:113–118

Backhed F, Manchester JK, Semenkovich CF et al (2007) Mechanisms underlying the resistance to diet-induced obesity in germ-free mice. Proc Natl Acad Sci USA 104:979–984

Candela M, Perna F, Carnevali P et al (2008) Interaction of probiotic Lactobacillus and Bifidobacterium strains with human intestinal epithelial cells: adhesion properties, competition against enteropathogens and modulation of IL-8 production. Int J Food Microbiol 125:286–292

Cani PD, Amar J, Iglesias MA et al (2007) Metabolic endotoxemia initiates obesity and insulin resistance. Diabetes 56:1761–1772

Cash HL, Whitham CV, Behrendt CL et al (2006) Symbiotic bacteria direct expression of an intestinal bactericidal lectin. Science 313:1126–1130

Dethlefsen L, McFall-Ngai M, Relman DA (2007) An ecological and evolutionary perspective on human-microbe mutualism and disease. Nature 449:811–818

Frank DN, St Amand AL, Feldman RA et al (2007) Molecular-phylogenetic characterization of microbial community imbalances in human inflammatory bowel diseases. Proc Natl Acad Sci USA 104:13780–13785

Fujihashi K, Kato H, van Ginkel FW et al (2001) A revisit of mucosal IgA immunity and oral tolerance. Acta Odontol Scand 59:301–308

Fukata M, Michelsen KS, Eri R et al (2005) Toll-like receptor-4 is required for intestinal response to epithelial injury and limiting bacterial translocation in a murine model of acute colitis. Am J Physiol Gastrointest Liver Physiol 288:G1055–G1065

Garner CE, Smith S, de Lacy Costello B et al (2007) Volatile organic compounds from feces and their potential for diagnosis of gastrointestinal disease. FASEB J 21:1675–1688

Gill HS (2003) Probiotics to enhance anti-infective defences in the gastrointestinal tract. Best Pract Res Clin Gastroenterol 17:755–773

Gill SR, Pop M, Deboy RT et al (2006) Metagenomic analysis of the human distal gut microbiome. Science 312:1355–1359

Holmes E, Loo RL, Stamler J et al (2008) Human metabolic phenotype diversity and its association with diet and blood pressure. Nature 453:396–400

Holmes E, Wilson ID, Nicholson JK (2008) Metabolic phenotyping in health and disease. Cell 134:714–717

Kalliomaki M, Collado MC, Salminen S et al (2008) Early differences in fecal microbiota composition in children may predict overweight. Am J Clin Nutr 87:534–538

Kellam P, Weiss RA (2006) Infectogenomics: insights from the host genome into infectious diseases. Cell 124:695–697

Kelly D, Campbell JI, King TP et al (2004) Commensal anaerobic gut bacteria attenuate inflammation by regulating nuclear-cytoplasmic shuttling of PPAR-gamma and RelA. Nat Immunol 5:104–112

Kindon H, Pothoulakis C, Thim L et al (1995) Trefoil peptide protection of intestinal epithelial barrier function: cooperative interaction with mucin glycoprotein. Gastroenterology 109: 516–523

Kuningas M, May L, Tamm R et al (2009) Selection for genetic variation inducing proinflammatory responses under adverse environmental conditions in a Ghanaian population. PLoS One 4:e7795

Lamont JT (1992) Mucus: the front line of intestinal mucosal defense. Ann N Y Acad Sci 664: 190–201

Lazarus R, Vercelli D, Palmer LJ et al (2002) Single nucleotide polymorphisms in innate immunity genes: abundant variation and potential role in complex human disease. Immunol Rev 190:9–25

Ley RE, Backhed F, Turnbaugh P et al (2005) Obesity alters gut microbial ecology. Proc Natl Acad Sci USA 102:11070–11075

Ley RE, Peterson DA, Gordon JI (2006a) Ecological and evolutionary forces shaping microbial diversity in the human intestine. Cell 124:837–848

Ley RE, Turnbaugh PJ, Klein S et al (2006b) Microbial ecology: human gut microbes associated with obesity. Nature 444:1022–1023

Ley TJ, Mardis ER, Ding L et al (2008) DNA sequencing of a cytogenetically normal acute myeloid leukemia genome. Nature 456:66–72

Liu Z, Li Z, Mao K et al. (2009) Dec2 promotes Th2 cell differentiation by enhancing IL-2R signaling. J Immunol 183: 6320–6329.

Ma D, Forsythe P, Bienenstock J (2004) Live *Lactobacillus reuteri* is essential for the inhibitory effect on tumor necrosis factor alpha-induced interleukin-8 expression. Infect Immun 72: 5308–5314

Macpherson AJ, Uhr T (2004) Induction of protective IgA by intestinal dendritic cells carrying commensal bacteria. Science 303:1662–1665

Marchesi JR, Holmes E, Khan F et al (2007) Rapid and noninvasive metabonomic characterization of inflammatory bowel disease. J Proteome Res 6:546–551

Mardis ER (2009) New strategies and emerging technologies for massively parallel sequencing: applications in medical research. Genome Med 1:40

Martin FP, Dumas ME, Wang Y et al (2007a) A top-down systems biology view of microbiome-mammalian metabolic interactions in a mouse model. Mol Syst Biol 3:112

Martin FP, Wang Y, Sprenger N et al (2007b) Effects of probiotic *Lactobacillus paracasei* treatment on the host gut tissue metabolic profiles probed via magic-angle-spinning NMR spectroscopy. J Proteome Res 6:1471–1481

Martin FP, Wang Y, Sprenger N et al (2008) Probiotic modulation of symbiotic gut microbial-host metabolic interactions in a humanized microbiome mouse model. Mol Syst Biol 4:157

Mayer L (2003) Mucosal immunity. Pediatrics 111:1595–1600

Mazmanian SK, Liu CH, Tzianabos AO et al (2005) An immunomodulatory molecule of symbiotic bacteria directs maturation of the host immune system. Cell 122:107–118

McCracken VJ, Lorenz RG (2001) The gastrointestinal ecosystem: a precarious alliance among epithelium, immunity and microbiota. Cell Microbiol 3:1–11

Medzhitov R (2007) Recognition of microorganisms and activation of the immune response. Nature 449:819–826

Neutra MR, Pringault E, Kraehenbuhl JP (1996) Antigen sampling across epithelial barriers and induction of mucosal immune responses. Annu Rev Immunol 14:275–300

Nibali L, Donos N, Henderson B (2009) Periodontal infectogenomics. J Med Microbiol 58: 1269–1274

O'Hara AM, O'Regan P, Fanning A et al (2006) Functional modulation of human intestinal epithelial cell responses by *Bifidobacterium infantis* and *Lactobacillus salivarius*. Immunology 118:202–215

Ordovas JM, Mooser V (2006) Metagenomics: the role of the microbiome in cardiovascular diseases. Curr Opin Lipidol 17:157–161

Pridmore RD, Berger B, Desiere F et al (2004) The genome sequence of the probiotic intestinal bacterium *Lactobacillus johnsonii* NCC 533. Proc Natl Acad Sci USA 101:2512–2517

Rakoff-Nahoum S, Paglino J, Eslami-Varzaneh F et al (2004) Recognition of commensal microflora by toll-like receptors is required for intestinal homeostasis. Cell 118:229–241

Rezzi S, Ramadan Z, Fay LB et al (2007) Nutritional metabonomics: applications and perspectives. J Proteome Res 6:513–525

Roselli M, Finamore A, Nuccitelli S et al (2009) Prevention of TNBS-induced colitis by different *Lactobacillus* and *Bifidobacterium* strains is associated with an expansion of gammadelta T and regulatory T cells of intestinal intraepithelial lymphocytes. Inflamm Bowel Dis 15:1526–1536

Schell MA, Karmirantzou M, Snel B et al (2002) The genome sequence of *Bifidobacterium longum* reflects its adaptation to the human gastrointestinal tract. Proc Natl Acad Sci USA 99: 14422–14427

Schena M, Shalon D, Davis RW et al (1995) Quantitative monitoring of gene expression patterns with a complementary DNA microarray. Science 270:467–470

Schneeberger EE, Lynch RD (2004) The tight junction: a multifunctional complex. Am J Physiol Cell Physiol 286:C1213–1228

Shanahan F (2005) Physiological basis for novel drug therapies used to treat the inflammatory bowel diseases I. Pathophysiological basis and prospects for probiotic therapy in inflammatory bowel disease. Am J Physiol Gastrointest Liver Physiol 288:G417–421

Shi M, Deng W, Bi E et al (2008) TRIM30 alpha negatively regulates TLR-mediated NF-kappa B activation by targeting TAB2 and TAB3 for degradation. Nat Immunol 9:369–377

Slack E, Hapfelmeier S, Stecher B et al (2009) Innate and adaptive immunity cooperate flexibly to maintain host–microbiota mutualism. Science 325:617–620

Stecher B, Hardt WD (2008) The role of microbiota in infectious disease. Trends Microbiol 16:107–114

Tlaskalova-Hogenova H, Stepankova R, Hudcovic T et al (2004) Commensal bacteria (normal microflora), mucosal immunity and chronic inflammatory and autoimmune diseases. Immunol Lett 93:97–108

Turnbaugh PJ, Hamady M, Yatsunenko T et al (2009) A core gut microbiome in obese and lean twins. Nature 457:480–484

Turnbaugh PJ, Ley RE, Hamady M et al (2007) The human microbiome project. Nature 449: 804–810

Turnbaugh PJ, Ley RE, Mahowald MA et al (2006) An obesity-associated gut microbiome with increased capacity for energy harvest. Nature 444:1027–1031

Vijay-Kumar M, Aitken JD, Carvalho FA et al (2010) Metabolic syndrome and altered gut microbiota in mice lacking toll-like receptor 5. Science 328:228–231

Vitali B, Ndagijimana M, Cruciani F et al (2010) Impact of a synbiotic food on the gut microbial ecology and metabolic profiles. BMC Microbiol 10:4

Wang Y, Tang H, Nicholson JK et al (2005) A metabonomic strategy for the detection of the metabolic effects of chamomile (*Matricaria recutita* L.) ingestion. J Agric Food Chem 53: 191–196

Wen L, Ley RE, Volchkov PY et al (2008) Innate immunity and intestinal microbiota in the development of type 1 diabetes. Nature 455:1109–1113

Xu J, Bjursell MK, Himrod J et al (2003) A genomic view of the human–*Bacteroides thetaiotaomicron* symbiosis. Science 299:2074–2076

Zeuzem S (2000) Gut–liver axis. Int J Colorectal Dis 15:59–82

Chapter 12
Autoimmune Disease and the Human Metagenome

Amy D. Proal, Paul J. Albert, and Trevor G. Marshall

Background

In 1922, Ernst Almquist – a colleague of Louis Pasteur – commented, "Nobody can pretend to know the complete life cycle and all the varieties of even a single bacterial species. It would be an assumption to think so" (Mattman, 2000). While Almquist's work on idiopathic bacteria in chronic disease never received the plaudits accorded to Pasteur's work, Almquist foresaw the complexity that would later be inherent to the field of metagenomics – a field that today forces us to examine how countless microbial genomes interact with the human genome across disease states.

Yet in the decades before novel genomic technology made a metagenomic understanding of disease possible, bacteria could only be cultured in vitro on a limited range of growth media. As most major diseases of the time – tuberculosis, pneumonia, leprosy, and others – were linked to the presence of a handful of acute pathogens able to grow under these constraints, a "game over" attitude toward infectious agents dominated the thinking of much of the medical community. Little consideration was given to the possible role of these pathogens in autoimmune and inflammatory disease states. Instead, for most of the twentieth century, the predominant feeling about the treatment, control, and prevention of diseases with a possible infectious etiology was optimism (Cohen, 2000).

Between 1940 and 1960, the development and successes of antibiotics and immunizations added to this optimism and, in 1969, Surgeon General William H. Stewart told the US Congress that it was time to "close the book on infectious diseases" (Avila et al., 2008). With "victory" declared, increasing emphasis was directed at the "noninfectious" diseases such as cancer and heart disease. In many cases, research on infectious disease or activities on their prevention and control were de-emphasized and resources were reduced or eliminated. As recently as the 1980s, pharmaceutical companies, believing that there were already enough antibiotics, began reducing the development of new drugs or redirecting it away from antibiotics.

A.D. Proal (✉)
Murdoch University, Murdoch, WA 6150, Australia
e-mail: amy.proal@gmail.com

K.E. Nelson (ed.), *Metagenomics of the Human Body*,
DOI 10.1007/978-1-4419-7089-3_12,

Despite this rosy narrative, some microbiologists were never convinced that drugs like penicillin had ended the war between man and microbe. In 1932, Razumov noted a large discrepancy between the viable plate count and total direct microscopic count of bacteria taken from aquatic habitats (Razumov, 1932). He found higher numbers (by several orders of magnitude) by direct microscopic counting than by the plating procedure. In 1949, Winogradsky confirmed Razumov's assessment and noted that many microbes are not satisfied with laboratory cultivation conditions. He remarked that readily cultivated bacteria in natural microbial communities "draw importance to themselves, whereas the other forms, being less docile, or even resistant, escape attention" (Relman, 1998). In 1985, Staley and Konopka pointed to Razumov's discrepancy and called it the "Great Plate Count Anomaly" (Grice et al., 2008). Their review describes work in which they compared the efficacy of a fluorescent dye versus standard plating procedures in detecting bacterial species in samples of water collected from Lake Washington. They found that only approximately 0.1–1.0% of the total bacteria present in any given sample could be enumerated by the plating procedure – causing them to conclude that, unless new methods for detecting bacteria were employed, "No breakthrough in determining species diversity seems likely in the near future."

Meanwhile, some microbiologists continued their best efforts to alter the pH and growth medium of their samples in an effort to look for previously undetected bacteria in chronic disease states. Over the course of a career spanning almost 50 years, Lida Mattman of Wayne State University cultured wall-less forms of bacteria from the blood samples of patients with over 20 inflammatory diagnoses including multiple sclerosis and sarcoidosis (Almenoff et al., 1996). She authored an entire textbook on novel approaches for in vitro cultivation of bacteria (Mattman, 2000). Over his 39-year career at Tulane University, Gerald Domingue published dozens of papers and book chapters devoted to the role of chronic forms of bacteria in inflammatory disease. "It is unwise to dismiss the pathogenic capacities of any microbe in a patient with a mysterious disease," he wrote. "Clearly, any patient with a history of recurrent infection and persistent disability is sending the signal that the phenomenon [infection with chronic bacteria or viruses] could be occurring. The so-called autoimmune diseases in which no organism can be identified by routine testing techniques are particularly suspect" (Domingue and Woody, 1997).

Yet, scientists like Mattman and Domingue faced serious challenges in trying to convince the medical community that their work was valid. Other research teams using less rigorous techniques often failed to duplicate their findings. Many of their observations were dismissed on the premise that their samples could have been contaminated. However, the greatest impediment toward the acceptance of this work was a set of rules set in motion by the nineteenth-century German physician Robert Koch. These rules, known as "Koch's postulates," stipulate that in order for a microbe to be deemed a causative agent of a disease, certain criteria must be met. The same microbe must be identified in every person with a given disease; the specific microbe must be able to be grown on pure culture medium in the lab; and, when reintroduced into a healthy animal or person, must produce the disease again.

While Koch's postulates may have offered certain clarity during the formative stages of the field of microbiology, the rules distracted scientists from considering the possibility that multiple species could be responsible for the onset of a single disease state. Even today, Koch's notions about disease are regularly invoked (Monaco et al., 2005) despite the emergence of a number of counter-examples. Neither *Mycobacterium leprae*, which is implicated in leprosy, nor *Treponema pallidum*, which causes syphilis, fulfill Koch's postulates, because these microbes cannot be grown in conventional culture media. Viruses further invalidate Koch's postulates because most require another living cell in order to replicate (Walker et al., 2006).

In the absence of clear connections between a single microbe and a single disease, most microbiologists necessarily assumed that the body was a sterile compartment and that inflammation, which might well suggest the presence of microbes, was attributed to an idiopathic causation. Unable to grow all but a fraction of bacteria found in the human body in the confines of a Petri dish, and constrained by a lack of technology with which to detect new microbes, the theory of autoimmune disease, in which the immune system loses tolerance and generates antibodies that target self gained momentum in the 1960s.

Yet over the past decade, the role of infectious agents in autoimmune disease has once again gained momentum. The 2004 International Congress on Autoimmunity in Budapest was themed "Autoimmunity and Infection" with many subsequent conferences and papers in the same vein. However, nearly all speakers discussed the role of viruses in autoimmune disease, whereas only a few contemplated bacteria. Autoimmune conditions were repeatedly attributed to easily cultured viruses such as Epstein–Barr and Herpes 6. Where bacteria were discussed, most reports centered on select pathogens such as *Chlamydia pneumoniae*. Yet because none of these pathogens has ever been detected in any one autoimmune disease state 100% of the time, such researchers continue to paint autoimmune diseases as a mosaic – in which the hallmarks of infection are continually present in bits and pieces but cannot be drawn into a fully cohesive picture. Yet the emerging science of metagenomics is beginning to unmask entirely new populations of microbes whose genomes allow for a means by which to bridge these gaps. The following chapter examines how this metagenomic microbiota can cause the dysfunction seen in a wide range of autoimmune conditions.

Culture-Independent Methods for Identifying Microbes

In 2007, a study orchestrated by NASA announced that the surfaces of the supposedly sterile "clean rooms," in which technicians assemble spacecraft, host an abundance of hardy bacteria (Moissl et al., 2007). Samples taken from clean rooms at the Jet Propulsion Laboratory in California, the Kennedy Space Flight Center in Florida, and the Johnson Space Center in Houston revealed the presence of almost 100 types of bacteria representing all the major bacterial phyla; 45% of the species identified were previously unknown to science. The findings came as a shock to

NASA officials, who were left to wonder exactly how many unknown microbes might have been taken to the moon and Mars.

These clean room bacteria had not been previously detected because they could not be characterized by standard cultivation techniques. To find them, the research team had used a genomic approach – RNA gene sequence analysis – to characterize the genetic material of the bacterial species in the rooms previously touted as sterile.

Similar culture-independent tools are beginning to revolutionize our understanding of autoimmune disease by allowing for a vastly more comprehensive understanding of the microbes that persist in *Homo sapiens*, microbes that may cause the generation of autoantibodies. Genomic sequencing techniques, including 16S RNA sequencing, PCR and, more recently, pyrosequencing, have made it clear that only a fraction of those microbes that persist in the human body will grow on the limited medium of a Petri dish. With the advent of these technologies, the field of metagenomics was born. Rather than focusing on the study of single microbes and their genomes, metagenomics provides a means of analyzing aspects of microbial communities through their underpinning genetics. The amount of novel microbial genetic information that is generated on a daily basis by metagenomic analysis is so great that multidisciplinary approaches that integrate statistics, bioinformatics, and mathematical methods are required to assess it effectively.

Today, the National Institutes of Health (NIH) estimates that a mere 10% of the cells that comprise *Homo sapiens* are human cells. The remaining 90% are bacterial in origin. The number of *Escherichia coli* in a single human is comparable to the entire human global population – approximately six billion people (Staley, 1997). Such knowledge has forever changed the manner in which the human organism is perceived. We may best describe the human being as a super-organism in which communities of different organisms flourish in symbiosis with the host. Yet, even with the availability of technology to explore the microbial world in depth, to date, only a fraction of the human bacterial microbiota has been genetically identified and characterized. As of late 2009, approximately 1,100 published complete bacterial genomes had been identified with 6,000 more under review (Liolios et al., 2008). Nevertheless, there are still huge gaps in our understanding of how the microbiota contributes to human health and disease.

Viruses (the virome) and phages are also key components of the microbiota. Like bacteria, many of these microbes have yet to be fully characterized by high-throughput genome sequencing. However, molecular analysis has revealed that nearly all humans acquire multiple viruses, usually within the first years of life, viruses that generally remain with them throughout life. Polyomaviruses infect between 72 and 98% of humans, surviving in the kidney, lung, and skin (Virgin et al., 2009). Similarly, human herpes viruses are extremely persistent. Anelioviruses, as well as adeno-associated virus, are now recognized to infect most humans by the end of childhood. The role of these viruses is unknown, but a significant number of people who harbor them become symptomatic later in life, suggesting that they may be capable of virulence under conditions of immune dysfunction. According to Herbert Virgin of Washington University, "We carry, for good or for ill, many lifelong [viral] passengers" (Virgin et al., 2009).

In the next 5 years, researchers associated with the NIH Human Microbiome Project (HMP) plan to use molecular genetic sequencing in an effort to catalog the bacterial component of the human microbiome. This initiative promises to increase our knowledge of bacterial diversity. The NIH has funded many more HMP projects, with the goal that the diagnosis, treatment, and prevention of many inflammatory diagnoses can be improved by examining how the microbiota differs between those people with a disease and their healthy counterparts. Thus far, targeted conditions include Crohn's disease (CD), inflammatory bowel disease, vaginosis, psoriasis, and other conditions now considered to be autoimmune. Early work has already demonstrated fundamental discrepancies in microbial composition between health and disease. Swidsinksi et al. found that patients with irritable bowel syndrome have more bacteria from diverse genera attached to their epithelial gut surfaces than do healthy controls. Some of these microbes, such as *Bacteriodes*, were found to penetrate the epithelial layer, at times intracellularly (Swidsinski et al., 2002). Enck et al. found that irritable bowel syndrome manifests with a relative decrease in populations of bifidobacteria and significant differences in a variety of other microbes, including those that cause the production of gas (Enck et al., 2009).

Medicine has become comfortable acknowledging that bacterial populations exist in the areas of the body in contact with the external environment, such as the airways, gastrointestinal tract, mouth, skin, and vagina/penis. For example, analysis of the human oral cavity by Nasidze et al. identified 101 bacterial genera in the mouth as well as an additional 64 genera previously unknown to science (Nasidze et al., 2009). Yet, microbes have also been shown to persist in many other body tissues, including joints and blood vessels. Some of the same bacteria identified in the salivary microbiome, such as *Actinobacillus actinomycetemcomitans* and *Porphyromonas gingivalis* – both of which cause tooth decay (Lamell et al., 2000) – have also been identified in atherosclerotic plaque (Kozarov et al., 2005). Bacterial DNA has been detected in the blood (Nikkari et al., 2001). Recently, 18 different bacterial taxa were detected in the amniotic fluid, which was previously believed to be completely sterile (DiGiulio et al., 2008). Analysis using 16S rRNA sequencing detected 28 distinct phylotypes on biofilm removed from prosthetic hip joints during revision arthroplasties – joints also removed from a body compartment also thought to be sterile. The prevalence of hydrothermal vent eubacteria, which were previously thought to persist only in the depths of the ocean since they were found at temperatures well above 176°F (80°C), was higher than the prevalence of *Staphylococcus aureus*, a common biofilm species (Fig. 12.1).

It is now more prudent to assume that tissues that become inflamed in disease most probably do so because of the actions of microscopic pathogens, rather than idiopathic causation. Different microbial populations have been identified in many nongastrointestinal autoimmune conditions including sarcoidosis (el-Zaatari et al., 1996), ankylosing spondylitis (Liu et al., 2001), chronic fatigue syndrome (Lombardi et al., 2009), rheumatoid arthritis, multiple sclerosis, Hashimoto's

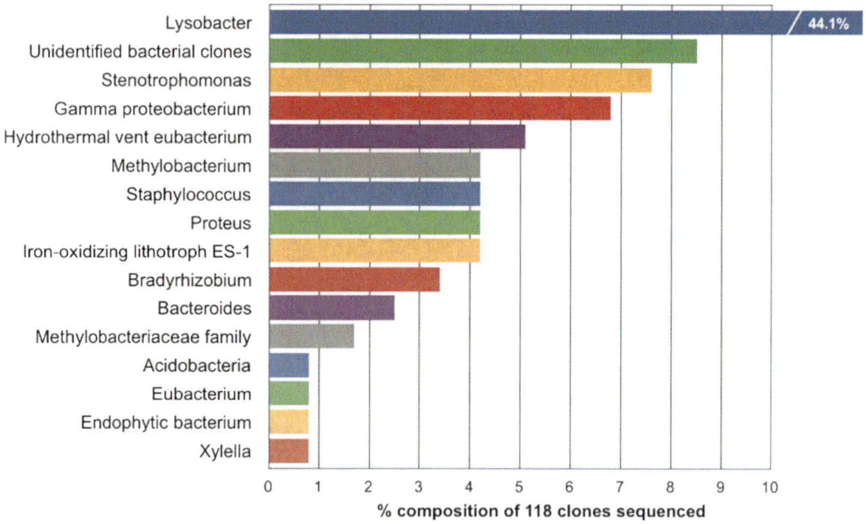

Fig. 12.1 Bacterial species identified by 16S rRNA gene sequencing of clones from 10 prosthetic hip joints (Dempsey et al., 2007)

thyroiditis, and others (Pordeus et al., 2008). These diseases share features of microbial infection including widespread inflammation and periods of relapse. Sarcoidosis and CD' are characterized by granuloma. In more than a dozen infectious diseases, granuloma is widely acknowledged to be a host-protective structure and to occur when acute inflammatory processes cannot destroy invading agents (Zumla and James, 1996).

The Human Metagenome

At only approximately 23,000 genes, the human genome is dwarfed by the thousands of genomes of the bacteria, viruses, and phages that persist in and on humans. Given the sheer number of microbial genes, it is no longer possible to study the human genome in isolation. Rather, the human genome is only one of myriad genomes that influence the *Homo sapiens* experience. Humans are controlled by a metagenome – a tremendous number of different genomes working in tandem. Because they are so small, thousands of microbial cells can persist inside a single infected human cell (Wirostko et al., 1989). The combined genetic contributions of these microbes invariably provide a vast number of gene products not encoded by our own relatively small genomes.

There is considerable similarity between the functions of the bacterial organisms and the human organisms. For example, humans and *E. coli* metabolize glucose-6-phosphate in a similar fashion, producing almost identical metabolites (Kuroki et al., 1993). Thus, the transgenomic interaction between an *E. coli* genome and the human genome, as they exchange nutrients and toxins, increases the complexity of transcription and translation for both species. The dihydrofolate reductase (DHFR) antagonist trimethoprim is such an effective antibiotic because, like humans, bacterial species possess a folate metabolism. Bacteria in the distal intestinal tract of mice have been shown to significantly alter the composition of human blood metabolites, including amino acids, indole-3-propionic acid (IPA), and organic acids containing phenol groups, providing another example of the significant interplay between bacterial and human metabolism. A broad, drug-like phase II metabolic response of the host to metabolites generated by the gut microbiota was observed (Wikoff et al., 2009), suggesting that the gut microbiome has a direct impact on the host's capacity for drug metabolism.

In the pre-genomic era, diseases were classified largely on the basis of symptom presentation; while in recent decades, researchers have attempted to categorize them based on common genes. Yet metagenomics dictates that we must also consider how the many microbial metabolites affect expression of these genes. Some genes and their related pathways have already been shown to influence the pathogenesis of autoimmune disease. For example, Goh et al. has shown that PTPN22 is related to rheumatoid arthritis, lupus, and diabetes mellitus (Goh et al., 2007). Yet expression of PTPN22 is also modified by the bacterial metagenome – it has been shown to be upregulated as part of the innate immune response to mycobacteria (Lykouras et al., 2008). The importance of understanding how microbes affect PTPN22 across multiple disease states has special impetus, given the increased rate of latent tuberculosis in the global population as well as studies showing high rates of infection by *Mycobacterium avium* among autoimmune patients (Bentley et al., 2008).

Many of the most well-studied persistent pathogenic bacteria have evolved mechanisms to evade the immune response and survive inside macrophages and other phagocytic cells. These include *Francisella tularensis* (Hazlett et al., 2008), *Mycobacterium tuberculosis* (Domingue and Woody, 1997), *Rickettsia massiliae* (Monaco et al., 2005), *Brucella* spp. (Baldwin and Goenka, 2006), *Listeria monocytogenes* (Birmingham et al., 2007), *Salmonella typhimurium* (Kuijl et al., 2007), among others. This suggests that other disease-causing components of the microbiota may also persist in the cytoplasm of nucleated cells, where they have access to both human DNA transcription and protein translation machinery (Hall et al., 2008). When *Shigella* persists within a macrophage it modulates numerous host signaling pathways, including those that inactivate mitogen-activated protein kinases (Lutjen-Drecoll, 1992). *Brucella* spp. downregulates genes involved in cell growth and metabolism, but upregulates those associated with the inflammatory response and the complement system upon infecting a macrophage.

Additionally, there appears to be an entire intra-cytoplasmic microbiota within phagocytic cells. Wirostko's team at Columbia University in the 1980s and 1990s used electron microscopy to identify entities within the cytoplasm of phagocytes from patients with juvenile rheumatoid arthritis, sarcoidosis (Wirostko et al., 1989), Crohn's, and other inflammatory diseases (Wirostko et al., 1987). The wide variety of elongated and globular formations, together with both the existence and absence of exoskeletons around the microbiota, would imply that the observed communities are metagenomic, rather than due to any one single obligate phagocytic pathogen.

Microbial Complexity

The HIV genome consists of a single strand of RNA comprising nine genes, from which are transcribed 19 proteins. Transcription is noncontiguous, and variations abound. For example, "Tat" is transcribed in multiple pieces that are subsequently joined. Yet 1,443 direct interactions (3,300 total interactions) have been identified between just these 19 proteins and the human metabolome (Fu et al., 2009). Consider that the average bacterial genome codes for hundreds or sometimes thousands of proteins. According to one recent estimate, the average human gut microbiota codes for 9 million unique genes (Yang et al., 2009). Factor in the proteins created by viruses and phages, and efforts to understand how these proteins affect the metabolome leave an observer with little more than stochastic noise, particularly since biological systems are replete with components showing nonlinear dynamic behavior.

Subsequently, interaction between the metagenome and the human genome introduces a new level of complexity to the study of autoimmune disease – complexity that renders it nearly impossible to fully comprehend the vast number of the interactions between the human genome and those microbial genomes capable of influencing the pathogenesis of autoimmune disease. According to Bunge, the size of a gene pool for a given environmental sample can be estimated by mathematical modeling, but the size of the gene pool for a microbial biosphere, such as the human body, may be beyond any current credible model (Bunge, 2009). While this complexity poses a significant challenge to systems biology and to Koch's simplistic one gene–one disease model, it does not impede the emergence of a better understanding of the human super-organism and the processes that cause disease.

Lifelong symbiosis between the human genome and persistent components of the metagenome has shifted the focus of microbiology away from the search for a single pathogen in a disease state. Many research teams are now striving to understand how components of the microbiota may cause disease. For example, researchers with the European MetaHIT Initiative are studying how bacteria in the gut may contribute to obesity and inflammatory bowel disease. The goal of the project is simply to examine associations between bacterial genes and human phenotypes. "We don't care if the name of the bacteria is *Enterobacter* or *Salmonella*. We want to know if there is an enzyme producing carbohydrates, an enzyme producing gas or an enzyme degrading proteins," explains Francisco Guarner of the project.

Studies focused on enzymes, proteins, and carbohydrates are studies of the metabolome. Metabolomic approaches can be used to characterize entire components of the microbiome that cannot easily be seen or studied directly. Because the downstream results of gene expression manifest in the human metabolome, the metabolome can be analyzed for the presence of those unique metabolites created under the influence of the microbiota. Dumas et al. used mass spectroscopy to identify the nonhuman metabolites present in the urine of subjects living in three distinct populations – the United States, China, and Japan (Dumas et al., 2006). He found that subjects in each population produced very different nonhuman metabolites. Thus, genetic makeup, healthcare, nutrition, and external toxins, factors associated with the acquisition of a particular microbiota, caused the three populations to become significantly different. Moreover, when five of the Japanese subjects moved to the United States, their metabolomes changed to resemble those of the American population. This suggests that the metagenome is indeed the product of its environment, and that the composition of the microbiota is far more important than regional variations in the human genome itself.

In another study, the INTERMAP epidemiological study used an ^{1}H NMR-based metabonomics approach to examine differences in the urine metabolite profiles for each of 4,630 participants from 17 populations in the USA, UK, Japan, and China (Stamler et al., 2003). Elevated blood pressure was associated with high levels of the bacterial co-metabolite formate. Interestingly, low levels of hippurate and alanine, which reflected gut microbial activities, were also found in subjects with high blood pressure (Holmes et al., 2008). This suggests that certain microbial metabolites may serve as useful biomarkers for a disease state.

The fact that components of the microbiota are seldom found as single entities further complicates the complexity of transgenomic control in *Homo sapiens*. While just a few decades ago, most of the bacteria in *Homo sapiens* were assumed to persist on their own in a planktonic form, it is now understood that large components of the microbiota persist in communities commonly called biofilms; they are sheltered by a self-created polymeric matrix that better protects them from the immune response. Hundreds of different microbes can persist in a single biofilm community, and individual bacteria often form a niche inside the biofilm that allows them to promote their own survival and the chronic nature of the infection. For example, more virulent bacteria may protect the biofilm from outside intrusion whereas other less innocuous species inside the biofilm may focus on obtaining nutrients for the community. As the biofilm forms and then develops, the collective genetic expression of microbes in the biofilm is altered dramatically. For example, the expression of 800 genes has been shown to be altered when a single bacterial species joins a biofilm (Sauer et al., 2002). Biofilms are increasingly being detected in autoimmune diseases where they were not known to previously exist. For example, Wolcott recently used pyrosequencing to demonstrate that the infectious agents that drive the development of diabetic leg, foot, and pressure ulcers are almost all in a biofilm state (Dowd et al., 2008).

Bacteria in biofilm, their planktonic counterparts, viruses, and other microbes rapidly and frequently share their DNA with other species, even distantly related species, through horizontal gene transfer. Genomic coherence is further muddled by homologous recombination. This further diversifies the variability present in the human microbiome. Horizontal gene transfer is now believed by many to occur so frequently that it has been proposed as a means by which species can acquire new genetic traits. Some argue that the number of microbes created through homologous recombination is so high that the concept of distinct bacterial species may become obsolete (Doolittle and Papke, 2006).

Thus, the concept that a single pathogen could cause the human metabolism to fail in the myriad of ways necessary to result in an advanced, systemic autoimmune disease is increasingly recognized as an outdated nineteenth-century concept. The postulates of Koch are no longer relevant in the era of the metagenome. Brock contends in his profile of Koch that attempts to rigidly apply Koch's postulates to the diagnosis of viral diseases may have significantly impeded the early development of the field of virology (Brock, 1988). The same can be said for the field of bacteriology, where the postulates have long impeded researchers from considering that the genomes of many different bacteria and other pathogens interact together to cause the range of symptoms we associate with autoimmune diagnoses.

Toward a More Nuanced View of the Human Microbiota

In *New science of metagenomics: revealing the secrets of our microbial planet*, the National Research Council writes, "The billions of benign microbes that live in the human gut help us to digest food, break down toxins, and fight off disease-causing microbes" (Committee on Metagenomics and National Research Council, 2007). While certain components of the microbiota clearly aid humans in these and other ways, strictly classifying microbes as either commensal or pathogenic may suggest too categorical a distinction. Emerging research suggests that bacteria are no more "good" or "bad" than their human counterparts, particularly when a commensal microbe can easily acquire a plasmid or virulence factor from another microbe. According to Fredricks and Relman, "The mobile nature of virulence-associated gene islands, transported between bacteria via plasmids or phages, creates the potential for acquired virulence in previously innocuous microbes" (Fredricks and Relman, 1998).

In September 2009, Malcolm Casadaban, an infectious disease researcher at the University of Chicago, died suddenly. An autopsy showed no obvious cause of death except *Yersinia* in his bloodstream. Dr. Casadaban, an associate professor at the university, was studying the bacteria to create a better vaccine for plague. Yet Casadaban was working with a strain of *Yersinia* that was supposed to be less virulent that those strains considered lethal. Researchers postulated that there must have been something unusual about the bacterium that caused it to be dangerous, such as a mutation. The so-called "innocuous" strain of *Yersinia* may have acquired a plasmid or gene that endowed it with newfound virulence.

Acquired virulence via horizontal gene transfer has been studied in anthrax. Although *Bacillus anthracis*, which causes fatal poisoning, and *B. cereus*, which causes nonlethal opportunistic infections, are generally classified as separate bacterial species, Hoffmaster discovered a *B. cereus* mutant that also causes a deadly form of pneumonia. Analysis revealed that the *B. cereus* mutant (*B. cereus* G9241) had acquired a plasmid with 99.6% sequence homology to pX01, *B. anthracis*' most virulent, toxin-encoding plasmid. Indeed, *B cereus* G9241 killed mice more quickly than *B. anthracis*. *Bacillus cereus* G9241 was deemed the product of horizontal gene transfer, causing Hoffmaster to note that, depending on the extent of horizontal gene transfer, nature could produce an unlimited number of variations and combinations of any given pathogen.

The distinction between commensalism and pathogenicity is further blurred by host-specific factors. For example, if a species of bacteria aids in the met

These play a vital role in allowing the innate immune system to target intracellular pathogens. For example, vitamin D-mediated human antimicrobial activity against *Mycobacterium tuberculosis* is dependent on the induction of cathelicidin (Liu et al., 2007). The VDR also transcribes Toll-like-receptor 2 (TLR2), which recognizes bacterial polysaccharides.

The TACO gene, when expressed, inhibits mycobacterial entry as well as survival. *Mycobacterium tuberculosis* (Mtb) downregulates the VDR, and thus expression of TACO in order to survive. Xu et al. showed that the VDR was downregulated 3.3 times in monocytic cell lines infected with Mtb (Xu et al., 2003). *Borrelia*, as assessed by BeadChip microarray, has been shown capable of downregulating VDR activity by a factor of 50-fold, with lysed *Borrelia* downregulating the receptor by a factor of 8 (Salazar et al., 2009). We have previously shown that at least one bacterial metabolite produced by gliding biofilm bacteria is also a strong VDR antagonist (Marshall, 2008). The HIV "tat" protein binds to the VDR in order to use this receptor to recognize its long terminal repeat (LTR) promoter region (Nevado et al., 2007). Thus, tat takes over the human VDR in order to transcribe HIV's own genome, so the HIV LTR can be recognized and express new HIV RNA. Tat also recruits histone acetyltransferase activity, including the CREB binding protein (CBP)/p300 complex, to acetylate the HIV LTR promoter region (Romani et al., 2009).

Slowing the ability of the VDR to express elements of the innate immune function is such a logical survival mechanism that it is almost certain that the other less studied components of the microbiota would have also evolved ways to dysregulate the VDR, and the other nuclear receptors orchestrating the innate immune response. Eukaryotic cells respond to the presence of the microbiota by activating signaling cascades such as the NF-kappaB pathway. Induction of such pathways leads to the upregulation of gene expression mediating pro-inflammatory and anti-apoptotic effector proteins. Thus, in order for pathogens (and potentially, symbionts) to continue their life cycle, it is necessary to evade or repress these cellular responses. This is especially true because acquisition of resistance to AMPs by a sensitive microbial strain is surprisingly improbable. Furthermore, the extension of human life during the past century now offers additional opportunity for microbes to evolve their specialization in order to survive in man.

Indeed, Yenamandra et al. recently showed that Epstein–Barr virus (EBV) also slows VDR activity (Yenamandra et al., 2009). Infection of human B cells with EBV induces metabolic activation, morphological transformation, cell proliferation, and eventual immortalization by altering the expression of a number of key nuclear receptors. The team found that the expression of 12 nuclear receptors was downregulated in the longer-lasting, younger lymphoblastoid cells. Among them were the VDR and the estrogen receptor beta (ERB), both downregulated by a factor of about 15 times (Fig. 12.2).

EBV is found in many common chronic disease states. Indeed, EBV has been detected in a subset of patients with nearly every autoimmune diagnosis, although it has rarely been detected in 100% of patients with any given condition. In some cases, infection with the virus is described as a "precipitating factor" for

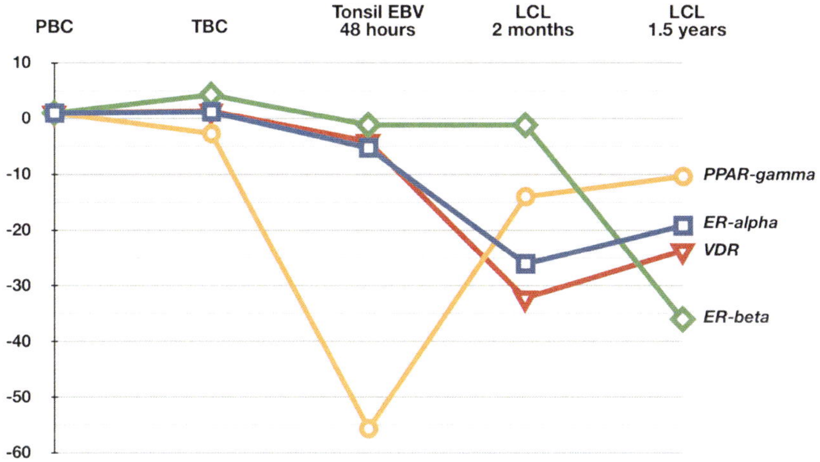

Fig. 12.2 Nuclear receptors mRNA expression is downregulated upon infection of B cells with EBV (Yenamandra et al., 2009)

autoimmune disease. That EBV downregulates VDR and ERB expression may explain this phenomenon. If a patient acquires EBV, the virus slows innate immune activity to the point where the endogenous microbiota can become dominant.

This is particularly true because, in addition to reducing expression of cathelicidin and beta-defensin, VDR dysregulation opens a number of other pathways that also influence immune activity and hormonal regulation. Blockage of the VDR prevents transcription of CYP24A1, an enzyme that normally breaks down excess levels of the active vitamin D metabolite 1,25-dihydroxyvitamin-D (1,25-D). Activation of protein kinase A (PKA) by bacterial cytokines also causes increased production of the enzyme CYP27B1, resulting in increased conversion of 25-hydroxyvitamin-D (25-D) into 1,25-D. Both processes result in a rise in 1,25-D.

High levels of 1,25-D in autoimmune disease have been confirmed in a clinical setting. Mawer et al. found that 1,25-D levels were particularly elevated in the synovial fluid surrounding the joints of patients with rheumatoid arthritis (Mawer et al., 1991). Abreu et al. found that in a cohort of 88 CD' patients, 35 patients or 40% had elevated levels of 1,25-D, which the authors defined as above 60 pg/ml (Abreu et al., 2004). Bell noted that patients with tuberculosis, pneumonia, AIDS, disseminated candidiasis, leprosy, rheumatoid arthritis, silicone-induced granulomas, Wegerner's granulomatosis, Hodgkin's disease, lymphoma, histocytic lymphoma, T-cell leukemia, plasma cell granuloma, leiomyoblastoma, seminoma, and subcutaneous fat necrosis all tend to manifest with higher than normal levels of 1,25-D (Bell, 1998). Blaney et al. found that of 100 patients with various autoimmune diagnoses, 85% had 1,25-D above the normal range (Fig. 12.3) (Blaney et al., 2009). Yoshizawa et al. reported that in VDR knockout mice, a circumstance that closely mimics extreme VDR dysregulation, 1,25-D levels increase by a factor of 10 (Yoshizawa et al., 1997). However, understanding 1,25-D's role in various inflammatory disease states is complicated by the fact that most researchers determining

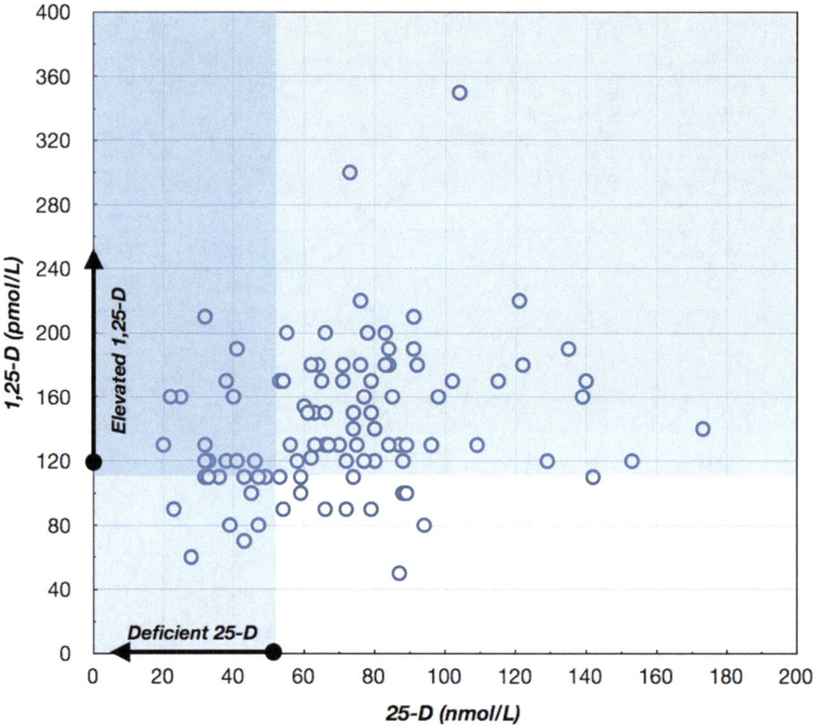

Fig. 12.3 25-D vs. 1,25-D in a cohort of 100 autoimmune patients (Blaney et al., 2009)

vitamin D status test subjects only for levels of the inactive vitamin D metabolite, 25-D.

In silico research indicates that 1,25-D has a high affinity for not just the VDR, but many of the body's other nuclear receptors (Proal et al., 2009). This suggests that at high concentrations it will displace their exogenous ligands. Those receptors affected by elevated 1,25-D include alpha thyroid, beta thyroid, the glucocorticoid (adrenal) receptor, and the progesterone receptor (Fig. 12.4). For example, 1,25-D has a very high affinity for the thyroid beta, suggesting that it can displace T3 and T4 from the binding pocket (Table 12.1) (Proal et al., 2009).

If 1,25-D prevents T3 from activating thyroid beta, then genes with thyroid beta promoters will be less energetically transcribed. This would result in thyroid disease and explain why increasing levels of thyroid hormone are necessary to maintain thyroid homeostasis as chronic disease progresses. Furthermore, since the functions of type 1 nuclear receptors are largely interdependent, if transcription by thyroid beta is dysregulated, system-wide transcription is also affected.

This leads to disruption of system-wide AMP production. Just as the VDR expresses cathelicidin and beta-defensin, other nuclear receptors also express AMPs. Brahmachary et al. have shown that the glucocorticoid receptor, the androgen receptor, and the VDR are, respectively, in control of 20, 17, and 16 families

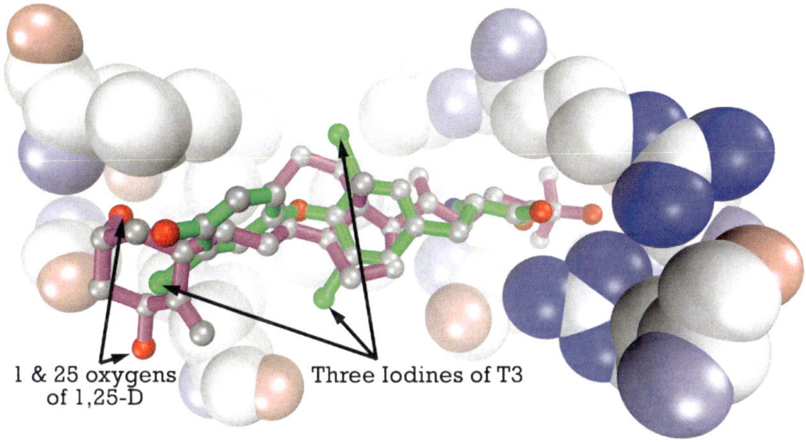

Fig. 12.4 The Thyroid-alpha nuclear receptor and T3, its native ligand [PDB:2H77], with the bound conformation of 1,25-D superimposed. Since the XSCORE Kd for 1,25-D is 8.4, and for T3 is 7.2, it is apparent that 1,25-D is capable of displacing T3 from binding to key receptor residues (shown here are Arg228, Asn179, Gly290, Leu292, Leu276, Ser277, Thr275, Ala263, Leu287, Ala180, Phe218, and Arg162) (Proal et al., 2009)

Table 12.1 Affinities of native ligands and 1,25-D for various nuclear receptors (Proal et al., 2009)

Nuclear receptor	Native ligand	Native ligand (Kd)	1,25-D (Kd)
Thyroid alpha	T3	7.20	8.41
Thyroid beta	T3	7.18	8.44
Glucocorticoid	Cortisol	7.36	8.12
Androgen	Testosterone	7.38	8.05
Progesterone	Progesterone	7.53	8.09

out of the 22 analyzed (Brahmachary et al., 2006). Thus, dysregulating VDR activity yields flow-on effects that potentially disable the bulk of the body's AMPs. A patient affected in this manner would become increasingly immunocompromised, allowing disease-causing components of the microbiota to proliferate with even greater ease.

This supports a disease model in which key components of the microbiota responsible for autoimmune conditions gradually shut down the innate immune response over a person's lifetime as bacteria, and other pathogens, incrementally accumulate into the microbiota. CD' is already characterized by diminishing functional antimicrobial activity, particularly when it comes to expression of cathelicidin and the beta-defensins (Nuding et al., 2007). Eventually, genes from the accumulating microbial metagenome may instigate a clinical disease symptomology, such as one of the autoimmune diagnoses, or simply drive the inflammation associated with the aches and pains of aging. Indeed, the lifelong accumulation of an increasingly diverse microbiota directly correlates with an age-related increase in diseases and symptoms associated with inflammation. The term "inflammaging" has been coined to explain "the now widely accepted phenomenon that aging is accompanied by a low-grade chronic, systemic up-regulation of the inflammatory response, and that the underlying inflammatory changes are common to most age-associated diseases" (Giunta, 2006).

Because 1,25-D is expressed in the human cycling endometrium and rises by 40% during early pregnancy, women may be disproportionately affected by the potential drop in AMP expression associated with VDR dysregulation (Viganò et al., 2006). This implies that females may more easily accumulate a more diverse microbiota than their male counterparts, which could help explain why women suffer from a higher risk of most autoimmune diagnoses.

Successive Infection and Variability in Disease Onset and Presentation

The makeup of a person's microbiota is unique: humans may share as little as 1% of the same species (Eckburg et al., 2005). Given that the human microbiome may play the principal causative role in autoimmune disease, it may not be by accident that the uniqueness with which patients' autoimmune disease symptoms develop parallels the incredible variability of the human microbiome. Traditionally, diseases have been understood to be discrete and have their own respective and distinct pathologies, hence the emphasis on diagnosis. However, if the spectrum of autoimmune disease were driven by a common factor – namely a person's microbial inhabitants – variability in disease could be explained by accounting for how the human microbiota accumulates and develops in any one person. Enck et al. recently analyzed fecal flora of stool samples from 35,292 adults whose ages ranged from 18 to 96 years of age in order to gauge the relative abundance and composition of various bacterial species over time (Enck et al., 2009). He found that while the number of bacteria in the fecal microbiota remained stable with age, the composition of the microbiota diversified as subjects became older, with the oldest subjects measured (over 60 years of age) representing the most profound changes. Older subjects were much more likely to have higher prevalence of microbes associated with chronic disease such as *Enterococcus* and *E. coli*.

A number of microbes that slow immune activity have already been identified indicating that bacteria/viral-driven suppression of innate immune activity may occur on a much larger scale than previously imagined. Each pathogen that decreases immune activity makes it easier for the host to pick up other pathogens, which themselves may further slow immune activity, creating a snowball effect. This process is known as successive infection and offers us a framework for understanding how not only diseases of the gastrointestinal tract develop, but also any number of other autoimmune and inflammatory diseases. As human genes are upregulated or downregulated by components of the microbiota, the body shifts farther and farther away from its natural state of homeostasis. Infected cells increasingly struggle to correctly produce human metabolites in the presence of numerous proteins and enzymes being created by the pathogenic genomes.

The ease with which a person acquires a pathogen from the environment, or from another person, depends largely on the state of their immune system. Those people who harbor low pathogenic loads and still have an active innate immune system could be expected to kill the acute and chronic pathogens they encounter. Conversely, those people with a compromised immune system will accumulate pathogens over time. We have previously discussed how VDR dysfunction, along with adrenal and androgen dysfunction, predispose to a weakened innate immune system, but there are many other factors in play. For example, Bukholm and team found that when the measles virus infects cell cultures, those cells are more susceptible to a secondary bacterial invasion (Bukholm et al., 1986).

Stress has also been shown to impede immune function, by inhibiting natural killer cell activity, lymphocyte populations, lymphocyte proliferation, antibody production, and reactivation of latent viral infections (Webster Marketon and Glaser, 2008). Already identified consequences on health include delayed wound healing, impaired responses to vaccination, and development and progression of cancer (Boscarino, 2004). Depending on the variety of stressful events that occur over a lifetime, people may be more susceptible to picking up microbes at different times. The immune response could be expected to be particularly weak after traumatic events such as surgery, a car accident, or even a pregnancy (McLean et al., 2005).

People accumulate microbiota-altering pathogens in myriad different ways, the most obvious being social contact. People in close proximity, particularly spouses and children, inevitably pick up components of each other's microbiomes (Wilhoite et al., 1993). Healthcare workers have higher rates of certain autoimmune and inflammatory conditions including breast cancer and malignant melanoma (Lie et al., 2007). Merely shaking hands causes the transfer of numerous microbes (Fierer et al., 2008). Genomic analysis of the bacteria on the hands of students leaving an exam room contained 332,000 genetically distinct bacteria belonging to 4,742 different species. Forty-five percent of the species detected were considered rare. This marked a 100-fold increase in the number of bacterial species detected over previous studies that had relied on purely culture-based methods to characterize the human hand microbiota.

Obesity is not currently accepted as an autoimmune condition, but Christakis and Fowler recently used quantitative analysis of a densely interconnected social network to conclude that obesity is transmitted among people (Christakis and Fowler, 2007). A person's risk of becoming obese increases by 57% if they have a friend who becomes obese, and by 37% if their spouse becomes obese. Although, as the team concludes, people may mimic the behavior of friends or family in ways that could cause them to gain or lose weight, it is also possible that the close proximity among many of the subjects in the study would have allowed them to directly exchange microbes. Since the composition of bacteria in the gut has, in several instances, been linked to the development of obesity (Kinross et al., 2008) – perhaps, in some cases, obesity is literally contagious. It seems likely the same could be said for any autoimmune condition with an infectious etiology.

In some cases, pathogens may be acquired in the womb, particularly if the mother already suffers from one or more autoimmune or inflammatory diagnoses.

Similarly, bacterial species including *Staphylococcus epidermidis*, *Streptococcus viridans*, *E. coli*, *Staphylococcus aureus*, *Streptococcus faecalis*, *Proteus*, and others have been detected in sperm (Merino et al., 1995). *Mycobacterium tuberculosis* and influenza HSN1 have been shown to cross the placental barrier. Already implicated in implantation failure, spontaneous abortion, and preterm birth, infection with *Shigella* is now proposed to cause endometriosis (Kodati et al., 2008). DiGiulio studied ribosomal DNA (rDNA) of bacteria, fungi, and archaea from amniotic fluid of 166 women in preterm labor with intact membranes. Fifteen percent of subjects harbored microbes that together belonged to 18 different taxa – including *Sneathia sanguinegens*, *Leptotrichia amnionii,* and an unassigned, uncultivated, and previously uncharacterized bacterium. A positive PCR was associated with histologic chorioamnionitis and funisitis. The correlation between positive PCR and preterm delivery was 100%.

Pathogens can also pass from mother to child during breast-feeding. For example, Human papillomavirus type 16 (also called high-risk HPV-16), which has been linked to cervical cancer, has been detected in human breast milk collected during the early period after a woman delivers her baby (Sarkola et al., 2008). Pathogens can also be transmitted from person to person through bodily fluids released during coughing, sneezing, and other intimate contact and are found nearly everywhere in our environment. For example, nontuberculosis *Mycobacteria* and other opportunistic human pathogens are enriched to high levels in many showerhead biofilms, >100-fold above background water contents. Catheters used to treat urinary tract infections and other conditions have, in some cases, been shown to harbor copious amounts of biofilm.

Early Infections Predispose a Person to Later Chronic Disease

Most of the bacteria implicated in autoimmune disease are slow-growing pathogens whose effects will take decades to manifest (Davenport et al., 2009). In this sense, bacteria acquired earlier in life can alter the ultimate microbiota in ways that may not be recognized for decades. According to Merkler et al., "In genetically susceptible individuals, early childhood infections seem to predispose them to [such disease as] multiple sclerosis or Type 1 diabetes years or even decades before clinical onset" (Merkler et al., 2006). A 2006 report by the Centers for Disease Control (CDC) echoes this sentiment: "A person's age at the time of infection – from intrauterine or perinatal (the time period surrounding birth), through childhood and adolescence, to adulthood and the elder years – may further influence the risk for chronic outcome. For example, perinatal herpes virus infection dramatically increases the risk of developing adult or pediatric chronic liver disease. Recurrent infections or perhaps serial infections with certain agents might also determine a person's risk for chronic outcome" (O'Connor et al., 2006).

Thus, while medicine generally assumes that once a person has recovered from an acute illness, they return to a state of complete health – the so-called "sterilizing immunity" – in truth, the long-term consequences of acute infection are somewhat poorly understood. Newborns who harbor certain types of bacteria in their throats, including *Streptococcus pneumoniae* and *Haemophilus influenzae*, are at increased risk for developing recurrent wheeze or asthma early in life (Bisgaard et al., 2007). Approximately two-thirds of patients with Guillain–Barré syndrome, a suspected autoimmune condition, have a history of an antecedent respiratory tract or gastrointestinal infection (Kuroki et al., 1993). Prenatal infections such as rubella, influenza, and toxoplasmosis are all associated with higher incidence of schizophrenia – with the children of those mothers exposed to influenza in the first trimester of gestation showing a seven-fold increased risk of schizophrenia (Brown, 2006). Reactive arthritis (Reiter's syndrome) is classically seen following infection with enteric pathogens such as *Yersinia*, *Salmonella*, *Campylobacter*, and *Shigella* (Hill Gaston and Lillicrap, 2003). Acute gastroenteritis, resulting from infection with the same pathogens, causes approximately 6–17% of patients to develop chronic irritable bowel syndrome.

In an especially provocative experiment, a team including Doron Merkler and Nobel Laureate Rolf Zinkernagel injected cytomegalovirus (CMV) into the brains of mice that were only a few days old (Merkler et al., 2006). The innate immune systems of the mice were able to eliminate CMV from most of the tissues except for those of the central nervous system. As a result, the virus persisted in the brains of the mice. Later in life, when the same mice were challenged by infection with a similar virus, they developed a condition resembling a type of autoimmune disease and died. The team referred to this concept as "viral *déjà vu*."

Incidents of food poisoning also point to unresolved features of acute infections. Siegler et al. noted that 10% of people who suffered from *E. coli* food poisoning later developed a relatively infrequent life-threatening complication called hemolytic uremic syndrome (HUS) where their kidneys and other organs fail (Siegler et al., 1994). According to the study, 10–20 years after patients recover, between 30 and 50% of *E. coli* survivors will have some kidney-related problem, conditions that include high blood pressure caused by scarred kidneys, slowly failing kidneys, or even end-stage kidney failure requiring dialysis.

Microbes can also be transmitted by donation of blood, bone marrow transplants, or organ donation, which, if pathogenic, can greatly disrupt the composition of the microbiota over time. The term "donor-acquired sarcoidosis" refers to the development of sarcoidosis in presumably naïve (nonsarcoidosis) transplant recipients who have received tissues or organs from donors who were not known or suspected to have active sarcoidosis (Padilla et al., 2002). Murphy studied over 8,500 people in the United Kingdom who underwent heart surgery between 1996 and 2003 (Murphy et al., 2007). Patients who had received red blood cell transfusions were about three times more likely to suffer a heart attack or stroke and were at a higher risk for infection, readmission to hospital, and death compared with heart patients who did not receive blood. The risks associated with blood transfusions were not influenced by a patient's age, hemoglobin levels, or the extent of their disability at the time of transfusion. Writing in the journal *Circulation*, Murphy et al. concluded: "Red blood cell transfusion appears to be harmful for almost all cardiac surgery patients and wastes a scarce commodity and other health service resources" (Murphy et al., 2007).

Comorbidity

Thus the catastrophic failure of the human metabolism we see in autoimmune disease – which at first glance appears so diverse and so different among different diagnoses – appears to be due to a single underlying mechanism: a ubiquitous microbiota, much of which has evolved to persist in the cytoplasm of nucleated cells. What differ among individuals as they gradually acquire a unique microbiota over time are the virulence, location, and combination of those pathogenic species. The high rate of comorbidity among inflammatory diagnoses (Anderson and Horvath, 2004) lends support for this explanation. Such comorbidity between seemingly unrelated diseases cannot be explained by laws of average – the risk of autoimmune disease is not evenly distributed. Figure 12.5 demonstrates the degree of comorbidity seen among various inflammatory diagnoses. Each "spoke" represents a study

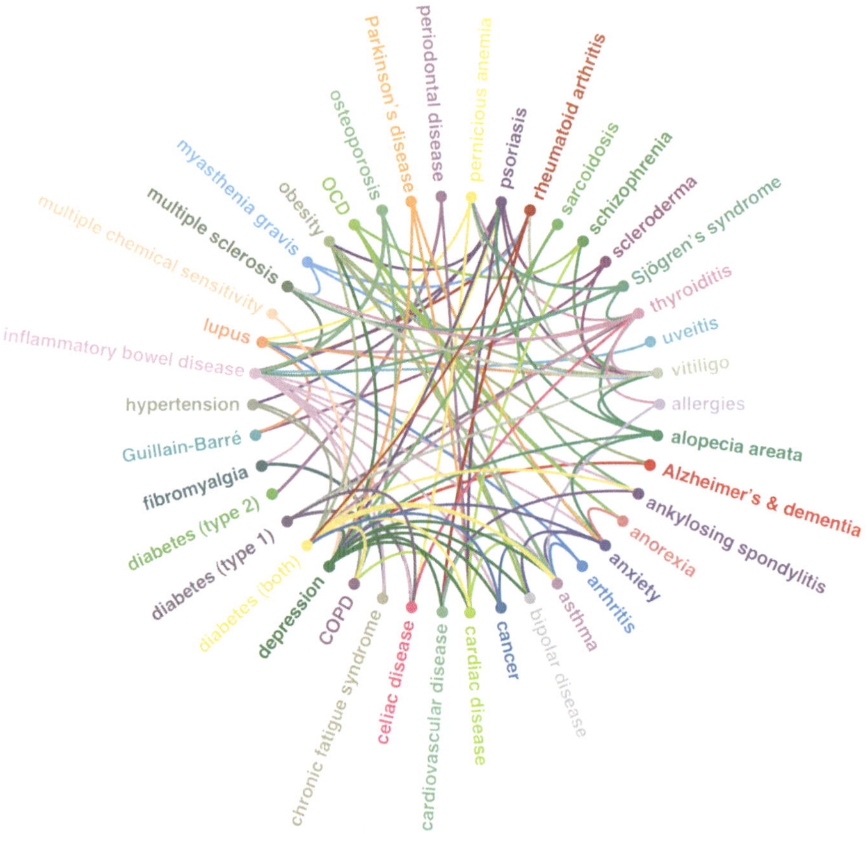

Fig. 12.5 Comorbidities among common inflammatory diseases. Each "spoke" of this wheel represents a published study appearing in MEDLINE, which shows a significant statistical relationship between one disease and another

from PubMed, which has demonstrated a significant statistical relationship between patients suffering from one inflammatory disease and the next.

In the case of multiple sclerosis, Barcellos et al. identified coexisting autoimmune phenotypes in patients with multiple sclerosis from families with several members with the disease and in their first-degree relatives (Barcellos et al., 2006). A total of 176 families (386 individuals and 1,107 first-degree relatives) were examined for a history of other autoimmune disorders. Forty-six (26%) index cases reported at least one coexisting autoimmune disorder. The most common were Hashimoto's thyroiditis (10%), psoriasis (6%), inflammatory bowel disease (3%), and rheumatoid arthritis (2%). One hundred and twelve (64%) families with a history of multiple sclerosis reported autoimmune disorders (excluding multiple sclerosis) in one or more first-degree relatives. Hashimoto's thyroiditis, psoriasis, and inflammatory bowel disease were also the most common diagnoses occurring in these family members. Such high rates of comorbidity support a model for autoimmune conditions in which no two people with the same diagnosis ever develop the exact same disease presentation; the interactions between an individual's genome and their unique metagenome are so varied that they are rarely identical.

Note that Fig. 12.1 suggests that patients with autoimmune diagnoses are also much more likely to suffer from mental conditions such as depression and anxiety. Increasing clinical evidence, including that from our own study (Perez, 2008), confirms the involvement of microbiota in neurological disease states. This suggests that both autoimmune and neurological diagnoses, which are currently balkanized into separate medical specialties, most probably result from the same underlying dysregulation of microbial populations.

Autoimmune and inflammatory conditions also suffer from specialty delineation. For example, VDR dysregulation does not just impact the autoimmune disease state. Researchers have reported epigenetic repression of VDR gene expression and activity in choriocarcinoma cell lines (Pospechova et al., 2009). Furthermore, the VDR expresses genes involved in both autoimmune and inflammatory processes. It transcribes insulin-like growth factor (IGFBP-3) (Wang et al., 2005), which influences the development of diabetes, yet also expresses metastasis suppressor protein 1 (MTSS1), which plays a vital role in repressing the cell cycle and promoting apoptosis in cancerous cells (Wang et al., 2005). Drawing a line between autoimmune and inflammatory disease makes these and other common mechanisms harder to recognize and study.

Causation vs. Association

If most autoimmune and inflammatory conditions do indeed arise from the same underlying disease process, then we must re-examine some of the cause and effect relationships postulated to exist among inflammatory conditions. For example, it is commonly believed that obesity is a causative factor in the development of diabetes (Hibbert-Jones et al., 2004). In fact, patients with Type 2 diabetes are so likely to become morbidly obese that the two conditions are sometimes collectively referred to as "diabesity" (Bailey, 2009). Obesity has been tied to microbial composition in the gut (Turnbaugh et al., 2006), the result of a microbial process. Roesch et al. found that the onset of Type 1 diabetes was tied to the presence of specific bacteria in the murine gut (Roesch et al., 2009). Additionally, at least one microbial species, *Streptomyces achromogenes*, secretes a substance, streptozocin, which can directly induce Type 1 diabetes (Bolzan and Bianchi, 2002). The diabetes disease process would also make it substantially harder for the immune system to regulate microbial gut composition. In particular, species that are extremely effective at extracting calories from food may thrive while their innocuous counterparts may find themselves out-competed. The expression of hormones that regulate appetite, such as leptin or ghrelin, could also become dysregulated by the bacterial microbiota (Fetissov et al., 2008). For example, *H. pylori* infection leads to a decrease in circulating ghrelin through a reduction in ghrelin-producing cells in the gastric mucosa (Weigt and Malfertheiner, 2009). In some cases, this could cause weight gain even in the absence of excess calorie consumption (English et al., 2002). In light of the above, obesity and diabetes might better be described as developing simultaneously. Treatments aimed at addressing those underlying factors contributing to both disease states might well prove the most effective.

The same dichotomy is found in other sets of parallel conditions such as tooth decay and dementia, rheumatoid arthritis and uveitis, high cholesterol and heart disease, and others. It is far more likely that both conditions arise from a common metagenomic microbiota than that one condition is causal for the other.

Microbial Interaction and Disease

One of the more striking characteristics of nonobese diabetic (NOD) mice is that exposure to *Mycobacteria* can prevent the onset of diabetes while precipitating lupus in the same animal (Harada et al., 1990; Hawke et al., 2003). While this phenomenon is difficult to interpret by studying the murine genome alone, it may help to consider the murine metagenome. If, as in humans, the murine metagenome causes disease as it accumulates over time, then the interactions between various microbial species may be telling. Even within the context of the ultimate example of symbiotic behavior, the biofilm, bacteria have been shown to compete with one another, sometimes even "cheating" to do so (Dunny et al., 2008). We would not have many antibiotics if it were not for competition among bacterial species. For example, the early tetracycline antibiotics were derived from species of *Streptomyces*, and are toxic to a number of its competitors.

With the NOD mice, introduction of a new species of bacteria into the microbiota, *Mycobacteria*, alters the microbiota in such a way as to wipe out, or at least diminish, the diabetes disease state. At the same time, the microbiota allows lupus to proliferate or dominate. Similar competition between microbes may also explain why lupus has been shown to inhibit the development of malaria (*Plasmodium falciparum*) (Zanini et al., 2009).

Autism, an inflammatory condition that has been associated with several unique microbial populations (Nicolson et al., 2007) may have a comparable dynamic at work. In children diagnosed with autism spectrum disorder, fever associated with intercurrent bacterial or viral infections – such as upper respiratory infections – has been shown to temporarily decrease aberrant behavior such as irritability and inappropriate speech (Curran et al., 2007).

Gastric surgery invariably alters the composition of the gastrointestinal microbiota. DePaula et al. found that after 39 diabetic type 2 patients in Brazil underwent bariatric surgery all subjects no longer required insulin therapy (DePaula et al., 2008). All subjects also experienced normalization of their cholesterol levels, 95.8% had their hypertension controlled, and 71% achieved targeted triglyceride levels. This correlates with data showing that the intestinal bacterial populations of normal weight individuals, morbidly obese individuals, and people who have undergone gastric bypass surgery are distinctly different. For example, *Firmicutes* were dominant in normal-weight and obese individuals but significantly decreased in post-gastric-bypass individuals, who had a proportional increase of gammaproteobacteria (Zhang et al., 2009).

Other microbial interactions can alter the pathogenicity of one or more species involved. The pathogenic potential of *Helicobacter hepaticus* in a mammalian colitis model is altered by the presence of different strains of *Bacteroides fragilis*. When the bacterial polysaccharide PSA is expressed on the microbial cell surface of *B. fragilis*, it suppresses pro-inflammatory interleukin-17 production to *H. hepaticus* (Mazmanian et al., 2008). Hoffman et al. found that when the bacterial species *Pseudomonas aeruginosa* and *S. aureus* were incubated together, *P. aeruginosa* created a protein, HQNO, which protected *S. aureus* from eradication by commonly used aminoglycoside antibiotics such as tobramycin (Hoffman et al., 2006). Besides, in cases of *P. aeruginosa* and *S. aureus* co-infection in the presence of HQNO, small-colony variants of *S. aureus* are selected for, making *S. aureus* more difficult for the immune system to target. Although we are far from understanding the full nature of these microbial interactions, it is clear that a microbiota constantly evolves so that the symptoms of any given disease are seldom static.

Familial Aggregation

The common disease–common variant hypothesis suggests that chronic diseases are the product of anywhere from one to thousands of disease-causing alleles. The HapMap single nucleotide polymorphisms (SNPs) cataloging project has identified over 3.1 million SNPs, with many more expected to be found as the project continues. However, only a fraction of these SNPs confers any more than a minimal statistically increased risk for disease (Chung et al., 2010). For example, in cancer, for nearly all regions conclusively identified by genome-wide association studies (GWAS), the per allele effect sizes estimated are less than 1.3. While over 85 regions have been conclusively associated in over a dozen different cancers, no more than five regions have been associated with more than one distinct cancer type (Chung et al., 2010). According to Stephen Chanock of NIH, "Nearly every candidate SNP [associated with cancer] has failed in the long run – maybe five or six are real by rigorous standards" (personal communication).

There appear to be factors at work other than just Mendelian inheritance. The increased risk of chronic disease among nonrelations in close proximity – the so-called "case clusters" – strongly implies an infectious dynamic at work. The evidence that the autoimmune disease sarcoidosis is communicable is particularly strong. A study of 215 sarcoidosis patients found that five husband-and-wife couples both had the disease – a rate 1,000 times greater than could be expected by chance (Rossman and Kreider, 2007). The NIH ACCESS research team also noted that the risk for sarcoidosis increased nearly five-fold in parents and siblings of people with the disease. A case-controlled study of residents of the Isle of Man found that 40% of people with sarcoidosis had been in contact with a person known to have the disease, compared with 1–2% of the control subjects (Gribbin et al., 2006). Another study reported three cases of sarcoidosis among 10 firefighters who apprenticed together (Kern et al., 1993).

The literature contains many examples of unexpected familial associations among seemingly distinct disease pathologies. For example, a 2008 study of parents of children with autism found they were more likely to have been hospitalized for a mental disorder than parents of control subjects, with schizophrenia being more common among case mothers and fathers compared with respective control parents (Daniels et al., 2008). In the case of schizophrenia and autism, both have been associated with prenatal viral infection (Fatemi et al., 2008). While a fetus can acquire these and many other pathogens directly, successive infection dictates that as children age they will manifest with inflammatory symptoms that may differ from those of their parents. Major factors that would influence the development of a discrete inflammatory diagnosis include the mix of species acquired, the sequence in which the pathogens are acquired, the subsequent changes in gene expression caused by the pathogens, and the profound effect on the body's proteins, enzymes, and metabolites caused by these changes. Because the adaptive immune response in infants takes several weeks to develop, infants are particularly prone to picking up pathogens during the first weeks of life (Bisgaard et al., 2007). Such pathogens could be acquired from any family or friends in contact with the child, especially the grandparents, who probably harbor some of the highest pathogenic loads. Palmer et al. found that infants pick up many of the species that make up their gut flora from family members within just a few weeks of birth, suggesting that nongut bacteria may easily be acquired during this time as well (Palmer et al., 2007).

Is Autoimmune Disease Predisposition Mendelian?

Two decades ago, the attention of the research community shifted toward a new source in an attempt to explain the etiology of autoimmune disease: the human genome. Begun formally in 1990, the US Human Genome Project was a 13-year effort coordinated by the US Department of Energy and the NIH. Its primary goal was to determine the sequence of chemical base pairs that make up DNA and to identify the genes of the human genome from both a physical and a functional standpoint. A working draft of the genome was released in 2000 and a complete version in 2003, with further analyses yet to be completed and published (Collins et al., 2003). Meanwhile, the private company Celera Genomics conducted a parallel project (Venter et al., 2001).

Early in the aftermath of the sequencing of the human genome, many geneticists advocated the common disease–common variant hypothesis, expressing certainty that the field would quickly determine genetic haplotypes that would correlate with and explain the bulk of chronic diseases. Dr. Francis Collins' 2001 statement was typical: "It should be possible to identify disease gene associations for many common illnesses in the next – 7 years" (Collins and McKusick, 2001). Researchers hoped that by dissecting the human genome, patients could be informed that they had "the gene" for breast cancer, sarcoidosis, rheumatoid arthritis, or any of the other autoimmune diagnoses. Targeted gene therapies could then be developed to effectively eradicate these conditions.

It may be too early to call human genomic research an unqualified failure (Buchanan et al., 2006), but it is difficult to ignore a lack of utility in identification of disease. Recently, the limited progress in the genetic analysis of common diseases has begun to be acknowledged (Davey Smith et al., 2005; Risch, 2000). Certainly there have been no widely successful gene therapies to date, and genome-driven personalized medicine has yet to live up to its early promise. To identify what some researchers refer to as the "missing heritability," geneticists have proposed GWA studies with historically unprecedented sample sizes. In the past year, researchers have publicly contemplated "daunting" sample sizes exceeding 500,000 subjects in concert with studies that would be conducted over periods as long as 45 years (Burton et al., 2009).

Ewald et al. argue that evolutionary forces that would cause a serious disease to be weeded from the population would also cause those people whose immune systems are prone to self-attack to be eliminated from the population (Cochran et al., 2000). An exception would occur if the disease offers a survival advantage. For example, the genetic disorder cystic fibrosis may confer resistance to tuberculosis (Poolman and Galvani, 2007). The Mendelian disorder sickle cell anemia is common in tropical countries because it confers resistance to malaria. With malaria, researchers can quantify the rate by generation at which the gene for sickle cell anemia is dropped from the population in the absence of an evolutionary advantage – as is the case when people migrate away from malaria-infested areas. However, no autoimmune diagnosis has been shown to confer any sort of beneficial survival trait. Under these circumstances, one would expect any faulty gene or network of genes associated with an autoimmune condition to be selected against, especially since many autoimmune conditions strike during the reproductive years. Chronic diseases have existed for thousands of years with manifestations of both arteriosclerosis (Azer, 1999) and cardiac disease observed in mummies of ancient Egypt (Miller et al., 2000). Ötzi the Neolithic Iceman who lived around 3300 BC had arthritis, allowing ample time for any alleles associated with autoimmune disease to be eliminated via natural selection (Dickson et al., 2003). Instead, the prevalence of autoimmune conditions seems to have remained essentially constant until quite recently.

SNPs and Autoimmune Disease

After noting that among his cohort of 31 patients with abdominal aortic aneurysm, SNPs in the gene BAK1 were different in aortic tissue than in blood samples from the same patients (Gottlieb et al., 2009) Gottlieb remarked, "Genome-wide association studies were introduced with enormous hype several years ago, and people expected tremendous breakthroughs. Unfortunately, the reality of these studies has been very disappointing, and our [own] discovery certainly could explain at least one of the reasons why." The conundrum that Gottlieb's study has exposed is that the human genome appears to vary between the tissue and plasma compartments. Medicine has always assumed that human DNA is homogeneous throughout the human body. We now need to explore the mechanisms whereby these different genetic sequences could arise through selective pressure in different tissues such as would exist if the tissue harbored a microbiota.

One of the mechanisms proposed for genetic predisposition states that genetic haplotypes predispose for disease processes. Because it is a highly polymorphic genomic region, MHC has served as the preferred axis for studying susceptibility to immune diseases. Major changes have been detected within the HLA class I and class II genes related to various populations across the globe. For example, in Type 1 diabetes, the most common haplotype in the Western world is AH8.1 (HLA-A1-B8 DR3-SC01). However, this haplotype is almost nonexistent in the Indian population and has been supplanted by the variant AH8.1v, which differs from the Caucasian AH8.1 at several gene loci (Mehra et al., 2007). Moreover, there are additional HLA-DR3 haplotypes HLA-A26-B8-DR3, HLA-A24-B8 DR3 (AH8.3), A2-B8-DR3 (AH8.4), and A31-B8-DR3 (AH8.5) that occur largely in the Indian population alone.

Similarly, the FCRL3-169T-C polymorphism, which is significantly associated with rheumatoid arthritis (RA) in East Asian populations is not associated with RA in Caucasians of European decent (Begovich et al., 2007). Interestingly, the frequency of the rs7528684 minor allele associated with FCRL3- varies as much within each of the two ethnic groups as it does between them. Furthermore, a recent large case-controlled study found that FCRL3-169T-C was not significantly associated with RA in Korean patients (Begovich et al., 2007).

Thus, no diagnostic certainty can be obtained by measuring genes on the HLA axis. None of the HLA haplotypes causes disease 100% of the time and none causes any one immune disease consistently. Patterns of haplotype variation are more suggestive of a regional infectious model rather than a model in which an illness is caused by widespread inherited variation of HLA haplotypes.

Potential Systematic Errors in the Interpretation of the Metagenome

Primers selected for most epidemiological studies are chosen without consideration for whether they might amplify DNA from the genomes of any intracellular microbes. As artist Pablo Picasso once remarked, "Computers are useless. They can only give you answers." If a software program fails to make provision for the possibility that a metagenome might also be present, the chances of a false-positive increase significantly during the process of genomic analysis. Similarities between bacterial and human genes will likely cause the analysis software to not assemble the genomic data properly. The likelihood of error is not minuscule as there is growing evidence of molecular mimicry, homology between bacterial and human proteins. For example, significant sequence homology exists between human carbonic anhydrase II and alpha-carbonic anhydrase of *H. pylori* (Guarneri et al., 2005). Moreover, the homologous segments contain the binding motif of the HLA molecule DRB1∗0405. The group A streptococcal carbohydrate antigen *N*-acetyl-glucosamine is able to cross react with cardiac myosin (Cunningham, 2003). Microbes including *E. coli*, *H. pylori*, *P. aeruginosa*, Cytomegalovirus, and *H. influenzae* share sequence homology with human pyruvate dehydrogenase complex-E2, which has been tied to the development of primary biliary cirrhosis (Bogdanos et al., 2004). The core oligosaccharides of low-M(r) LPSs of *C. jejuni* serotypes that are associated with the development of Guillain–Barré syndrome are homologous to neural gangliosides.

Before we can be certain that all measured SNPs and HLA haplotypes are a product of only the human genome and not the metagenome, researchers must begin to actively choose PCR primer pairs that are unlikely to amplify microbial DNA. Primers need to be certified not only to amplify a unique sequence in the human genome, but also as not likely to amplify genes from any of the thousands of bacterial and viral genomes in the metagenomic databases. Although PCR amplification usually involves more than one stage of genomic selectivity, the increasing use of arrays of RNA probes increases the likelihood that a fragment of metagenomic RNA will unexpectedly match a probe, and increases the possibility of a false-positive being signaled for the particular SNP being sought.

Antibodies in Response to Microbial DNA

Autoimmune diseases are characterized largely by the presence of autoantibodies. Although autoantibodies were reported over a century ago, many scientists at the time were unwilling to accept the possibility that the immune system attacks its own cells. Ehrlich argued that autoimmunity was not possible and proposed the theory of *horror autotoxicus* to describe the body's innate aversion to immunological self-destruction by the production of autoantibodies. Now that humans are understood to be the product of multiple genomes, increasing evidence supports Ehrlich's view. When an innate immune system is forced to respond to a chronic microbiota, the resulting cascade of chemokines and cytokines will also stimulate an adaptive response. Antibodies are notoriously polyspecific, and the likelihood that antibodies generated to target metagenomic fragments will also target human proteins (target "self") is finite.

A litany of research implies a re-evaluation of the "autoantibody." Recently researchers have shown that certain autoantibodies are created in response to several well-studied pathogens. "Lupus-specific autoantibodies" such as RO, La, or dsDNA are often generated in response to EBV (Barzilai et al., 2007). Similarly, anti-EBNA-1 antibodies are able to bind lupus-specific autoantigens such as Sm or Ro (Harley and James, 2006). Casali and Slaughter found that, in humans, EBV is a polyclonal B cell activator, and in vitro transformation with EBV results in production of rheumatoid factor (RF) (Casali et al., 1987; Slaughter et al., 1978). Possnett et al. argues that high titers of RF are not only associated with severe rheumatoid arthritis but also appear in a number of other diseases including viral, bacterial, and parasitic infections (Posnett and Edinger, 1997). Maturation of RF can be initiated by chronic infections (Djavad et al., 1996). For example, patients with subacute bacterial endocarditis, which is frequently tied to the presence of *Streptococcus*, also often present with high levels of RF (Russell et al., 1992). Williams et al. showed that once the offending infectious agent is removed with antibiotic therapy, the RF disappears (Williams and Kunkel, 1962). Similarly, the autoimmune disease thrombocytopenic purpura (ITP) is mediated by what are considered to be anti-platelet autoantibodies. However, Asahi et al. found that eradication of *H. pylori* is effective in increasing platelet count in nearly half of ITP patients infected with the bacterium (Asahi et al., 2006). Barzilai and team also found that Hepatitis B shares amino acid sequences with different autoantigens, further suggesting that the so-called autoantibodies may actually be created in response to pathogens (Barzilai et al., 2007). Autoantibodies have been detected in patients without autoimmune disease during periods of infection. Berlin et al. collected sera from 88 patients with acute infections (41 bacterial, 23 viral, 17 parasitic, and 7 rikettsial (Berlin et al., 2007)). Elevated titers of autoantibodies including annexin-V, prothrombin, ASCA, ANA, or antiphospholipid antibodies were detected in approximately half of the subjects, with 34 individuals harboring elevated titers of at least two "autoantibodies."

EBV, *E. coli*, *Salmonella,* and other pathogens discussed above are easily detected by culture-based methods that may explain why their presence has already been tied to "autoantibody" production. Yet the vast majority of the human microbiota is understudied. This means that what we now consider to be autoantibodies in many autoimmune diagnoses may also indicate the presence of pathogens, but pathogens that have yet to be fully characterized and named. Thus, in addition to looking for antibodies to well-characterized pathogens, it is also important that we look for antibodies indicating the presence of the underlying chronic microbiota, some of which we may also be mistaking for autoantibodies. Like the pathogens that may create them, many of these antibodies may not yet be detected by standard testing. If this is the case, hundreds of pathogen-induced antibodies may exist and impact the autoimmune disease state, but the possible detection and correlation of such antibodies with specific components of the microbiota remains difficult until a much larger portion of the microbiota has been characterized.

Because many antibodies demonstrate a high degree of polyspecificity, it is possible that in some cases, antibodies initially directed against pathogens could also attack human tissue (Christen et al., 2010). According to Bozic, oxidative alterations, affecting either the hypervariable region or the receptor site of IgGs, may influence their functions (Bozic et al., 2007). Similarly, McIntyre reported the appearance and disappearance of antiphospholipid antibodies subsequent to oxidation reactions in human blood (McIntyre, 2004). Dimitrov et al. have shown that a fraction of antibodies present in all healthy individuals begin to recognize large number of self-antigens only after a transient exposure to certain protein-destabilizing conditions, including low or high pH, high salt concentration, chaotropic factors, and redox-active agents (Dimitrov et al., 2008). This points to at least one mechanism whereby the oxidative stress that accumulates in inflamed tissue could be at least partly responsible for the apparent polyspecificity of antibodies and autoantibodies.

Molecular mimicry, in which peptides from pathogens share sequence or structural similarities with self-antigens, may also contribute to autoantibody production. Lekakh et al. found that autoantibodies with polyspecific activity in the serum of healthy donors were able to cross react with DNA and lipopolysaccharides (LPSs) of widespread species of bacteria including *E. coli*, *P. aeruginosa*, *Shigella boydii*, and *Salmonella* (Lekakh et al., 1991). CD' is classified as an autoimmune condition based largely on the presence of perinuclear anti-nuclear cytoplasmic antibodies (pANCAs) in patients with the disease. Yet recently two major species of proteins immunoreactive to pANCA were detected in bacteria from anaerobic libraries, implicating colonic bacteria as a possible trigger for the disease-associated immune response.

We previously discussed how factors other than calorie consumption may contribute to the weight gain often associated with autoimmune or inflammatory conditions. Fetissov et al. studied healthy women for the presence of IgG or IgA autoantibodies directed against 14 key regulatory peptides and neuropeptides, including ghrelin, leptin, vasopressin, and insulin (Fetissov et al., 2008). They found numerous cases of sequence homology among these peptides and the protein structures of over 30 microbes including *Lactobacilli*, *H. pylori*, *E. coli*, *Yersinia pseudotuberculosis*, and *Listeria monocytogenes*, suggesting that the "autoantibodies" were actually the result of molecular mimicry. In the presence of certain pathogenic bacterial species, the production of IgG autoantibodies directed against ghrelin was upregulated, suggesting a complex interplay between autoantibody levels and microbial antigens. This suggested that these so-called "autoantibodies" might not only have physiologic implications in pathways that regulate hunger and satiety but also represent a key link between the gut and the brain.

An increasing number of studies also show that what are currently perceived as autoantibodies can often be detected in the so-called healthy individuals years before the full presentation of an autoimmune disease state. Many researchers now espouse that early detection of these antibodies can help predict whether or not such a "healthy" person will develop an autoimmune disease. For example, in an 8-year prospective study, Swaak et al. examined the diagnostic significance of anti-double-stranded deoxyribonucleic acid (anti-dsDNA) determination in a group of 441 patients without systemic lupus erythematosus (SLE) whose sera were found to contain antibodies to dsDNA on routine screening (Swaak and Smeenk, 1985). Within 1 year, 69% (304) of these patients fulfilled the preliminary American Rheumatism Association (ARA) criteria for SLE. Eighty-two of the remaining 137 patients were followed up for several years. At the end of the study, 52% of these patients had also developed SLE. The team concluded that about 85% of patients without SLE with anti-dsDNA in the circulation would develop SLE within a few years.

Another recent study of blood from 441 healthy Portuguese blood donors found autoantibodies for rheumatoid factor, anticyclic citrunillated peptides, anti-mitochondria, anti-*Sacharomyces cerevisiae*, ANA, anti-TTG, and anti-Beta2-glycoprotein (Tavares-Ratado et al., 2009). More than 30% of the blood contained one or more of the antibodies, 4% exhibited two antibodies, and nearly 1% had three or more antibodies present. It is clear that sub-clinical autoimmune disease is much more common than previously thought.

This gradual presentation of an increasing number of the so-called "autoantibodies" in the years before a patient meets the official criteria for an autoimmune diagnosis supports the model of successive infection described earlier – pathogenic components of the microbiota gradually accumulate over the course of a lifetime until bacterial, viral, and phage load reaches a level at which a diagnosis can be made. It also supports the contention that individuals perceived as "healthy" may still harbor and accumulate pathogenic microbes that will eventually lead to an inflammatory diagnosis, or a process associated with "aging." Indeed, it is possible that any antibodies that damage "self" do so as an unintended polyspecific consequence of their activity against the metagenomic pathogens.

Therapies in the Era of the Metagenome

At the 2008 International Conference on Metagenomics in La Jolla, CA, James Kinross of the Imperial College of London began his speech with the following statement: "We surgeons have been operating on the gut for literally thousands of years and the microbiota has just been this extraordinary elephant in the room. We seem to have completely ignored the fact that we've co-evolved with thousands of bacteria over millions of years and that they somehow may be important to our health. As doctors, we routinely do terrible things to the microbiota and I'm sure this has implications for our health."

Although most physicians are undoubtedly well intentioned, Kinross is correct in that many clinicians are generally not offered training that would keep them up to date with advances in metagenomics. The result is that many doctors still believe that nonmucosal surfaces of the body are largely sterile and that bacteria and other pathogens are not driving factors in the autoimmune processes. Instead, the standard of care for patients with autoimmune disease continues to be corticosteroids and TNF-alpha blocking medications. According to a 2008 report, TNF-alpha inhibitors accounted for 80% of RA drug sales in the United States, France, Germany, Italy, Spain, the United Kingdom, and Japan. Use of these immunosuppressants is still grounded in the theory that autoimmune disease results from an overly exuberant immune response and these drugs are administered without consideration for the presence of a metagenome. Whether helpful or harmful, there is no question that by dramatically slowing the immune response, such therapies must necessarily and profoundly affect the composition, development, and stability of the human microbiota.

Despite the copious use of these immunosuppressant drugs in autoimmune conditions, they provide, at best, short-term palliation. Gottlieb et al. showed that steroid use causes relapse in sarcoidosis (Gottlieb et al., 1997). Additionally, there are no definitive studies showing corticosteroids improve long-term prognosis in the treatment of chronic inflammatory illness, nor is there any demonstrated reduction in mortality. Van den Bosch and Grutters write, "Remarkably, despite over 50 years of use, there is no proof of long-term (survival) benefit from corticosteroid treatment" (Grutters and van den Bosch, 2006). On the contrary, one of the side effects of TNF-alpha inhibitors is an increased risk of tuberculosis. Several studies have shown that TNF-alpha production is required for the proper expression of acquired specific resistance following infection with *M. tuberculosis* (Allie et al., 2008; Arend et al., 2003). So if we inhibit TNF-alpha expression, we would expect a long-term increase in the prevalence of not only tuberculosis, but also in any of the autoimmune or inflammatory diseases already associated with chronic forms of mycobacteria and other bacteria (Bull et al., 2003; Burnham et al., 1978).

The failure of these first-line therapies to cure "autoimmunity," and the range of detrimental side effects associated with their use, suggests that slowing the immune response of patients with autoimmune disease is counterproductive, allowing microbial populations to develop unchecked. Now that autoimmune conditions are more widely understood as illnesses in which myriad pathogens may trigger or drive the disease process, efforts to target the root cause of autoimmune disease should instead be targeted toward activating the innate immune response, not suppressing it.

Our own work (Perez, 2008) offers an example of the results of stimulating rather than suppressing the innate immune response of patients with autoimmune disease. Over the past 7 years, we have observed the effects of an experimental therapy for autoimmune disease that uses the VDR agonist olmesartan to reverse pathogen-induced VDR dysregulation. Subjects are also administered sub-inhibitory bacteriostatic antibiotics, which weaken bacterial ribosomes so that pathogens can more easily be targeted by the reactivated immune system. Nearly all of the hundreds of patients to start the therapy reported the predicted increase in specific symptoms of their autoimmune diagnosis. After months, or sometimes years, of dealing with these symptomatic flares, the very symptoms that waxed and waned in synchronism with antibiotic administration began to disappear, resulting in improvement and, in many cases, eventual resolution of the disease process. This response has been noted in the widely varying diagnoses of sarcoidosis, rheumatoid arthritis, lupus, Type 2 diabetes, uveitis, Hashimoto's thyroiditis, ankylosing spondylitis, chronic fatigue syndrome, and fibromyalgia among others. The often dramatic elevations in disease activity observed among study subjects – particularly during the early stages of therapy – cannot be attributed to side effects of the protocol medications, as individually the drugs are well known and unremarkable (Schwocho and Masonson, 2001). Additionally, when healthy individuals have been administered the same medications they do not suffer any similar symptoms.

The most viable hypothesis for these temporary surges in disease symptoms and inflammatory markers is that treatment medications allow the immune system to mount an effective attack on an intracellular microbiota, such as the microbiota observed by Wirostko et al. It is reasonable to expect that when intraphagocytic pathogens are killed, some of the host cells will also undergo apoptosis, phagocytosis, or simply disintegration, leading to an increase in inflammation. For over 100 years, researchers have noted that the death of acute and persistent pathogens is accompanied by a surge in inflammation. They have attributed the temporary rise in inflammation to an increase in endotoxin and cytokine release upon bacterial death. Known as the Jarisch–Herxheimer reaction, or immunopathology, this phenomenon has been previously demonstrated after antibiotic administration in diseases including tuberculosis (Cheung and Chee, 2009), borreliosis (Vidal et al., 1998), tick-borne relapsing fever (Mitiku and Mengistu, 2002), multiple sclerosis (Kissler, 2001), Whipple disease (Peschard et al., 2001), and syphilitic alopecia (Pareek, 1977), among others. Zinkernagel also observed immunopathology in the mice he had infected with a persistent neuro-active virus (Zinkernagel et al., 2009). Similarly, immune reconstitution inflammatory syndrome (IRIS) is a condition seen in some cases of AIDS following the use of antiretroviral drugs. As the immune system begins to recover, it responds to previously acquired opportunistic infections with an overwhelming inflammatory response that, like the immunopathological reaction we observe, makes the symptoms of the infection temporarily worse (Shelburne et al., 2002). At this point in time, the exact species or forms of bacteria potentially killed by any one subject in our own study cohort remain unknown. As the focus of the HMP moves beyond the mucosal surfaces, and catalogs L-forms and other intracellular species within body tissues, a clearer picture of disease pathogenesis will emerge. However, as long as patients continue to report improvement and recovery, determining the exact nature of pathogens being targeted by the therapy has not been a high priority, given the limited resources currently allocated to this research team.

Some subjects in the cohort have reported drops in viral titers, suggesting that once the immune system is no longer burdened by the pathogenic components of the bacterial microbiota, it may regain the ability to target chronic viruses as well. This suggests that treatments that reverse immunosuppression caused by the bacterial microbiota might also prove useful in mitigating viral virulence.

Our research suggests that while some people report being "allergic" to certain bacteriostatic antibiotics, what they perceive as an "allergy" may actually be immunopathological reactions. For example, there are reports of minocycline "inducing lupus" (Geddes, 2007). A more logical explanation may be that certain patients harbor persistent bacterial species that predispose for sub-clinical lupus. When minocycline is administered, some of these bacteria are killed, resulting in immunopathological reactions that are mistakenly interpreted as clinical manifestation of the disease.

What we have initiated needs further testing. However, the reports of profound immunopathological reactions in autoimmune subjects imply the need to re-examine whether palliative drugs actually provide long-term benefit for patients with autoimmune disease. Whether at the doctor's office or the health food store, patients with autoimmune conditions continually seek out palliative drugs or supplements that successfully reduce symptoms by lowering inflammation. Yet, if bacteria drive the pathogenesis of autoimmune inflammation, and chronic bacterial death invariably results in temporary increases in discomfort, then treatments that mitigate symptoms may well do so at the expense of proliferation in pathogenic components of the microbiota. Commonly used immunosuppressive compounds include vitamin D, which, although its immunosuppressive properties have now been identified (Arnson et al., 2007), is now viewed as the ultimate inexpensive wonder drug (Holick, 2008). Frequent use of vitamin D, as well as other substances that slow immune activity, could at least partially account for the recently increased prevalence of nearly every autoimmune disease (Luque et al., 2006).

L-Form Bacteria: An Often Overlooked Component of the Microbiota

Certain stages of the bacterial life cycle result in the loss of the cell wall. L-form bacteria are often less than 0.2 μm in diameter (Domingue and Woody, 1997) and are therefore difficult to view with a standard optical microscope. Not only do these L-form variants fail to succumb to antibiotics that target the bacterial cell wall, but those antibiotics also encourage the formation of L-forms. "Treatment with penicillin does not merely select for L-forms (which are penicillin-resistant) but actually induces L-form growth," states Josep Casadesus of the University of Sevilla (Casadesus, 2007). In fact, researchers deliberately culture classical forms of bacteria in conjunction with various beta-lactam antibiotics in order to create L-forms (Mattman, 2000). The ability of the L-form to flourish in the face of treatment with the beta-lactam antibiotics points to a mechanism by which acute bacterial forms can mutate into latent mutants that may cause disease at a later time. Some researchers have deemed the conversion into the L-form state to be a universal property of bacteria (Gumpert and Taubeneck, 1983).

Joseleau-Petit et al. showed that classical forms of bacteria transform into the L-form only if they are denied the ability to form a normal cell wall (Joseleau-Petit et al., 2007). The beta-lactam antibiotics work toward this end by blocking the creation of penicillin-binding proteins (PBPs) – proteins responsible for forming the cross-linked chains associated with a peptidoglycan-derived cell wall. When the ability of the PBPs to create a full cell wall is blocked, the cells also become spherical and osmosensitive. Recently, Glover et al. performed the first systematic genetic evaluation of genes and pathways involved in the formation and survival of unstable L-form bacteria (Glover et al., 2009). Microarray analysis of L-form versus classical bacterial colonies revealed many upregulated genes of unknown function as well as multiple overexpressed stress pathways shared in common with persister cells and biofilms. Dell'Era et al. also observed cell division and changes in gene expression in stable *L. monocytogenes* L-forms (Dell'Era et al., 2009).

Since the discovery of the L-forms in 1935 (Kleineberger-Nobel, 1951), they have been described in hundreds of publications. Yet because researchers are only just beginning to use molecular tools to study the L-form, they are still seldom factored into the mix of microbes that compose the human microbiome. However, over the years, L-forms have been implicated in dozens of diseases of unknown etiology, including RA, multiple sclerosis, sarcoidosis, glomerulonephritis, idiopathic hematuria, interstitial cystitis, rheumatic fever, and syphilis – as well as a large number of chronic and relapsing infections (Domingue and Woody, 1997; Mattman, 2000).

A Research Consideration: Men Are Not Tall Mice Without Tails

The emerging role of the human microbiota implies a reconsideration of certain long-standing and frequently invoked models of disease. According to Javier Mestas of University of California, Irvine, "There has been a tendency to ignore differences and in many cases, perhaps, make the assumption that what is true in mice is necessarily true in humans. By making such assumptions we run the risk of overlooking aspects of human immunology that do not occur, or cannot be modeled, in mice" (Mestas and Hughes, 2004). Murine models are still used in an effort to understand most autoimmune and inflammatory conditions, despite the obvious differences between the murine and human immune systems.

For example, there are major differences in the Toll-like receptors. TLR1-9 exists in both mouse and man, although TLR8 detects single-stranded RNA in man and has no known function in the mouse. TLR10 exists in humans only; it is a degenerative pseudo-gene in the mouse. TLR11, 12 and 13 in mice do not exist in man and their function is not yet well defined.

Analysis of the human and murine VDR offers other examples of discord between man and mouse. Marshall's molecular dynamics emulation showed that the drug olmesartan, a putative VDR agonist, binds into a different conformation in the murine VDR to that of *Homo sapiens* (Marshall, 2008), calling into question the whole concept of drug safety testing in murine models.

While the human VDR transcribes dozens of genes necessary for a robust innate immune response, including many key AMPs, the VDR does not similarly control the murine innate immune system.

The murine innate immune response is dependent on a cascade of nitric oxide functions in a manner yet to be fully understood (Bogdan, 2001). Although mice have VDRs, the homology differs, and they express different genes than the human

VDR. For example, the gene encoding the calcium-binding protein osteocalcin is "robustly" transcribed by the VDR in humans, but not in mice.

Brahmachary et al. showed that the rat VDR does not express the cathelicidin AMPs, marking an important difference in the way the two species target invading pathogens (Brahmachary et al., 2006). Gombart et al. recently expanded on the finding by providing evidence of an evolutionarily fixed, Alu-mediated divergence in steroid hormone nuclear receptor gene regulation between humans/primates and other mammals (Gombart et al., 2009). This divergence, which placed the cathelicidin pathway under VDR control only in humans and closely related primates, remained under purifying selection for the past 55–60 million years, and yet even cathelicidin in primates is not identical to that in man. Eventually, the pathway evolved to become a key component of a novel innate immune response unique to human infection. Because the murine VDR does not express cathelicidin, there is less of an evolutionary incentive for components of the murine microbiota to dysregulate its expression. This suggests that the survival mechanisms employed by the human and murine microbiotas may be very different. Thus, the intermingling of murine and human biologies in the literature hinders our ability to fully understand nuclear receptor control of the AMPs and other key aspects of innate immunity.

Discussion

The prevailing theory of autoimmune disease, which dictates that the body creates autoantibodies that attack its own cells, was developed during an era when culture-based methods vastly underestimated the number of microbes capable of persisting in and on *Homo sapiens*. The advent of culture-independent tools such as 16S RNA sequencing, single cell sampling, and pyrosequencing has opened the door to an era of discovery. Rather than a sterile compartment, the human body is now known to teem with thousands of species of bacteria, viruses, and phages. In addition to persisting on the body's external surfaces, these microbes survive in the blood and in many of the tissues, which become inflamed during autoimmune disease, suggesting that what were once thought to be "autoimmune" processes may instead result from the presence of persistent microbes. Metagenomics is allowing us to study these microbes in the tissues within which they naturally persist, where they can be examined in the context of other microbes in their community. A more exact understanding of how networks of microbes can interact to cause disease has superseded Koch's postulates, which stipulate that a single microbe causes a single disease.

While diseases were once categorized largely on the basis of symptom presentation, they can now be classified based on their underlying genetics. Yet the expression of key human genes is continually altered by a plethora of microbial metabolites through an almost imponderable number of interactions. These metabolites, some of which are created by bacteria considered to be "friendly" or innocuous, can directly drive the pathogenesis of autoimmune disease by altering the expression of genes such as ACE and PTN22, genes associated with diagnoses including rheumatoid arthritis, lupus, diabetes mellitus, myocardial infarction, renal tubular dysgenesis, and Alzheimer's. It is becoming apparent that autoimmune processes cannot be fully understood if the human genome is studied in isolation. An understanding of the interactions between the human genome and the metagenome calls for a more nuanced understanding of the microbiota. Classifying certain microbes as purely commensal may underrepresent the full spectrum of their actions. Indeed, harmless species of bacteria and viruses can easily acquire virulent plasmids via horizontal gene transfer or homologous recombination.

The microbiota has persisted in and on the human body for millennia. It has evolved to slow the host immune response in order to ensure microbial survival. Pathogens such as *M. tuberculosis*, *Borrelia*, EBV, and HIV have evolved to dysregulate the VDR nuclear receptor, inhibiting expression of the beta-defensin and cathelicidin AMPs along with TLR2. Flow-on effects from VDR dysregulation can further alter AMP expression via (at least) the alpha-thyroid, androgen, and glucocorticoid nuclear receptors. This may result in the immunosuppression and hormonal imbalances characteristic of many autoimmune diagnoses.

The bacteria that cause autoimmune disease likely accumulate over a lifetime, with individuals picking up pathogens with greater ease over time, as the immune response becomes increasingly constrained. Successive infection dictates that even people with the same autoimmune diagnosis are unlikely to present with identical clusters of symptoms and helps explain the high levels of comorbidity observed among these patients. Common autoimmune comorbidities include inflammatory conditions such as cardiovascular disease, along with mental diagnoses such as depression or anxiety, suggesting these conditions may also be driven by the microbiota. Thus, insights gained from studying microbial composition in autoimmune disease can accelerate research in other areas of medicine. Recently, several studies have shown the presence of "autoantibodies" in autism with antinuclear antibody seropositivity showing a significant positive association with disease severity, mental retardation, and electroencephalogram abnormalities. Rather than assign autism to the end of a growing list of autoimmune diagnoses, this knowledge might be better used as a basis on which to further explore the role that components of the microbiota may play in driving the pathogenesis of disease.

Analyzing autoimmune disease through the lens of metagenomics calls for a re-evaluation of the autoantibody. Polyspecific autoantibodies are increasingly being associated with elements of the microbiota, making it likely that the term "autoimmune" will soon lose its diagnostic utility. When a disabled immune system is forced to respond to the presence of a chronic microbiota, the resulting cascade of cytokines and chemokines will stimulate an adaptive immune response. The adaptive immune system will then proceed to generate antibodies to fragments of DNA generated by apoptosis or phagocytosis of infected cells. This is supported by studies showing that the so-called autoantibodies such as RO, La, dsDNA, and RF can be created in response to various bacterial and viral pathogens. Autoantibodies are often observed before a patient becomes fully symptomatic with an autoimmune diagnosis, reflecting the gradual accumulation of persistent microbes.

Rather than focusing on phenotypes and subsets of the metagenome, microbiome research may instead benefit from broader approaches geared toward understanding shared mechanisms of persistence. Translational medicine should aim at cutting through barriers among specialities, even between biologists and clinicians, so that more of the pieces of the emerging jigsaw of disease etiology can drop into place, and autoimmune disease patients can fully benefit from the insights gained from metagenomic science.

Acknowledgments The authors wish to acknowledge the assistance of Dr. Elena Kashuba for sharing her data and helping us prepare Fig. 12.2.

References

Abreu MT, Kantorovich V, Vasiliauskas EA et al (2004) Measurement of vitamin D levels in inflammatory bowel disease patients reveals a subset of Crohn's disease patients with elevated 1,25-dihydroxyvitamin D and low bone mineral density. Gut 53(8):1129–1136

Allie N, Alexopoulou L, Quesniaux VJF et al (2008) Protective role of membrane tumour necrosis factor in the host's resistance to mycobacterial infection. Immunology 125(4):522–534

Almenoff PL, Johnson A, Lesser M et al (1996) Growth of acid fast L-forms from the blood of patients with sarcoidosis. Thorax 51(5):530–533

Anderson G, Horvath J (2004) The growing burden of chronic disease in America. Public Health Rep 119(3):263–270

Arend SM, Breedveld FC, van Dissel JT (2003) TNF-alpha blockade and tuberculosis: better look before you leap. Neth J Med 61(4):111–119

Arnson Y, Amital H, Shoenfeld Y (2007) Vitamin D and autoimmunity: new aetiological and therapeutic considerations. Ann Rheum Dis 66(9):1137–1142

Asahi A, Kuwana M, Suzuki H et al (2006) Effects of a *Helicobacter pylori* eradication regimen on anti-platelet autoantibody response in infected and uninfected patients with idiopathic thrombocytopenic purpura. Haematologica 91(10):1436–1437

Avila M, Said N, Ojcius DM (2008) The book reopened on infectious diseases. Microbes Infect 10(9):942–947

Azer SA (1999) Arterial disease in antiquity. Med J Aust 171(5):280

Bailey CJ (2009) New therapies for diabesity. Curr Diab Rep 9(5):360–367

Baldwin CL, Goenka R (2006) Host immune responses to the intracellular bacteria *Brucella*: does the bacteria instruct the host to facilitate chronic infection? Crit Rev Immunol 26(5):407–442

Barcellos LF, Kamdar BB, Ramsay PP et al (2006) Clustering of autoimmune diseases in families with a high-risk for multiple sclerosis: a descriptive study. Lancet Neurol 5(11):924–931

Barzilai O, Ram M, Shoenfeld Y (2007) Viral infection can induce the production of autoantibodies. Curr Opin Rheumatol 19(6):636–643

Begovich AB, Chang M, Schrodi SJ (2007) Meta-analysis evidence of a differential risk of the FCRL3 -169T–>C polymorphism in white and East Asian rheumatoid arthritis patients. Arthritis Rheum 56(9):3168–3171

Bell NH (1998) Renal and nonrenal 25-hydroxyvitamin D-1alpha-hydroxylases and their clinical significance. J Bone Miner Res 13(3):350–353

Bentley RW, Keenan JI, Gearry RB et al (2008) Incidence of *Mycobacterium avium* subspecies paratuberculosis in a population-based cohort of patients with Crohn's disease and control subjects. Am J Gastroenterol 103(5):1168–1172

Berlin T, Zandman-Goddard G, Blank M et al (2007) Autoantibodies in nonautoimmune individuals during infections. Ann N Y Acad Sci 1108:584–593

Birmingham CL, Canadien V, Gouin E et al (2007) *Listeria monocytogenes* evades killing by autophagy during colonization of host cells. Autophagy 3(5):442–451

Bisgaard H, Hermansen MN, Buchvald F et al (2007) Childhood asthma after bacterial colonization of the airway in neonates. N Engl J Med 357(15):1487–1495
Blaney GP, Albert PJ, Proal AD (2009) Vitamin D metabolites as clinical markers in autoimmune and chronic disease. Ann N Y Acad Sci 1173:384–390
Bogdan C (2001) Nitric oxide and the immune response. Nat Immunol 2(10):907–916
Bogdanos DP, Baum H, Grasso A et al (2004) Microbial mimics are major targets of crossreactivity with human pyruvate dehydrogenase in primary biliary cirrhosis. J Hepatol 40(1):31–39
Bolzan AD, Bianchi MS (2002) Genotoxicity of streptozotocin. Mutat Res 512(2–3): 121–134
Boscarino JA (2004) Posttraumatic stress disorder and physical illness: results from clinical and epidemiologic studies. Ann N Y Acad Sci 1032:141–153
Bozic B, Cucnik S, Kveder T et al (2007) Autoimmune reactions after electro-oxidation of IgG from healthy persons: relevance of electric current and antioxidants. Ann N Y Acad Sci 1109:158–166
Brahmachary M, Schonbach C, Yang L et al (2006) Computational promoter analysis of mouse, rat and human antimicrobial peptide-coding genes. BMC Bioinformatics 7(Suppl 5):S8
Brock TD (1988) Robert Koch, a life in medicine and bacteriology. Science Tech Publishers, Springer, New York, NY
Brown AS (2006) Prenatal infection as a risk factor for schizophrenia. Schizophr Bull 32(2): 200–202
Buchanan AV, Weiss KM, Fullerton SM (2006) Dissecting complex disease: the quest for the Philosopher's Stone? Int J Epidemiol 35(3):562–571
Bukholm G, Modalsli K, Degre M (1986) Effect of measles-virus infection and interferon treatment on invasiveness of *Shigella flexneri* in HEp2-cell cultures. J Med Microbiol 22(4):335–341
Bull TJ, McMinn EJ, Sidi-Boumedine K et al (2003) Detection and verification of *Mycobacterium avium* subsp. paratuberculosis in fresh ileocolonic mucosal biopsy specimens from individuals with and without Crohn's disease. J Clin Microbiol 41(7):2915–2923
Bunge J (2009) Statistical estimation of uncultivated microbial diversity uncultivated microorganisms. Springer, New York, NY, pp 1–18
Burnham WR, Lennard-Jones JE, Stanford JL et al (1978) Mycobacteria as a possible cause of inflammatory bowel disease. Lancet 2(8092 Pt 1):693–696
Burton PR, Hansell AL, Fortier I et al (2009) Size matters: just how big is BIG?: quantifying realistic sample size requirements for human genome epidemiology. Int J Epidemiol 38(1): 263–273
Casadesus J (2007) Bacterial L-forms require peptidoglycan synthesis for cell division. Bioessays 29(12):1189–1191
Casali P, Burastero SE, Nakamura M et al (1987) Human lymphocytes making rheumatoid factor and antibody to ssDNA belong to Leu-1+ B-cell subset. Science 236(4797):77–81
Cheung CM, Chee SP (2009) Jarisch–Herxheimer reaction: paradoxical worsening of tuberculosis chorioretinitis following initiation of antituberculous therapy. Eye (Lond) 23(6):1472–1473
Christakis NA, Fowler JH (2007) The spread of obesity in a large social network over 32 years. N Engl J Med 357(4):370–379
Christen U, Hintermann E, Holdener M et al (2010) Viral triggers for autoimmunity: Is the 'glass of molecular mimicry' half full or half empty? J Autoimmun 34(1):38–44
Chung CC, Magalhaes W, Gonzalez-Bosquet J et al (2010) Genome-wide association studies in cancer – current and future directions. Carcinogenesis 31(1):111–120
Cochran GM, Ewald PW, Cochran KD (2000) Infectious causation of disease: an evolutionary perspective. Perspect Biol Med 43(3):406–448
Cohen ML (2000) Changing patterns of infectious disease. Nature 406(6797):762–767
Collins FS, McKusick VA (2001) Implications of the human genome project for medical science. JAMA 285(5):540–544
Collins FS, Morgan M, Patrinos A (2003) The human genome project: lessons from large-scale biology. Science 300(5617):286–290

Committee on Metagenomics, and National Research Council (2007) New science of metagenomics: revealing the secrets of our microbial planet. National Academies Press, Washington, DC

Cunningham MW (2003) Autoimmunity and molecular mimicry in the pathogenesis of post-streptococcal heart disease. Front Biosci 8:s533–543

Curran LK, Newschaffer CJ, Lee LC et al (2007) Behaviors associated with fever in children with autism spectrum disorders. Pediatrics 120(6):e1386–1392

Daniels JL, Forssen U, Hultman CM et al (2008) Parental psychiatric disorders associated with autism spectrum disorders in the offspring. Pediatrics 121(5):e1357–1362

Davenport MP, Belz GT, Ribeiro RM (2009) The race between infection and immunity: how do pathogens set the pace? Trends Immunol 30(2):61–66

Davey Smith G, Ebrahim S, Lewis S et al (2005) Genetic epidemiology and public health: hope, hype, and future prospects. Lancet 366(9495):1484–1498

Dell'Era S, Buchrieser C, Couve E et al (2009) *Listeria monocytogenes* L-forms respond to cell wall deficiency by modifying gene expression and the mode of division. Mol Microbiol 73(2):306–322

Dempsey KE, Riggio MP, Lennon A et al (2007) Identification of bacteria on the surface of clinically infected and non-infected prosthetic hip joints removed during revision arthroplasties by 16S rRNA gene sequencing and by microbiological culture. Arthritis Res Ther 9(3):R46

DePaula AL, Macedo AL, Rassi N et al (2008) Laparoscopic treatment of type 2 diabetes mellitus for patients with a body mass index less than 35. Surg Endosc 22(3):706–716

Dickson JH, Oeggl K, Handley LL (2003) The iceman reconsidered. Sci Am 288(5):70–79

DiGiulio DB, Romero R, Amogan HP et al (2008) Microbial prevalence, diversity and abundance in amniotic fluid during preterm labor: a molecular and culture-based investigation. PLoS One 3(8):e3056

Dimitrov JD, Lacroix-Desmazes S, Kaveri SV et al (2008) Insight into the mechanism of the acquired antibody auto-reactivity. Autoimmun Rev 7(6):410–414

Djavad N, Bas S, Shi X et al (1996) Comparison of rheumatoid factors of rheumatoid arthritis patients, of individuals with mycobacterial infections and of normal controls: evidence for maturation in the absence of an autoimmune response. Eur J Immunol 26(10): 2480–2486

Domingue GJ Sr, Woody HB (1997) Bacterial persistence and expression of disease. Clin Microbiol Rev 10(2):320–344

Doolittle WF, Papke RT (2006) Genomics and the bacterial species problem. Genome Biol 7(9):116

Dowd SE, Wolcott RD, Sun Y et al (2008) Polymicrobial nature of chronic diabetic foot ulcer biofilm infections determined using bacterial tag encoded FLX amplicon pyrosequencing (bTEFAP). PLoS One 3(10):e3326

Dumas ME, Maibaum EC, Teague C et al (2006) Assessment of analytical reproducibility of 1H NMR spectroscopy based metabonomics for large-scale epidemiological research: the INTERMAP Study. Anal Chem 78(7):2199–2208

Dunny GM, Brickman TJ, Dworkin M (2008) Multicellular behavior in bacteria: communication, cooperation, competition and cheating. Bioessays 30(4):296–298

Eckburg PB, Bik EM, Bernstein CN et al (2005) Diversity of the human intestinal microbial flora. Science 308(5728):1635–1638

el-Zaatari FA, Naser SA, Markesich DC et al (1996) Identification of *Mycobacterium avium* complex in sarcoidosis. J Clin Microbiol 34(9):2240–2245

Enck P, Zimmermann K, Rusch K et al (2009) The effects of ageing on the colonic bacterial microflora in adults. Z Gastroenterol 47(7):653–658

English PJ, Ghatei MA, Malik IA et al (2002) Food fails to suppress ghrelin levels in obese humans. J Clin Endocrinol Metab 87(6):2984

Fatemi SH, Reutiman TJ, Folsom TD et al (2008) The role of cerebellar genes in pathology of autism and schizophrenia. Cerebellum 7(3):279–294

Fetissov SO, Hamze Sinno M, Coeffier M et al (2008) Autoantibodies against appetite-regulating peptide hormones and neuropeptides: putative modulation by gut microflora. Nutrition 24(4):348–359

Fierer N, Hamady M, Lauber CL et al (2008) The influence of sex, handedness, and washing on the diversity of hand surface bacteria. Proc Natl Acad Sci USA 105(46):17994–17999

Fredricks DN, Relman DA (1998) Infectious agents and the etiology of chronic idiopathic diseases. Curr Clin Top Infect Dis 18:180–200

Fu W, Sanders-Beer BE, Katz KS et al (2009) Human immunodeficiency virus type 1, human protein interaction database at NCBI. Nucleic Acids Res 37(Database issue):D417–422

Geddes R (2007) Minocycline-induced lupus in adolescents: clinical implications for physical therapists. J Orthop Sports Phys Ther 37(2):65–71

Giunta S (2006) Is inflammaging an auto[innate]immunity subclinical syndrome? Immun Ageing 3:12

Glover WA, Yang Y, Zhang Y (2009) Insights into the molecular basis of L-form formation and survival in *Escherichia coli*. PLoS One 4(10):e7316

Goh KI, Cusick ME, Valle D et al (2007) The human disease network. Proc Natl Acad Sci USA 104(21):8685–8690

Gombart AF, Saito T, Koeffler HP (2009) Exaptation of an ancient Alu short interspersed element provides a highly conserved vitamin D-mediated innate immune response in humans and primates. BMC Genomics 10:321

Gottlieb B, Chalifour LE, Mitmaker B et al (2009) BAK1 gene variation and abdominal aortic aneurysms. Hum Mutat 30(7):1043–1047

Gottlieb JE, Israel HL, Steiner RM et al (1997) Outcome in sarcoidosis. The relationship of relapse to corticosteroid therapy. Chest 111(3):623–631

Gribbin J, Hubbard RB, Le Jeune I et al (2006) Incidence and mortality of idiopathic pulmonary fibrosis and sarcoidosis in the UK. Thorax 61(11):980–985

Grice EA, Kong HH, Renaud G et al (2008) A diversity profile of the human skin microbiota. Genome Res 18(7):1043–1050

Grutters JC, van den Bosch JM (2006) Corticosteroid treatment in sarcoidosis. Eur Respir J 28(3):627–636

Guarneri F, Guarneri C, Benvenga S (2005) *Helicobacter pylori* and autoimmune pancreatitis: role of carbonic anhydrase via molecular mimicry? J Cell Mol Med 9(3):741–744

Gumpert J, Taubeneck U (1983) Characteristic properties and biological significance of stable protoplast type L-forms. Experientia Suppl 46:227–241

Hall CB, Caserta MT, Schnabel K et al (2008) Chromosomal integration of human herpesvirus 6 is the major mode of congenital human herpesvirus 6 infection. Pediatrics 122(3): 513–520

Harada M, Kishimoto Y, Makino S (1990) Prevention of overt diabetes and insulitis in NOD mice by a single BCG vaccination. Diabetes Res Clin Pract 8(2):85–89

Harley JB, James JA (2006) Epstein-Barr virus infection induces lupus autoimmunity. Bull NYU Hosp Jt Dis 64(1–2):45–50

Hawke CG, Painter DM, Kirwan PD et al (2003) Mycobacteria, an environmental enhancer of lupus nephritis in a mouse model of systemic lupus erythematosus. Immunology 108(1):70–78

Hazlett KR, Caldon SD, McArthur DG et al (2008) Adaptation of *Francisella tularensis* to the mammalian environment is governed by cues which can be mimicked in vitro. Infect Immun 76(10):4479–4488

Hibbert-Jones E, Regan G, Bramwell J (2004) What do we know about diabetes and obesity in adults and children? J Fam Health Care 14(4):95–98

Hill Gaston JS, Lillicrap MS (2003) Arthritis associated with enteric infection. Best Pract Res Clin Rheumatol 17(2):219–239

Hoffman LR, Deziel E, D'Argenio DA et al (2006) Selection for *Staphylococcus aureus* small-colony variants due to growth in the presence of *Pseudomonas aeruginosa*. Proc Natl Acad Sci USA 103(52):19890–19895

Holick MF (2008) Vitamin D: a D-Lightful health perspective. Nutr Rev 66(10 Suppl 2): S182–194

Holmes E, Loo RL, Stamler J et al (2008) Human metabolic phenotype diversity and its association with diet and blood pressure. Nature 453(7193):396–400

Joseleau-Petit D, Liebart JC, Ayala JA et al (2007) Unstable *Escherichia coli* L forms revisited: growth requires peptidoglycan synthesis. J Bacteriol 189(18):6512–6520

Kern DG, Neill MA, Wrenn DS et al (1993) Investigation of a unique time-space cluster of sarcoidosis in firefighters. Am Rev Respir Dis 148(4 Pt 1):974–980

Kinross JM, von Roon AC, Holmes E et al (2008) The human gut microbiome: implications for future health care. Curr Gastroenterol Rep 10(4):396–403

Kissler H (2001) Is multiple sclerosis caused by a silent infection with malarial parasites? A historico-epidemiological approach: part II. Med Hypotheses 57(3):292–301

Kleineberger-Nobel E (1951) Filterable forms of bacteria. Bacteriol Rev 15(2):77–103

Knodler LA, Finlay BB (2001) *Salmonella* and apoptosis: to live or let die? Microbes Infect/Inst Pasteur 3(14–15):1321–1326

Kodati VL, Govindan S, Movva S et al (2008) Role of *Shigella* infection in endometriosis: a novel hypothesis. Med Hypotheses 70(2):239–243

Kozarov EV, Dorn BR, Shelburne CE et al (2005) Human atherosclerotic plaque contains viable invasive *Actinobacillus actinomycetemcomitans* and *Porphyromonas gingivalis*. Arterioscler Thromb Vasc Biol 25(3):e17–18

Kuijl C, Savage ND, Marsman M et al (2007) Intracellular bacterial growth is controlled by a kinase network around PKB/AKT1. Nature 450(7170):725–730

Kuroki S, Saida T, Nukina M et al (1993) *Campylobacter jejuni* strains from patients with Guillain–Barre syndrome belong mostly to Penner serogroup 19 and contain beta-N-acetylglucosamine residues. Ann Neurol 33(3):243–247

Lamell CW, Griffen AL, McClellan DL et al (2000) Acquisition and colonization stability of *Actinobacillus actinomycetemcomitans* and *Porphyromonas gingivalis* in children. J Clin Microbiol 38(3):1196–1199

Lekakh IV, Rott GM, Poverennyi AM (1991) "Masked" autoantibodies from the serum of healthy blood donors cross-reacting with DNA and bacterial lipopolysaccharides. Biull Eksp Biol Med 111(5):516–518

Lie JA, Andersen A, Kjaerheim K (2007) Cancer risk among 43000 Norwegian nurses. Scand J Work Environ Health 33(1):66–73

Liolios K, Mavromatis K, Tavernarakis N et al (2008) The Genomes On Line Database (GOLD) in 2007: status of genomic and metagenomic projects and their associated metadata. Nucleic Acids Res 36(Database issue):D475–479

Liu PT, Stenger S, Tang DH et al (2007) Cutting edge: vitamin D-mediated human antimicrobial activity against *Mycobacterium tuberculosis* is dependent on the induction of cathelicidin. J Immunol 179(4):2060–2063

Liu Y, Penttinen MA, Granfors K (2001) Insights into the role of infection in the spondyloarthropathies. Curr Rheumatol Rep 3(5):428–434

Lombardi VC, Ruscetti FW, Das Gupta J et al (2009) Detection of an infectious retrovirus, xmrv, in blood cells of patients with chronic fatigue syndrome. Science 326(5952):585–589

Luque C, Cisternas FA, Araya M (2006) Changes in the patterns of disease after the epidemiological transition in health in Chile, 1950–2003. Rev Med Chil 134(6):703–712

Lutjen-Drecoll E (1992) Morphology of the pars plana region. Dev Ophthalmol 23:50–59

Lykouras D, Sampsonas F, Karparianos A et al (2008) Human genes in TB infection: their role in immune response. Monaldi Arch Chest Dis 69(1):24–31

Machado AM, Figueiredo C, Touati E et al (2009) *Helicobacter pylori* infection induces genetic instability of nuclear and mitochondrial DNA in gastric cells. Clin Cancer Res 15(9): 2995–3002

Marshall TG (2008) VDR nuclear receptor is key to recovery from cognitive dysfunction. In: Days of molecular medicine. Stockholm, Sweden

Marshall TG (2008) Vitamin D discovery outpaces FDA decision making. BioEssays 30(2): 173–182

Mattman LH (2000) Cell wall deficient forms: stealth pathogens. CRC Press, Boca Raton, FL

Mawer EB, Hayes ME, Still PE et al (1991) Evidence for nonrenal synthesis of 1,25-dihydroxyvitamin D in patients with inflammatory arthritis. J Bone Miner Res 6(7): 733–739

Mazmanian SK, Round JL, Kasper DL (2008) A microbial symbiosis factor prevents intestinal inflammatory disease. Nature 453(7195):620–625

McIntyre JA (2004) The appearance and disappearance of antiphospholipid autoantibodies subsequent to oxidation–reduction reactions. Thromb Res 114(5–6):579–587

McLean SA, Williams DA, Clauw DJ (2005) Fibromyalgia after motor vehicle collision: evidence and implications. Traffic Inj Prev 6(2):97–104

Mehra NK, Kumar N, Kaur G et al (2007) Biomarkers of susceptibility to type 1 diabetes with special reference to the Indian population. Indian J Med Res 125(3):321–344

Merino G, Carranza-Lira S, Murrieta S et al (1995) Bacterial infection and semen characteristics in infertile men. Arch Androl 35(1):43–47

Merkler D, Horvath E, Bruck W et al (2006) "Viral deja vu" elicits organ-specific immune disease independent of reactivity to self. J Clin Invest 116(5):1254–1263

Mestas J, Hughes CC (2004) Of mice and not men: differences between mouse and human immunology. J Immunol 172(5):2731–2738

Miller R, Callas DD, Kahn SE et al (2000) Evidence of myocardial infarction in mummified human tissue. JAMA 284(7):831–832

Mitiku K, Mengistu G (2002) Relapsing fever in Gondar, Ethiopia. East Afr Med J 79(2):85–87

Moissl C, Osman S, La Duc MT et al (2007) Molecular bacterial community analysis of clean rooms where spacecraft are assembled. FEMS Microbiol Ecol 61(3):509–521

Monaco C, Mathur A, Martin JF (2005) What causes acute coronary syndromes? Applying Koch's postulates. Atherosclerosis 179(1):1–15

Muller M (2006) Premature cellular senescence induced by pyocyanin, a redox-active *Pseudomonas aeruginosa* toxin. Free Radic Biol Med 41(11):1670–1677

Murphy GJ, Reeves BC, Rogers CA et al (2007) Increased mortality, postoperative morbidity, and cost after red blood cell transfusion in patients having cardiac surgery. Circulation 116(22):2544–2552

Nasidze I, Li J, Quinque D et al (2009) Global diversity in the human salivary microbiome. Genome Res 19(4):636–643

Nevado J, Tenbaum SP, Castillo AI et al (2007) Activation of the human immunodeficiency virus type I long terminal repeat by 1 alpha,25-dihydroxyvitamin D3. J Mol Endocrinol 38(6): 587–601

Nicolson GL, Gan R, Nicolson NL et al (2007) Evidence for *Mycoplasma* ssp., *Chlamydia pneunomiae*, and human herpes virus-6 coinfections in the blood of patients with autistic spectrum disorders. J Neurosci Res 85(5):1143–1148

Nikkari S, McLaughlin IJ, Bi W et al (2001) Does blood of healthy subjects contain bacterial ribosomal DNA? J Clin Microbiol 39(5):1956–1959

Nuding S, Fellermann K, Wehkamp J et al (2007) Reduced mucosal antimicrobial activity in Crohn's disease of the colon. Gut 56(9):1240–1247

O'Connor SM, Taylor CE, Hughes JM (2006) Emerging infectious determinants of chronic diseases. Emerg Infect Dis 12(7):1051–1057

Padilla ML, Schilero GJ, Teirstein AS (2002) Donor-acquired sarcoidosis. Sarcoidosis Vasc Diffuse Lung Dis 19(1):18–24

Palmer C, Bik EM, DiGiulio DB et al (2007) Development of the human infant intestinal microbiota. PLoS Biol 5(7):e177

Pareek SS (1977) Syphilitic alopecia and Jarisch-Herxheimer reaction. Br J Vener Dis 53(6): 389–390

Perez TH (2008) Bacteria induced vitamin D receptor dysfunction in autoimmune disease: theoretical and practical implications for interpretation of serum vitamin D metabolite levels. In: 6th International Congress on Autoimmunity, Porto, Portugal, September 11 2008

Peschard S, Brinkane A, Bergheul S et al (2001) Whipple disease associated with pulmonary arterial hypertension. Jarisch–Herxheimer reaction after antibiotic therapy. Presse Med 30(31 Pt 1):1549–1551

Poolman EM, Galvani AP (2007) Evaluating candidate agents of selective pressure for cystic fibrosis. J R Soc Interface 4(12):91–98

Pordeus V, Szyper-Kravitz M, Levy RA et al (2008) Infections and autoimmunity: a panorama. Clin Rev Allergy Immunol 34(3):283–299

Posnett DN, Edinger J (1997) When do microbes stimulate rheumatoid factor? J Exp Med 185(10):1721–1723

Pospechova K, Rozehnal V, Stejskalova L et al (2009) Expression and activity of vitamin D receptor in the human placenta and in choriocarcinoma BeWo and JEG-3 cell lines. Mol Cell Endocrinol 299(2):178–187

Proal AD, Albert PJ, Marshall TG (2009) Dysregulation of the vitamin D nuclear receptor may contribute to the higher prevalence of some autoimmune diseases in women. Ann N Y Acad Sci 1173:252–259

Ramchandran L, Shah NP (2008) Proteolytic profiles and angiotensin-I converting enzyme and alpha-glucosidase inhibitory activities of selected lactic acid bacteria. J Food Sci 73(2): M75–81

Razumov AS (1932) The direct method of calculation of bacteria in water: comparison with the Koch method. Mikrobiologija 1:131–146

Relman DA (1998) Detection and identification of previously unrecognized microbial pathogens. Emerg Infect Dis 4(3):382–389

Risch NJ (2000) Searching for genetic determinants in the new millennium. Nature 405(6788):847–856

Roesch LF, Lorca GL, Casella G et al (2009) Culture-independent identification of gut bacteria correlated with the onset of diabetes in a rat model. ISME J 3(5):536–548

Romani B, Engelbrecht S, Glashoff RH (2009) Functions of tat: the versatile protein of human immunodeficiency virus type 1. J Gen Virol

Rossman MD, Kreider ME (2007) Lesson learned from ACCESS (A Case Controlled Etiologic Study of Sarcoidosis). Proc Am Thorac Soc 4(5):453–456

Russell MW, Wu HY, White PL et al (1992) Serum antibody responses to *Streptococcus mutans* antigens in humans systemically infected with oral streptococci. Oral Microbiol Immunol 7(6):321–325

Salazar JC, Duhnam-Ems S, La Vake C et al (2009) Activation of human monocytes by live *Borrelia burgdorferi* generates TLR2-dependent and -independent responses which include induction of IFN-beta. PLoS Pathog 5(5):e1000444

Sarkola M, Rintala M, Grenman S et al (2008) Human papillomavirus DNA detected in breast milk. Pediatr Infect Dis J 27(6):557–558

Sauer K, Camper AK, Ehrlich GD et al (2002) *Pseudomonas aeruginosa* displays multiple phenotypes during development as a biofilm. J Bacteriol 184(4):1140–1154

Schwocho LR, Masonson HN (2001) Pharmacokinetics of CS-866, a new angiotensin II receptor blocker, in healthy subjects. J Clin Pharmacol 41(5):515–527

Shelburne SA 3rd, Hamill RJ, Rodriguez-Barradas MC et al (2002) Immune reconstitution inflammatory syndrome: emergence of a unique syndrome during highly active antiretroviral therapy. Medicine (Baltimore) 81(3):213–227

Siegler RL, Pavia AT, Christofferson RD et al (1994) A 20-year population-based study of postdiarrheal hemolytic uremic syndrome in Utah. Pediatrics 94(1):35–40

Slaughter L, Carson DA, Jensen FC et al (1978) In vitro effects of Epstein–Barr virus on peripheral blood mononuclear cells from patients with rheumatoid arthritis and normal subjects. J Exp Med 148(5):1429–1434

Staley JT (1997) Biodiversity: are microbial species threatened? Curr Opin Biotechnol 8(3): 340–345

Stamler J, Elliott P, Dennis B et al (2003) INTERMAP: background, aims, design, methods, and descriptive statistics (nondietary). J Hum Hypertens 17(9):591–608

Swaak T, Smeenk R (1985) Detection of anti-dsDNA as a diagnostic tool: a prospective study in 441 non-systemic lupus erythematosus patients with anti-dsDNA antibody (anti-dsDNA). Ann Rheum Dis 44(4):245–251

Swidsinski A, Ladhoff A, Pernthaler A et al (2002) Mucosal flora in inflammatory bowel disease. Gastroenterology 122(1):44–54

Tavares-Ratado P, Geraldes A, Simões V et al (2009 of Conference, September 11–13). Prevalence of circulating autoantibodies in portugese blood donors. In: 4th Asian Congress on Autoimmunity, Singapore, September 11–13 2009

Turnbaugh PJ, Ley RE, Mahowald MA et al (2006) An obesity-associated gut microbiome with increased capacity for energy harvest. Nature 444(7122):1027–1031

Venter JC, Adams MD, Myers EW et al (2001) The sequence of the human genome. Science 291(5507):1304–1351

Vidal V, Scragg IG, Cutler SJ et al (1998) Variable major lipoprotein is a principal TNF-inducing factor of louse-borne relapsing fever. Nat Med 4(12):1416–1420

Viganò P, Lattuada D, Mangioni S et al (2006) Cycling and early pregnant endometrium as a site of regulated expression of the vitamin D system. J Mol Endocrinol 36(3):415–424

Virgin HW, Wherry EJ, Ahmed R (2009) Redefining chronic viral infection. Cell 138(1):30–50

Walker L, Levine H, Jucker M (2006) Koch's postulates and infectious proteins. Acta Neuropathol 112(1):1–4

Wang TT, Tavera-Mendoza LE, Laperriere D et al (2005) Large-scale in silico and microarray-based identification of direct 1,25-dihydroxyvitamin D3 target genes. Mol Endocrinol 19(11):2685–2695

Wang Y, Beydoun MA (2007) The obesity epidemic in the United States–gender, age, socioeconomic, racial/ethnic, and geographic characteristics: a systematic review and meta-regression analysis. Epidemiol Rev 29:6–28

Webster Marketon JI, Glaser R (2008) Stress hormones and immune function. Cell Immunol 252(1–2):16–26

Weigt J, Malfertheiner P (2009) Influence of *Helicobacter pylori* on gastric regulation of food intake. Curr Opin Clin Nutr Metab Care 12(5):522–525

Wikoff WR, Anfora AT, Liu J et al (2009) Metabolomics analysis reveals large effects of gut microflora on mammalian blood metabolites. Proc Natl Acad Sci USA 106(10): 3698–3703

Wilhoite SL, Ferguson DA Jr, Soike DR et al (1993) Increased prevalence of *Helicobacter pylori* antibodies among nurses. Arch Intern Med 153(6):708–712

Williams RC Jr, Kunkel HG (1962) Rheumatoid factor, complement, and conglutinin aberrations in patients with subacute bacterial endocarditis. J Clin Invest 41:666–675

Wirostko E, Johnson L, Wirostko B (1987) Crohn's disease. Rifampin treatment of the ocular and gut disease. Hepatogastroenterology 34(2):90–93

Wirostko E, Johnson L, Wirostko B (1989) Sarcoidosis associated uveitis. Parasitization of vitreous leucocytes by mollicute-like organisms. Acta Ophthalmol (Copenh) 67(4):415–424

Xu Y, Xie J, Li Y et al (2003) Using a cDNA microarray to study cellular gene expression altered by *Mycobacterium tuberculosis*. Chin Med J 116(7):1070–1073

Yang X, Xie L, Li Y et al (2009) More than 9,000,000 unique genes in human gut bacterial community: estimating gene numbers inside a human body. PLoS One 4(6):e6074

Yenamandra SP, Lundin A, Arulampalam V et al (2009) Expression profile of nuclear receptors upon Epstein–Barr virus-induced B cell transformation. Exp Oncol 31(2):92–96

Yilmaz O, Yao L, Maeda K et al (2008) ATP scavenging by the intracellular pathogen *Porphyromonas gingivalis* inhibits P2X7-mediated host-cell apoptosis. Cell Microbiol 10(4):863–875

Yoshizawa T, Handa Y, Uematsu Y et al (1997) Mice lacking the vitamin D receptor exhibit impaired bone formation, uterine hypoplasia and growth retardation after weaning. Nat Genet 16(4):391–396

Zanini GM, De Moura Carvalho LJ, Brahimi K et al (2009) Sera of patients with systemic lupus erythematosus react with plasmodial antigens and can inhibit the in vitro growth of *Plasmodium falciparum*. Autoimmunity 42(6):545–552

Zhang H, DiBaise JK, Zuccolo A et al (2009) Human gut microbiota in obesity and after gastric bypass. Proc Natl Acad Sci USA 106(7):2365–2370

Zinkernagel MS, Bolinger B, Krebs P et al (2009) Immunopathological basis of lymphocytic choriomeningitis virus-induced chorioretinitis and keratitis. J Virol 83(1):159–166

Zumla A, James DG (1996) Granulomatous infections: etiology and classification. Clin Infect Dis 23(1):146–158

Chapter 13
Metagenomic Applications and the Potential for Understanding Chronic Liver Disease

Bernd Schnabl

Introduction

Liver Fibrosis and Cirrhosis

Liver fibrosis is the common consequence of chronic hepatic injury of any cause. It consists of a progressive accumulation of extracellular matrix proteins in the liver parenchyma that distorts the normal hepatic architecture by forming a fibrous scar. In early stages of fibrosis, the liver functions relatively well and few people experience symptoms (Bataller and Brenner, 2005). But as the inflammation and liver injury continue, scar tissue builds up and connects with existing scar tissue, which can eventually disrupt the metabolic functions of the liver. The main clinical consequences of advanced hepatic fibrosis are impaired hepatocellular function and increased intrahepatic resistance to blood flow, resulting in hepatic insufficiency and portal hypertension, respectively (Gines et al., 2004). To date, the most effective therapy is to remove the causative agent (e.g., alcohol consumption, obesity, hepatitis B and C viral infections, and biliary obstruction). However, if this cannot be achieved, there are no other effective treatment options for hepatic fibrosis. Chronic liver disease may progress to cirrhosis, an end-stage disease and a major cause of morbidity and mortality worldwide. Patients who progress to liver cirrhosis have a poor prognosis, and liver transplantation is often indicated. Cirrhosis causes more than 27,000 deaths per year in the United States (Gines et al., 2004). Even those who survive cirrhosis face a high risk of developing long-term morbidity and hepatocellular carcinoma.

Half of cirrhotic patients die within 2 years after diagnosis, and among the leading causes of death are infections (30–50%). The mortality of infections is more than 20 times increased in patients with cirrhosis. In-hospital mortality of cirrhotic patients with infection is approximately 15%, more than twice

B. Schnabl (✉)
Department of Medicine, University of California San Diego, MC0702, 9500 Gilman Drive, La Jolla, CA 92093, USA
e-mail: beschnabl@ucsd.edu

K.E. Nelson (ed.), *Metagenomics of the Human Body*,
DOI 10.1007/978-1-4419-7089-3_13,

that of patients without infection (Wong et al., 2005). Infections in cirrhosis are mainly caused by bacteria and the main sites of infection are ascites (spontaneous bacterial peritonitis), blood, respiratory, and urinary tract. The most common organisms are *Staphylococcus aureus, Enterococcus faecalis, Streptococcus pneumoniae, Klebsiella,* and *Escherichia coli* (Frances et al., 2008).

Bacterial Translocation

Bacterial translocation is defined as the passage of viable indigenous bacteria or their products from the intestinal tract through the epithelial mucosa to the mesenteric lymph nodes (MLNs), systemic circulation, or extraintestinal organs. It is considered an important mechanism for the incidence of infections, including spontaneous bacterial peritonitis and sepsis in cirrhotic patients (Wiest and Garcia-Tsao, 2005). Factors that prevent bacterial translocation include prevention of bacterial overgrowth, integrity of the mucosal barrier, and immunocompetence as bacteria breaching the epithelial barrier are killed in the mesenteric lymph nodes.

Intestinal Symbiosis and Dysbiosis

Human microbiota research was largely dependent on conventional culture techniques of bacteria. Only a minority of the intestinal microbiota is culturable (approximately 20%) (Eckburg et al., 2005). However, recent technical advances have helped to characterize the gastrointestinal microflora in its deep biodiversity and functional contribution to the host biology. Molecular methods sequence the bacterial 16S ribosomal (r)RNA used in taxonomical classification of bacteria. This culture-independent strategy using massive parallel sequencing and comparative genomics provides us with new information about the composition and the architecture of both human and animal microflora. Metagenomic analysis applies sequencing to the total genomic DNA from a microbial community and gives us more insight into the metabolic function of the bacterial population rather than providing us with the 16S rRNA identities only.

The intestinal tract is estimated to contain up to 10^{14} microbes; a microbial population that has developed various mechanisms to influence its eukaryotic host, in general, in a beneficial and symbiotic fashion. The gastrointestinal tract harbors all three domains of life (Bacteria, Archaea, and Eukarya), 10 bacterial phyla (Firmicutes, Bacteroidetes, Proteobacteria, Actinobacteria, Fusobacteria, Verrucomicrobia, Cyanobacteria, TM7, Spirochaetes, and VadinBE97), and more than 15,000 species-level bacterial phylotypes (Camp et al., 2009). The human gastrointestinal tract is dominated by only two bacterial phyla, Firmicutes and Bacteroidetes, which make up >99% of the identified phylotypes (Eckburg et al., 2005; Ley et al., 2006). There is a symbiotic relationship between eukaryotic and prokaryotic cells in the human intestine. The significant number of inflammatory cells in the lamina propria of the colonized intestinal tract and the preservation of an intact epithelial lining reflect this delicate and symbiotic relationship between the host and the microbiota.

Dysbiosis, defined as disruption of this mutual crosstalk between the microflora and the gastrointestinal tract, has been associated with disease. Dysbiosis has been largely studied as changes in bacterial species. However, it has become evident recently (using metagenomics approaches) that changes in the genome of intestinal bacteria are more important than changes in the phyla that are present (Turnbaugh et al., 2009). A dysbiotic enteric microflora has been implicated in the pathogenesis of inflammatory bowel disease (Garrett et al., 2007; Frank et al., 2007).

The immune system (both innate and adaptive immunity) is responsible for guarding the number and composition of bacteria, whereas the intestinal microflora has their own means to modify host processes. Intestinal epithelial cells are the main interface between the host and the intestinal microflora. Epithelial cells have many important functions in protecting the host from pathogen invasion as well as in maintaining a symbiotic relationship between the host and the microflora. The innate immune system has developed a series of phylogenetically conserved receptors including Toll-like receptors (TLRs). TLRs have the ability to recognize specific microbial products, known as pathogen-associated molecular patterns (PAMPs), such as lipopolysaccharide (LPS), the principal glycolipid component of the outer membrane of Gram-negative bacteria, and also peptidoglycan, double-stranded RNA, and unmethylated DNA (Akira et al., 2006). Ligand binding to pattern recognition receptors activates signaling cascades that control transcription of defensive or proinflammatory genes. Antimicrobial effector molecules that are secreted by enterocytes or Paneth cells play an important role in the maintenance of the host–microbe symbiotic relationship, and they are also necessary for defense against overt pathogens (Akira et al., 2006; Medzhitov, 2007). Pattern recognition receptor signaling stimulated by commensal bacteria does not activate tissue responses that is interpreted as disease state. It rather contributes to maintaining intestinal homeostasis.

Requirement of Bacterial Products for Liver Fibrogenesis

The innate immune system is important not only for intestinal homeostasis, but also for liver injury and fibrogenesis. Several studies have addressed a role for LPS signaling in experimental liver fibrosis. Mice deficient in CD14 or LBP as LPS binding and presenting molecules are resistant to liver injury and fibrosis induced by bile duct ligation (Isayama et al., 2006). The most convincing evidence supporting a role for TLR4 in liver fibrosis are studies using Toll-like receptor 4 (TLR4) mutant C3H/Hej and TLR4-deficient mice. Hepatic inflammation and fibrosis are strongly decreased in the TLR4 mutant C3H/Hej strain and TLR4-deficient mice following bile duct ligation or carbon tetrachloride (CCl_4) administration (Seki et al., 2007). TLR4 has also been shown in patients as important modulator of liver fibrosis. TLR4 as the cellular LPS receptor was identified as one of seven genes associated with increased risk of developing cirrhosis in patients with chronic hepatitis C (Huang et al., 2007). TLR4 D299G and T399I SNPs that are associated with protection from hepatic fibrosis reduce TLR4-mediated inflammatory and fibrogenic signaling (Guo et al., 2009).

Toll-like receptor 9 (TLR9) is a pattern recognition receptor that is activated by CpG motifs present specifically in bacterial DNA. We have shown that TLR9-deficient mice are resistant to experimental liver fibrosis (Gabele et al., 2008). Together, these results indicate that bacterial translocation with a subsequent increase of bacterial products is important not only for infections in patients with chronic liver disease, but also in promoting hepatic fibrogenesis.

Nonalcoholic Fatty Liver Disease

Obesity is a major health burden in Western countries and is closely associated with several metabolic disorders such as diabetes, dyslipidemia, and heart disease. It is also linked to an epidemic in nonalcoholic fatty liver disease (NAFLD) as the hepatic manifestation of the metabolic syndrome. NAFLD spans a spectrum of hepatic disorders from simple steatosis to nonalcoholic steatohepatitis (NASH). A "two-hit" model has been proposed for the development of NAFLD (Schenk et al., 2008). The first hit is the initial hepatic lipid accumulation, but a second hit is required for liver injury, inflammation, and fibrosis. The metabolic syndrome is an abnormal energy imbalance between multiple organs. White adipose tissue is primarily important for storage of triglycerides, which are released as free fatty acids at times of energy need. Skeletal muscle and brown adipose tissue serve to decrease energy in times of excess through adaptive thermogenesis.

Obesity is closely correlated with changes in the microbiome and metagenome. Studies in obese mice with a mutation in the leptin gene (ob/ob) and their lean wild-type littermates revealed that obesity is associated with a division-wide increase and decrease in the relative abundance of the Firmicutes and Bacteroidetes, respectively (Ley et al., 2005). Furthermore, when wild-type germfree mice fed a standard chow are colonized with a microbiota harvested from ob/ob or lean donors, adiposity in recipients of the obese microbiota increased more than in recipients of a lean microbiota (Turnbaugh et al., 2006). This change involves several mechanisms, which were mostly identified by metagenomic techniques. One is the fermentation of indigestible polysaccharides by the microbial flora into monosaccharides and short-chain fatty acids. A higher amount of monosaccharides and short-chain fatty acids are subsequently absorbed from the gut lumen and transported to the liver with resulting induction of *de novo* hepatic lipogenesis (Backhed et al., 2004). In addition, fasting-induced adipocyte factor (FIAF) as a circulating lipoprotein lipase inhibitor is selectively suppressed in the intestinal epithelium of the host resulting in deposition of triglycerides in adipocytes (Backhed et al., 2004). FIAF also induces the expression of peroxisomal proliferator-activated receptor coactivator (Pgc-1α) resulting in increased fatty acid oxidation (Backhed et al., 2007). Together, the microbiota of obese individuals is more efficient at extracting energy from the intestine than the microbiota of lean individuals. The linkage between adiposity and changes in the gut microbial ecology also appears to apply to humans (Ley et al., 2005; Turnbaugh et al., 2006). Twin studies (obese and nonobese) revealed that the human gut microbiome is shared among family members but that each person's gut microbial community varies in the specific bacterial lineages present (Turnbaugh et al., 2009).

Metabolic profiling of genetic susceptibility to impaired glucose homeostasis and fatty liver disease demonstrated a disruption of choline metabolism. Conversion of choline into methylamines by microbiota in a susceptible mouse strain reduces the bioavailability of choline and mimics the effect of choline-deficient diets causing NAFLD (see below) (Dumas et al., 2006). These data indicate an additional active role for the intestinal microflora in the development of insulin resistance and fatty liver disease.

As 33% of the general population has fatty liver disease (Szczepaniak et al., 2005), the question arises whether changes observed in the microflora of obese individuals are also present in individuals with NAFLD or NASH. This question has been addressed by several studies. Small intestinal bacterial overgrowth was present in 50% of patients with NASH, which is significantly higher than in a sex- and age-matched healthy control group (Wigg et al., 2001). As endotoxin seems to be involved in the pathogenesis of hepatic fat deposition, one study investigated changes in gut leakiness. Patients with NAFLD show significantly increased intestinal permeability, small intestinal bacterial overgrowth, and alterations in the intestinal tight junctions as compared to healthy individuals. Gut permeability and small intestinal bacterial overgrowth correlated with severity of steatosis, but not fibrosis or hepatic inflammation on liver biopsy (Miele et al., 2009). The beneficial effect of probiotics on hepatic steatosis (Li et al., 2003; Ma et al., 2008; Velayudham et al., 2009) provides further evidence supporting a contribution of the intestinal microflora and a role of gut-derived endotoxemia in the onset of the hepatic steatosis.

Alcohol-Induced Liver Disease

Alcohol abuse is the most important cause of liver cirrhosis in industrialized countries. Alcoholic liver disease is characterized by fatty liver (steatosis), which might progress to alcoholic hepatitis and fibrosis. Chronic alcoholic liver disease may progress in 15–40% to cirrhosis, an end-stage disease and a major cause of morbidity and mortality worldwide (Bode et al., 2005). Patients with chronic alcohol abuse or alcoholic liver cirrhosis have more frequently intestinal bacterial overgrowth as compared to healthy individuals. Both aerobic and anaerobic microorganisms were higher in jejunal aspirates in patients with chronic alcohol abuse (Bode et al., 1984; Casafont Morencos et al., 1996).

Alcoholic and nonalcoholic steatohepatitis (ASH and NASH) apart from the underlying etiology share common pathologic and clinical features. One study explored the intestinal flora following alcohol administration and onset of steatohepatitis using length heterogeneity PCR (LH-PCR). After 10 weeks of alcohol feeding, mucosa-associated bacterial microbiota fingerprints in the colon were markedly different in the alcohol-fed rats as compared to the isocaloric (dextrose) fed rats, but no differences were observed in the ileum. Both probiotic *Lactobacillus GG* and prebiotic oats supplementation prevented alcohol-induced colonic dysbiosis (Mutlu et al., 2009). As LH-PCR is only able to identify changes in the bacterial profile, changes in microbiome and metagenome remain to be determined in future studies using sequencing. Interestingly, the onset of endotoxemia in the same model of alcohol feeding in rats precedes the onset of dysbiosis markedly. Endotoxemia was found already 2 weeks after alcohol feeding (Keshavarzian et al., 2009). Thus, alcohol-induced dysbiosis does not cause intestinal permeability with resulting endotoxemia, but it might increase the levels of endotoxin at later stages.

Increased intestinal permeability has been reported in patients with alcoholic liver disease (Parlesak et al., 2000). Translocation of enteric bacteria to mesenteric lymph nodes occurs in 31% of patients with decompensated cirrhosis and is reduced to the level found in noncirrhotic patients by selective intestinal decontamination (Cirera et al., 2001). The presence of bacterial DNA in serum as a molecular marker for bacterial translocation was 34% in noninfected patients with liver cirrhosis and in all cases of spontaneous bacterial peritonitis (Frances et al., 2008). There are conflicting results in animal models of alcohol-induced liver injury. Although reports found no significant difference in bacterial translocation to mesenteric lymph nodes or the systemic circulation following alcohol administration for 14 days as compared to control-fed animals (Mason et al., 1998; Fleming et al., 2001), translocation of viable intestinal flora has been reported in rats fed alcohol for 2 weeks (Napolitano et al., 1995).

Plasma levels of LPS or endotoxin, a major component of the Gram-negative bacterial outer membrane, increase with the severity of liver dysfunction in cirrhotic patients, and are also significantly higher in patients with chronic hepatitis than in healthy subjects (Lin et al., 1995). Endotoxemia in patients with alcoholic cirrhosis is significantly higher than in patients with nonalcoholic liver cirrhosis (Fukui et al., 1991; Bode et al., 1987). In addition, endotoxemia is also more frequent in patients with mild forms of alcoholic hepatitis without evidence of fibrosis or cirrhosis (Bode and Bode 2005; Parlesak et al., 2000). A correlation between alcohol ingestion and increased systemic levels of endotoxin has been also demonstrated in animal models of alcohol-induced liver injury (Adachi et al., 1995; Nanji et al., 1993; Tamai et al., 2000). It was also noted that increased intestinal permeability with subsequent endotoxemia occurs prior to the development of alcoholic steatohepatitis in rats (Keshavarzian et al., 2009). Plasma levels of peptidoglycan, which makes up about 70 and 20% of Gram-positive and Gram-negative bacterial cell walls, are also increased following acute alcohol exposure (Tabata et al., 2002).

Several studies investigated possible mechanisms for increased bacterial translocation. Acute ingestion of alcohol alters the epithelial barrier in the colon through ethanol oxidation into acetaldehyde by the colonic microflora and results in activation of mast cells (Ferrier et al., 2006). In agreement with this report, tissue oxidative stress was increased in jejunal, ileal, and colonic mucosa of rats as early as 4 weeks following alcohol feeding (Keshavarzian et al., 2009). Cytokines including interleukin (IL)-1β, IL-6 and tumor necrosis factor (TNF), and cell adhesion molecule inter-cellular adhesion molecule (ICAM)-1 were elevated in the ileum of mice fed ethanol for 14 days (Fleming et al., 2001).

As mentioned above, bacterial translocation not only causes severe infections in cirrhotic patients, but it also causes progression of alcohol-induced liver injury and fibrosis. Several studies addressed a role for LPS signaling in experimental alcohol-induced liver disease. Selective intestinal decontamination with antibiotics (polymyxin B and neomycin) decreases plasma endotoxin levels without affecting gut permeability and prevents alcoholic liver injury (Adachi et al., 1995; Enomoto et al., 1999, 2001). Mice deficient in CD14 as cellular co-receptor for LPS are resistant to alcohol-induced liver injury (Yin et al., 2001). The most convincing evidence supporting a role for LPS in alcoholic liver injury are studies using TLR4 mutant C3H/Hej mice. Hepatic steatosis, inflammation, and necrosis are strongly reduced in the TLR4 mutant C3H/Hej strain following ethanol administration as compared to wildtype mice (Uesugi et al., 2001). It was postulated that LPS binds via TLR4 to hepatic Kupffer cells with resulting induction of TNF. Indeed, TNFR-I-deficient mice are also resistant to alcohol-induced liver damage (Yin et al., 1999). Additional data supporting the crucial pathogenic role of endotoxin in alcoholic liver injury comes from interventional, therapeutic studies. Feeding a Gram-positive probiotic lactobacillus strain with concomitant displacement of Gram-negative bacteria protected mice from ethanol-induced liver injury with a concomitant decrease in systemic endotoxin levels (Nanji et al., 1994). Thus, animal studies emphasize the importance of bacterial translocation with resulting endotoxemia as a modifier for disease progression.

Chronic Liver Disease

Dysbiosis and Bacterial Translocation in Experimental Models of Chronic Liver Disease

Experimental animal models of bile duct ligation or CCl_4 induced liver cirrhosis as a model for end-stage liver disease, demonstrate intestinal bacterial overgrowth and translocation of bacteria and bacterial products including LPS and bacterial DNA (Parks et al., 1996; Guarner et al., 1997). Using conventional culture techniques and following bile duct ligation, the prevalence of mostly Gram-negative bacteria such as *E. coli, Enterococci, Klebsiella, Pseudomonas, Pasteurella, Shigella*, and *Proteus* increased in the cecum. Cirrhotic rats with bacterial translocation had a higher prevalence of intestinal bacterial overgrowth as compared to rats without translocation. The prevalence of bacterial overgrowth and translocation in cirrhotic animals was 67 and 47–78%, respectively (Sanchez et al., 2005; Guarner et al., 2006; Garcia-Tsao et al., 1995; Runyon et al., 1994). The majority of translocated bacteria are Gram-negative (71%), and Gram-positive cocci were isolated from mesenteric lymph nodes in 29% (Guarner et al., 2006).

Following ligation of the common bile duct in rats as early as 3 days post-intervention, there was an increase of cecal bacteria and translocation to the mesenteric lymph nodes, whereas blood cultures remained sterile. Bacteria recovered from the mesenteric lymph nodes included *E. coli, Enterobacter, Pseudomonas, Proteus*, and *Staphylococcus* (Ding et al., 1994). There is evidence that conjugated bile acids regulate expression of host genes to promote innate defense against luminal bacteria. Conjugated bile acids bind to the nuclear receptor for bile acids, farnesoid receptor X (FXR), in enterocytes to induce the antimicrobial proteins angiogenin 1 and RNase family member 4 that prevent bacterial overgrowth and promote epithelial cell integrity (Inagaki et al., 2006). As liver disease results in an impaired bile flow and a decrease of conjugated bile acids in the small intestine, this might contribute to small intestinal bacterial overgrowth frequently observed in patients with chronic liver disease.

As PCR for bacterial DNA is more sensitive than microbial culture, one study found bacterial DNA in the serum and mesenteric lymph nodes in 58 and 63% of cirrhotic rats, respectively (Guarner et al., 2006). Together, these data suggest that intestinal bacterial overgrowth is one of the factors predisposing to bacterial translocation.

Using jejunal self-filling blind loops to create small bowel bacterial overgrowth causes significant hepatic inflammation, leading to fibrosis in susceptible rat strains (Lichtman et al., 1990). Antibiotics and intestinal decontamination prevent hepatic injury and fibrosis induced by CCl_4 or choline-deficient diet, suggesting an important contributory role of the intestinal microflora to liver fibrosis progression (Rutenburg et al., 1957). As selective intestinal decontamination with antibiotics prevents spontaneous bacterial peritonitis, the port of entry for bacteria appears to be the gut (Chang et al., 1998; Singh et al., 1995; Runyon et al., 1995). Thus, animal studies emphasize the importance of intestinal bacterial overgrowth and bacterial translocation as a modifier for liver disease progression.

The exact reason for increased bacterial translocation is currently unknown. In cirrhotic rats there is an increased number of mucosal lymphocytes in the intestine of cirrhotic mice, although their interferon (IFN)-γ production seems to be impaired resulting in a disruption of effective mucosal defense against intraluminal microorganisms (Inamura et al., 2003). Another study suggested that changes in the glycosylation of the intestinal mucosa and increased hydrophobicity of luminal bacteria results in increased bacterial adherence to epithelial cells, which might facilitate translocation across the epithelial barrier. An increase of sialic acid, fucose, hexose, and hexosamine has been observed in the brush border membrane and in the surfactant layer. In vitro interaction assays of normal bacteria to the brush border membrane isolated from cirrhotic rats showed a decreased adherence with mannose, galactose, and fucose treatment, but not with glucose or glucosamine treatment. This is interesting as it remains to be determined whether these changes are secondary to differences in epithelial cell gene expression or in the bacterial metagenome. As oxidative stress can modulate glycosylation on epithelial cell surface and alter the viscosity of the mucus, both can increase bacterial binding. This is supported by the fact that antioxidants reversed oxidative stress, changes in glycosylation and bacterial adherence (Natarajan et al., 2006). Together, these data indicate that intestinal mucosal damage is associated with changes in mucosal surface glycosylation, which might facilitate bacterial adherence. Other studies have confirmed increased oxidative damage in ileal and cecal mucosa of cirrhotic rats (Chiva et al., 2003).

Dendritic cells are key effectors in bacterial trafficking from the intestinal lumen to mesenteric lymph nodes and are capable of sampling bacteria through adjacent epithelial cells and of carrying bacteria to the mesenteric lymph nodes. An immune response involving B cells and T cells is then induced in mesenteric lymph nodes (Guarner et al., 2006). B cells and T cells interact with antigen presenting cells (APCs) to initiate responses in the draining mesenteric lymph nodes. This is supported by a study investigating Th1 polarization and monocyte activation in mesenteric lymph nodes of cirrhotic rats. TNF levels produced by monocytes and IFN-γ secreted by T cells were found to be increased, indicative of an immune response by enteric bacteria initiated in mesenteric lymph nodes in cirrhosis (Munoz et al., 2005). TNF is increased in culture-positive mesenteric lymph nodes from patients with liver cirrhosis (Genesca et al., 2003). Bacterial translocation is reduced in cirrhotic rats treated with anti-TNF antibodies (Frances et al., 2007). These findings support a role for TNF in the development of a disturbed function of the intestinal barrier as it might contribute to intestinal leakiness.

Using enteric antibiotics to reduce the intestinal microflora and bacterial translocation to mesenteric lymph nodes was found to attenuate the expansion of activated T cells and monocytes in the mesenteric lymph nodes and to normalize TNF and IFN-γ production in the mesenteric lymph nodes and blood. These findings suggest a strong link between the ongoing pro-inflammatory immune response observed in the mesenteric lymph nodes and blood of rats with cirrhosis and gut bacterial antigens (Munoz et al., 2005).

Intestinal Bacterial Overgrowth in Patients with Chronic Liver Disease

There is ample evidence that patients with chronic liver disease of various etiologies have intestinal bacterial overgrowth. The prevalence of intestinal bacterial overgrowth was significantly higher in patients with advanced liver dysfunction (13–53% in Child–Pugh Class A, 27–69% in Child–Pugh Class B, and 48–75% in Child–Pugh Class C) (Morencos et al., 1995; Pande et al., 2009; Bauer et al., 2001; Jun et al., 2009). Overgrowth was predominantly due to oropharyngeal flora including *Streptococci, Veillonella* spp, *Corynebacterium* spp, *Stomatococcus*, and *Staphylococcus*, and also included *Bacteroidetes* spp., *E. coli*, and *Klebsiella* spp. *Bifidobacterium* in the fecal flora of patients with liver cirrhosis was decreased as compared to healthy controls (Zhao et al., 2004). The cause for increased intestinal bacterial overgrowth in patients with chronic liver disease has not been fully elucidated. A correlation between low gastric pH and increased bacteria in the jejunum was described in patients with liver cirrhosis (Bauer et al., 2001; Shindo et al.,

1993). A delayed transit time as characterized by an alteration of the migrating motor complex (MMC) in the small intestine has been observed in cirrhotic patients causing intestinal stasis and overgrowth (Chang et al., 1998; Madrid et al., 1997; Gunnarsdottir et al., 2003). This is supported by a clinical study in which a pro-motility agent cisapride reversed small intestinal dysmotility and bacterial overgrowth in patients with liver cirrhosis (Madrid et al., 2001). Interestingly, the patient cohort also showed an improvement in liver function as assessed by the Child–Pugh score and hepatic encephalopathy (Madrid et al., 2001). There are currently no studies assessing the microbiome and metagenome in early stages of chronic liver disease.

Dysbiotic gut microbiota might contribute to the initiation and/or progression of liver disease by several mechanisms, including an increase in intestinal permeability. Microbiota and their products modulate barrier function by affecting epithelial pro-inflammatory responses and mucosal repair functions. In addition, a dysbiotic microflora as source of profibrogenic mediators (e.g., endotoxin) can contribute to disease progression given that there is a coexisting increase in intestinal permeability. However, a number of questions remain. First, what is the cause for intestinal bacterial overgrowth? Second, is intestinal dysbiosis and overgrowth a prerequisite for bacterial translocation or is it just an innocent bystander? The intact mucosal barrier likely prevents detection of bacteria by TLR-bearing cells such as macrophages. In the case of epithelial cell damage and bacterial penetration and translocation, macrophages recognize and detect bacteria via TLRs. Bacterial translocation could thus drive a damage response with overexpression of pro-inflammatory cytokines, leading to disruption of epithelial tight junctions and increased intestinal permeability as observed in patients with chronic liver disease.

Bacterial Translocation in Cirrhotic Patients

Bacterial overgrowth might contribute to bacterial translocation and liver dysfunction. This is indeed shown as bacterial DNA in peripheral blood was more frequently found in patients with increased disease severity and in cirrhotic patients with small intestinal bacterial overgrowth as compared to patients without overgrowth (Jun et al., 2009). Furthermore, small intestinal bacterial overgrowth was an independent and a major risk factor for the presence of bacterial DNA in cirrhotic patients. In addition, patients with bacterial overgrowth of the small intestine develop spontaneous bacterial peritonitis (SBP) more frequently than patients without bacterial overgrowth (Chang et al., 1998; Morencos et al., 1995).

Intestinal permeability was significantly increased in patients with liver cirrhosis as compared to healthy controls and was more relevant with overt clinical complications as measured by urinary excretion of indigested carbohydrates (Pascual et al., 2003). Direct evidence for increased intestinal translocation comes from two studies demonstrating increased bacterial DNA in patients with cirrhosis. Bacterial DNA in peripheral blood was present in 32–34% of cirrhotic patients and in all cases of spontaneous bacterial peritonitis (Frances et al., 2008; Such et al., 2002). A study of patients with liver cirrhosis who underwent liver transplantation or liver resection showed that bacterial translocation to the mesenteric lymph nodes increased with advanced cirrhosis (3% in Child–Pugh Class A, 8% in Child–Pugh Class B, and 31% in Child–Pugh Class C). Selective intestinal decontamination decreased translocation to the mesenteric lymph nodes to the level of noncirrhotic patients. Organisms of enteric origin grown from mesenteric lymph nodes included *Clostridium perfingens, E. coli, S. viridans, Citrobacter* spp., *Enterococcus*, and *S. equinus* (Cirera et al., 2001). Interestingly, only two of these translocated enteric organisms were Gram-negative bacteria.

Changes in the gut microbiome may also affect levels of mediators for liver injury. Increases in plasma LPS have been observed in cirrhotic patients with small intestinal bacterial overgrowth (Bauer et al., 2002). Endotoxemia occurs frequently in humans with a variety of chronic liver diseases such as chronic hepatitis, and the severity of liver disease correlates with the degree of endotoxemia (Lin et al., 1995). Endotoxemia has been associated with decreased survival rates in patients with liver cirrhosis (Bauer et al., 2002).

Potential Interventions Using Antibiotics, Prebiotics, Probiotics, or Synbiotics

Nonabsorbable or poorly absorbable antibiotics are used for selective intestinal decontamination to reduce the enteric flora or eliminate specific bacteria species.

The use of probiotics is an emerging therapeutic strategy in chronic liver disease. Probiotics are defined as microorganisms (bacteria or yeast) that induce the growth of other microorganisms causing or supporting the balance of autochthonic microflora in the gastrointestinal tract, when consumed in adequate amounts. Probiotics could presumably exert the same beneficial effects as the commensal microflora does. Probiotics exert their beneficial health effects to the host via many different mechanisms, including remodeling of microbial communities, immunomodulation by upregulation of anti-inflammatory factors, and suppression of pro-inflammatory mediators, effects on epithelial cell differentiation and proliferation (Preidis and Versalovic, 2009). These also include or may result in augmentation of the intestinal barrier function and prevention of bacterial translocation, which are especially important in chronic liver disease (Zareie et al., 2006; Sheth and Garcia-Tsao, 2008).

Prebiotics are short-chain carbohydrates or more complex saccharides that cannot be digested by pancreatic and brush-border enzymes, but can be selectively used and fermented by the commensal microbiota. Prebiotics can enhance the proliferation of beneficial microbes by manipulation of the energy source in an attempt to remodel intestinal microbial communities. As this strategy emphasizes the functional significance of these dynamic bacterial communities, the therapeutic manipulation of the microflora has the potential to ameliorate disease.

Combination of the above three components (synbiotics) might provide synergistic and more effective therapies. Directed manipulation of the microbiome within the gastrointestinal tract may yield health benefits. Combining data from metagenomics and metabolomics will facilitate rational attempts to manipulate commensal microflora by dietary supplements.

Animal Studies

As discussed above, selective intestinal decontamination using antibiotics results in reduced hepatic fibrogenesis in various animal models of liver injury and fibrosis (Seki et al., 2007; Adachi et al., 1995; Enomoto et al., 1999, 2001; Rutenburg et al., 1957).

There are conflicting results whether probiotics can prevent bacterial translocation in rodent models of end-stage liver disease. *Lactobacillus* spp. or *Lactobacillus acidophilus* failed to prevent bacterial translocation to mesenteric lymph nodes or ascitic fluid (Bauer et al., 2002; Wiest et al., 2003). Probiotics in these animals did not sufficiently alter the fecal flora suggesting that alternative probiotics or higher dosages might be required (Sheth and Garcia-Tsao, 2008). *Lactobacillus johnsonii La1* in conjunction with antioxidants (vitamin C and glutamate) or antioxidants alone reduced bacterial translocation. As antioxdiants alone abrogated bacterial translocation, it is unclear what the effect of probiotics alone would have been (Chiva et al., 2002). A controlled study in ob/ob mice with fatty liver disease demonstrated improved liver histology with decreased fat deposition and liver enzymes in mice treated with a probiotic mixture containing *Streptococcus thermophilus, Bifidobacterium breve, Bifidobacterium longum, Bifidobacterium infantis, L. acidophilus, Lactobacillus plantarum, Lactobacillus casei*, and *Lactobacillus bulgaricus* (VSL#3) for 4 weeks. Insulin resistance and fatty acid oxidation were markedly improved in these mice (Li et al., 2003). Similar results were obtained using VSL#3 in mice fed a high-fat diet. Probiotics have been shown to increase hepatic NKT cells and reduce inflammatory signaling (Ma et al., 2008). In an animal model of NASH induced by methionine-choline deficient (MCD) diet feeding of VSL#3 ameliorated liver fibrosis, but not inflammation and steatosis (Velayudham et al., 2009).

Patients

Selective intestinal decontamination results in decreased endotoxin levels as surrogate markers of bacterial translocation after treatment of human cirrhotic patients with norfloxacin for 1 month (Rasaratnam et al., 2003). Antibiotic-induced changes in the composition of the gastrointestinal microflora by antibiotics can influence the susceptibility to specific enteric pathogens and can induce antibiotic resistance to other microorganisms. Antibiotics disturb the normal mechanisms of microbial community regulation and pathogen colonization resistance (Brandl et al., 2008). Results from long-term treatment of cirrhotic patients are currently lacking.

A randomized controlled trial using *E. coli* Nissle (Mutaflor) versus placebo for 42 days in 39 patients with liver cirrhosis reduced intestinal dysbiosis with a restoration of normal colonic colonization. This was characterized by increased numbers of *Lactobacilli* and *Bifidobacteria*. Although not statistically significant, there was a trend toward lower endotoxin concentrations and improvement in Child–Pugh classification (Lata et al., 2007). Synbiotic treatment of cirrhotic patients with lactic acid bacteria and fiber supplementation increased the fecal content of non-urease-producing *Lactobacillus* species at the expense of other bacterial species, which was associated with a significant improvement of hepatic encephalopathy, a common neurological complication in patients with liver cirrhosis. Synbiotic treatment was also associated with a significant reduction in endotoxemia and improvement of the Child–Pugh score in nearly 50% of cases. Improvement was significantly lower in patients receiving fiber alone (29%) or placebo (8%) (Liu et al., 2004). Probiotic treatment (*Bifidobacterium bifidum* and *L. plantarum*) for 5 days resulted in restoration of colonic bowel flora and improvement of liver tests in patients with alcoholic liver injury (Kirpich et al., 2008). A possible mechanism is by restoring phagocytic neutrophil activity (Stadlbauer et al., 2008). A third open pilot study enrolled patients with chronic liver disease secondary to NAFLD or chronic hepatitis C infection and end-stage liver disease patients with alcoholic or hepatitis C-related cirrhosis. This nonrandomized and non-placebo-controlled study showed an improvement in liver function tests (AST, ALT, albumin, and bilirubin) and reduced levels of inflammatory serum cytokines after treatment with the probiotic mixture VSL#3 for 3 months (Loguercio et al., 2005). Despite these promising data from human trials, a recent Cochrane systematic review concluded that the lack of randomized control trials makes it impossible to support or refute probiotics for patients with NAFLD or NASH (Lirussi et al., 2007). Metagenomic analysis of the intestinal microbiome in patients with liver disease will provide us with a better understanding of probiotic interactions in the intestine and their possible combination with prebiotics.

Conclusion

Various alterations in the gastrointestinal tract of patients with chronic liver disease might not be as clinically evident as other clinical features, such as ascites, esophageal varices, and hepatic encephalopathy. However, changes in the intestinal microbiome and metagenome might cause increased intestinal permeability with resulting bacterial translocation and endotoxemia. Increased plasma levels of endotoxin drives hepatic fibrogenesis, and bacterial translocation is a major risk factor for infections and hence mortality in this patient population. Ongoing research initiatives are expected to expand our knowledge in the near future, providing data that will allow us to gain more insight into metagenomics and metabolomics of these patients. Appropriate use of knowledge uncovered by the gut microbiome and metagenome will help identifying key substrates and signaling mediators that help explain microbe–microbe and microbe–host interactions. Therapeutic approaches with probiotic, antibiotic, prebiotic, and synbiotic treatment will target specific microorganisms and metabolites and with a better understanding of the host–microbe interaction, the directed use of "biotics" will remodel microbial communities in a sophisticated and successful way. This strategy represents an attractive therapeutic approach for patients with chronic liver disease.

Acknowledgments I thank Dr. David A. Brenner, University of California, San Diego, for critically reading this manuscript.

References

Adachi Y, Moore LE, Bradford BU, Gao W, Thurman RG (1995) Antibiotics prevent liver injury in rats following long-term exposure to ethanol. Gastroenterology 108:218–224

Akira S, Uematsu S, Takeuchi O (2006) Pathogen recognition and innate immunity. Cell 124: 783–801

Backhed F, Ding H, Wang T, Hooper LV, Koh GY, Nagy A, Semenkovich CF, Gordon JI (2004) The gut microbiota as an environmental factor that regulates fat storage. Proc Natl Acad Sci USA 101:15718–15723

Backhed F, Manchester JK, Semenkovich CF, Gordon JI (2007) Mechanisms underlying the resistance to diet-induced obesity in germ-free mice. Proc Natl Acad Sci USA 104:979–984

Bataller R, Brenner DA (2005) Liver fibrosis. J Clin Invest 115:209–218

Bauer TM, Fernandez J, Navasa M, Vila J, Rodes J (2002) Failure of *Lactobacillus* spp. to prevent bacterial translocation in a rat model of experimental cirrhosis. J Hepatol 36:501–506

Bauer TM, Schwacha H, Steinbruckner B, Brinkmann FE, Ditzen AK, Aponte JJ, Pelz K, Berger D, Kist M, Blum HE (2002) Small intestinal bacterial overgrowth in human cirrhosis is associated with systemic endotoxemia. Am J Gastroenterol 97:2364–2370

Bauer TM, Steinbruckner B, Brinkmann FE, Ditzen AK, Schwacha H, Aponte JJ, Pelz K, Kist M, Blum HE (2001) Small intestinal bacterial overgrowth in patients with cirrhosis: prevalence and relation with spontaneous bacterial peritonitis. Am J Gastroenterol 96:2962–2967

Bode C, Bode JC (2005) Activation of the innate immune system and alcoholic liver disease: effects of ethanol per se or enhanced intestinal translocation of bacterial toxins induced by ethanol? Alcohol Clin Exp Res 29:166S–171S

Bode C, Kugler V, Bode JC (1987) Endotoxemia in patients with alcoholic and nonalcoholic cirrhosis and in subjects with no evidence of chronic liver disease following acute alcohol excess. J Hepatol 4:8–14

Bode JC, Bode C, Heidelbach R, Durr HK, Martini GA (1984) Jejunal microflora in patients with chronic alcohol abuse. Hepatogastroenterology 31:30–34

Brandl K, Plitas G, Mihu CN, Ubeda C, Jia T, Fleisher M, Schnabl B, DeMatteo RP, Pamer EG (2008) Vancomycin-resistant enterococci exploit antibiotic-induced innate immune deficits. Nature 455:804–807

Camp JG, Kanther M, Semova I, Rawls JF (2009) Patterns and scales in gastrointestinal microbial ecology. Gastroenterology 136:1989–2002

Casafont Morencos F, de las Heras Castano G, Martin Ramos L, Lopez Arias MJ, Ledesma F, Pons Romero F (1996) Small bowel bacterial overgrowth in patients with alcoholic cirrhosis. Dig Dis Sci 41:552–556

Chang CS, Chen GH, Lien HC, Yeh HZ (1998) Small intestine dysmotility and bacterial overgrowth in cirrhotic patients with spontaneous bacterial peritonitis. Hepatology 28: 1187–1190

Chiva M, Guarner C, Peralta C, Llovet T, Gomez G, Soriano G, Balanzo J (2003) Intestinal mucosal oxidative damage and bacterial translocation in cirrhotic rats. Eur J Gastroenterol Hepatol 15:145–150

Chiva M, Soriano G, Rochat I, Peralta C, Rochat F, Llovet T, Mirelis B, Schiffrin EJ, Guarner C, Balanzo J (2002) Effect of *Lactobacillus johnsonii La1* and antioxidants on intestinal flora and bacterial translocation in rats with experimental cirrhosis. J Hepatol 37:456–462

Cirera I, Bauer TM, Navasa M, Vila J, Grande L, Taura P, Fuster J, Garcia-Valdecasas JC, Lacy A, Suarez MJ, Rimola A, Rodes J (2001) Bacterial translocation of enteric organisms in patients with cirrhosis. J Hepatol 34:32–37

Ding JW, Andersson R, Soltesz V, Willen R, Bengmark S (1994) Obstructive jaundice impairs reticuloendothelial function and promotes bacterial translocation in the rat. J Surg Res 57: 238–245

Dumas ME, Barton RH, Toye A, Cloarec O, Blancher C, Rothwell A, Fearnside J, Tatoud R, Blanc V, Lindon JC, Mitchell SC, Holmes E, McCarthy MI, Scott J, Gauguier D, Nicholson JK (2006) Metabolic profiling reveals a contribution of gut microbiota to fatty liver phenotype in insulin-resistant mice. Proc Natl Acad Sci USA 103:12511–12516

Eckburg PB, Bik EM, Bernstein CN, Purdom E, Dethlefsen L, Sargent M, Gill SR, Nelson KE, Relman DA (2005) Diversity of the human intestinal microbial flora. Science 308:1635–1638

Enomoto N, Ikejima K, Yamashina S, Hirose M, Shimizu H, Kitamura T, Takei Y, Sato, Thurman RG (2001) Kupffer cell sensitization by alcohol involves increased permeability to gut-derived endotoxin. Alcohol Clin Exp Res 25:51S–54S

Enomoto N, Yamashina S, Kono H, Schemmer P, Rivera CA, Enomoto A, Nishiura T, Nishimura T, Brenner DA, Thurman RG (1999) Development of a new, simple rat model of early alcohol-induced liver injury based on sensitization of Kupffer cells. Hepatology 29:1680–1689

Ferrier L, Berard F, Debrauwer L, Chabo C, Langella P, Bueno L, Fioramonti J (2006) Impairment of the intestinal barrier by ethanol involves enteric microflora and mast cell activation in rodents. Am J Pathol 168:1148–1154

Fleming S, Toratani S, Shea-Donohue T, Kashiwabara Y, Vogel SN, Metcalf ES (2001) Pro- and anti-inflammatory gene expression in the murine small intestine and liver after chronic exposure to alcohol. Alcohol Clin Exp Res 25:579–589

Frances R, Chiva M, Sanchez E, Gonzalez-Navajas JM, Llovet T, Zapater P, Soriano G, Munoz C, Balanzo J, Perez-Mateo M, Song XY, Guarner C, Such J (2007) Bacterial translocation is downregulated by anti-TNF-alpha monoclonal antibody administration in rats with cirrhosis and ascites. J Hepatol 46:797–803

Frances R, Zapater P, Gonzalez-Navajas JM, Munoz C, Cano R, Moreu R, Pascual S, Bellot P, Perez-Mateo M, Such J (2008) Bacterial DNA in patients with cirrhosis and noninfected ascites mimics the soluble immune response established in patients with spontaneous bacterial peritonitis. Hepatology 47:978–985

Frank DN, St Amand AL, Feldman RA, Boedeker EC, Harpaz N, Pace NR (2007) Molecular–phylogenetic characterization of microbial community imbalances in human inflammatory bowel diseases. Proc Natl Acad Sci USA 104:13780–13785

Fukui H, Brauner B, Bode JC, Bode C (1991) Plasma endotoxin concentrations in patients with alcoholic and non-alcoholic liver disease: reevaluation with an improved chromogenic assay. J Hepatol 12:162–169

Gabele E, Muhlbauer M, Dorn C, Weiss TS, Froh M, Schnabl B, Wiest R, Scholmerich J, Obermeier F, Hellerbrand C (2008) Role of TLR9 in hepatic stellate cells and experimental liver fibrosis. Biochem Biophys Res Commun 376:271–276

Garcia-Tsao G, Lee FY, Barden GE, Cartun R, West AB (1995) Bacterial translocation to mesenteric lymph nodes is increased in cirrhotic rats with ascites. Gastroenterology 108: 1835–1841

Garrett WS, Lord GM, Punit S, Lugo-Villarino G, Mazmanian SK, Ito S, Glickman JN, Glimcher LH (2007) Communicable ulcerative colitis induced by T-bet deficiency in the innate immune system. Cell 131:33–45

Genesca J, Marti R, Rojo F, Campos F, Peribanez V, Gonzalez A, Castells L, Ruiz-Marcellan C, Margarit C, Esteban R, Guardia J, Segura R (2003) Increased tumour necrosis factor alpha production in mesenteric lymph nodes of cirrhotic patients with ascites. Gut 52:1054–1059

Gines P, Cardenas A, Arroyo V, Rodes J (2004) Management of cirrhosis and ascites. N Engl J Med 350:1646–1654

Guarner C, Gonzalez-Navajas JM, Sanchez E, Soriando G, Frances R, Chiva M, Zapater P, Benlloch S, Munoz C, Pascual S, Balanzo J, Perez-Mateo M, Such J (2006) The detection of bacterial DNA in blood of rats with CCl4-induced cirrhosis with ascites represents episodes of bacterial translocation. Hepatology 44:633–639

Guarner C, Runyon BA, Young S, Heck M, Sheikh MY (1997) Intestinal bacterial overgrowth and bacterial translocation in cirrhotic rats with ascites. J Hepatol 26:1372–1378

Gunnarsdottir SA, Sadik R, Shev S, Simren M, Sjovall H, Stotzer PO, Abrahamsson H, Olsson R, Bjornsson ES (2003) Small intestinal motility disturbances and bacterial overgrowth in patients with liver cirrhosis and portal hypertension. Am J Gastroenterol 98:1362–1370

Guo J, Loke J, Zheng F, Hong F, Yea S, Fukata M, Tarocchi M, Abar OT, Huang H, Sninsky JJ, Friedman SL (2009) Functional linkage of cirrhosis-predictive single nucleotide polymorphisms of Toll-like receptor 4 to hepatic stellate cell responses. Hepatology 49: 960–968

Huang H, Shiffman ML, Friedman S, Venkatesh R, Bzowej N, Abar OT, Rowland CM, Catanese JJ, Leong DU, Sninsky JJ, Layden TJ, Wright TL, White T, Cheung RC (2007) A 7 gene signature identifies the risk of developing cirrhosis in patients with chronic hepatitis C. Hepatology 46:297–306

Inagaki T, Moschetta A, Lee YK, Peng L, Zhao G, Downes M, Yu RT, Shelton JM, Richardson JA, Repa JJ, Mangelsdorf DJ, Kliewer SA (2006) Regulation of antibacterial defense in the small intestine by the nuclear bile acid receptor. Proc Natl Acad Sci USA 103:3920–3925

Inamura T, Miura S, Tsuzuki Y, Hara Y, Hokari R, Ogawa T, Teramoto K, Watanabe C, Kobayashi H, Nagata H, Ishii H (2003) Alteration of intestinal intraepithelial lymphocytes and increased bacterial translocation in a murine model of cirrhosis. Immunol Lett 90:3–11

Isayama F, Hines IN, Kremer M, Milton RJ, Byrd CL, Perry AW, McKim SE, Parsons C, Rippe RA, Wheeler MD (2006) LPS signaling enhances hepatic fibrogenesis caused by experimental cholestasis in mice. Am J Physiol Gastrointest Liver Physiol 290:G1318-1328

Jun DW, Kim KT, Lee OY, Chae JD, Son BK, Kim SH, Jo YJ, Park YS (2009) Association between small intestinal bacterial overgrowth and peripheral bacterial DNA in cirrhotic patients. Dig Dis Sci

Keshavarzian A, Farhadi A, Forsyth CB, Rangan J, Jakate S, Shaikh M, Banan A, Fields JZ (2009) Evidence that chronic alcohol exposure promotes intestinal oxidative stress, intestinal hyperpermeability and endotoxemia prior to development of alcoholic steatohepatitis in rats. J Hepatol 50:538–547

Kirpich IA, Solovieva NV, Leikhter SN, Shidakova NA, Lebedeva OV, Sidorov PI, Bazhukova TA, Soloviev AG, Barve SS, McClain CJ, Cave M (2008) Probiotics restore bowel flora and improve liver enzymes in human alcohol-induced liver injury: a pilot study. Alcohol 42:675–682

Lata J, Novotny I, Pribramska V, Jurankova J, Fric P, Kroupa R, Stiburek O (2007) The effect of probiotics on gut flora, level of endotoxin and Child–Pugh score in cirrhotic patients: results of a double-blind randomized study. Eur J Gastroenterol Hepatol 19:1111–1113

Ley RE, Backhed F, Turnbaugh P, Lozupone CA, Knight RD, Gordon JI (2005) Obesity alters gut microbial ecology. Proc Natl Acad Sci USA 102:11070–11075

Ley RE, Turnbaugh PJ, Klein S, Gordon JI (2006) Microbial ecology: human gut microbes associated with obesity. Nature 444:1022–1023

Li Z, Yang S, Lin H, Huang J, Watkins PA, Moser AB, Desimone C, Song XY, Diehl AM (2003) Probiotics and antibodies to TNF inhibit inflammatory activity and improve nonalcoholic fatty liver disease. Hepatology 37:343–350

Lichtman SN, Sartor RB, Keku J, Schwab JH (1990) Hepatic inflammation in rats with experimental small intestinal bacterial overgrowth. Gastroenterology 98:414–423

Lin RS, Lee FY, Lee SD, Tsai YT, Lin HC, Lu RH, Hsu WC, Huang CC, Wang SS, Lo KJ (1995) Endotoxemia in patients with chronic liver diseases: relationship to severity of liver diseases, presence of esophageal varices, and hyperdynamic circulation. J Hepatol 22: 165–172

Lirussi F, Mastropasqua E, Orando S, Orlando R (2007) Probiotics for non-alcoholic fatty liver disease and/or steatohepatitis. Cochrane Database Syst Rev (1): CD005165

Liu Q, Duan ZP, Ha DK, Bengmark S, Kurtovic J, Riordan SM (2004) Synbiotic modulation of gut flora: effect on minimal hepatic encephalopathy in patients with cirrhosis. Hepatology 39: 1441–1449

Loguercio C, Federico A, Tuccillo C, Terracciano F, D'Auria MV, De Simone C, Del Vecchio Blanco C (2005) Beneficial effects of a probiotic VSL#3 on parameters of liver dysfunction in chronic liver diseases. J Clin Gastroenterol 39:540–543

Ma X, Hua J, Li Z (2008) Probiotics improve high fat diet-induced hepatic steatosis and insulin resistance by increasing hepatic NKT cells. J Hepatol 49:821–830

Madrid AM, Cumsille F, Defilippi C (1997) Altered small bowel motility in patients with liver cirrhosis depends on severity of liver disease. Dig Dis Sci 42:738–742

Madrid AM, Hurtado C, Venegas M, Cumsille F, Defilippi C (2001) Long-term treatment with cisapride and antibiotics in liver cirrhosis: effect on small intestinal motility, bacterial overgrowth, and liver function. Am J Gastroenterol 96:1251–1255

Mason CM, Dobard E, Kolls J, Nelson S (1998) Effect of alcohol on bacterial translocation in rats. Alcohol Clin Exp Res 22:1640–1645

Medzhitov R (2007) Recognition of microorganisms and activation of the immune response. Nature 449:819–826

Miele L, Valenza V, La Torre G, Montalto M, Cammarota G, Ricci R, Masciana R, Forgione A, Gabrieli ML, Perotti G, Vecchio FM, Rapaccini G, Gasbarrini G, Day CP, Grieco A (2009) Increased intestinal permeability and tight junction alterations in nonalcoholic fatty liver disease. Hepatology 49:1877–1887

Morencos FC, de las Heras Castano G, Martin Ramos L, Lopez Arias MJ, Ledesma F, Pons Romero F (1995) Small bowel bacterial overgrowth in patients with alcoholic cirrhosis. Dig Dis Sci 40:1252–1256

Munoz L, Albillos A, Nieto M, Reyes E, Lledo L, Monserrat J, Sanz E, de la Hera A, Alvarez-Mon M (2005) Mesenteric Th1 polarization and monocyte TNF-alpha production: first steps to systemic inflammation in rats with cirrhosis. Hepatology 42:411–419

Mutlu E, Keshavarzian A, Engen P, Forsyth CB, Sikaroodi M, Gillevet P (2009) Intestinal dysbiosis: a possible mechanism of alcohol-induced endotoxemia and alcoholic steatohepatitis in rats. Alcohol Clin Exp Res 33:1836–1846

Nanji AA, Khettry U, Sadrzadeh SM, Yamanaka T (1993) Severity of liver injury in experimental alcoholic liver disease. Correlation with plasma endotoxin, prostaglandin E2, leukotriene B4, and thromboxane B2. Am J Pathol 142:367–373

Nanji AA, Khettry U, Sadrzadeh SM (1994) *Lactobacillus* feeding reduces endotoxemia and severity of experimental alcoholic liver (disease). Proc Soc Exp Biol Med 205: 243–247

Napolitano LM, Koruda MJ, Zimmerman K, McCowan K, Chang J, Meyer AA (1995) Chronic ethanol intake and burn injury: evidence for synergistic alteration in gut and immune integrity. J Trauma 38:198–207

Natarajan SK, Ramamoorthy P, Thomas S, Basivireddy J, Kang G, Ramachandran A, Pulimood AB, Balasubramanian KA (2006) Intestinal mucosal alterations in rats with carbon tetrachloride-induced cirrhosis: changes in glycosylation and luminal bacteria. Hepatology 43:837–846

Pande C, Kumar A, Sarin SK (2009) Small-intestinal bacterial overgrowth in cirrhosis is related to the severity of liver disease. Aliment Pharmacol Ther 29:1273–1281

Parks RW, Clements WD, Pope C, Halliday MI, Rowlands BJ, Diamond T (1996) Bacterial translocation and gut microflora in obstructive jaundice. J Anat 189(Pt 3):561–565

Parlesak A, Schafer C, Schutz T, Bode JC, Bode C (2000) Increased intestinal permeability to macromolecules and endotoxemia in patients with chronic alcohol abuse in different stages of alcohol-induced liver disease. J Hepatol 32:742–747

Pascual S, Such J, Esteban A, Zapater P, Casellas JA, Aparicio JR, Girona E, Gutierrez A, Carnices F, Palazon JM, Sola-Vera J, Perez-Mateo M (2003) Intestinal permeability is increased in patients with advanced cirrhosis. Hepatogastroenterology 50:1482–1486

Preidis GA, Versalovic J (2009) Targeting the human microbiome with antibiotics, probiotics, and prebiotics: gastroenterology enters the metagenomics era. Gastroenterology 136: 2015–2031

Rasaratnam B, Kaye D, Jennings G, Dudley F, Chin-Dusting J (2003) The effect of selective intestinal decontamination on the hyperdynamic circulatory state in cirrhosis. A randomized trial. Ann Intern Med 139:186–193

Runyon BA, Borzio M, Young S, Squier SU, Guarner C, Runyon MA (1995) Effect of selective bowel decontamination with norfloxacin on spontaneous bacterial peritonitis, translocation, and survival in an animal model of cirrhosis. Hepatology 21:1719–1724

Runyon BA, Squier S, Borzio M (1994) Translocation of gut bacteria in rats with cirrhosis to mesenteric lymph nodes partially explains the pathogenesis of spontaneous bacterial peritonitis. J Hepatol 21:792–796

Rutenburg AM, Sonnenblick E, Koven I, Aprahamian HA, Reiner L, Fine J (1957) The role of intestinal bacteria in the development of dietary cirrhosis in rats. J Exp Med 106:1–14

Sanchez E, Casafont F, Guerra A, de Benito I, Pons-Romero F (2005) Role of intestinal bacterial overgrowth and intestinal motility in bacterial translocation in experimental cirrhosis. Rev Esp Enferm Dig 97:805–814

Schenk S, Saberi M, Olefsky JM (2008) Insulin sensitivity: modulation by nutrients and inflammation. J Clin Invest 118:2992–3002

Seki E, De Minicis S, Osterreicher CH, Kluwe J, Osawa Y, Brenner DA, Schwabe RF (2007) TLR4 enhances TGF-beta signaling and hepatic fibrosis. Nat Med 13:1324–1332

Sheth AA, Garcia-Tsao G (2008) Probiotics and liver disease. J Clin Gastroenterol 42(Suppl 2):S80–84

Shindo K, Machida M, Miyakawa K, Fukumura M (1993) A syndrome of cirrhosis, achlorhydria, small intestinal bacterial overgrowth, and fat malabsorption. Am J Gastroenterol 88: 2084–2091

Singh N, Gayowski T, Yu VL, Wagener MM (1995) Trimethoprim-sulfamethoxazole for the prevention of spontaneous bacterial peritonitis in cirrhosis: a randomized trial. Ann Intern Med 122:595–598

Stadlbauer V, Mookerjee RP, Hodges S, Wright GA, Davies NA, Jalan R (2008) Effect of probiotic treatment on deranged neutrophil function and cytokine responses in patients with compensated alcoholic cirrhosis. J Hepatol 48:945–951

Such J, Frances R, Munoz C, Zapater P, Casellas JA, Cifuentes A, Rodriguez-Valera F, Pascual S, Sola-Vera J, Carnicer F, Uceda F, Palazon JM, Perez-Mateo M (2002) Detection and identification of bacterial DNA in patients with cirrhosis and culture-negative, nonneutrocytic ascites. Hepatology 36:135–141

Szczepaniak LS, Nurenberg P, Leonard D, Browning JD, Reingold JS, Grundy S, Hobbs HH, Dobbins RL (2005) Magnetic resonance spectroscopy to measure hepatic triglyceride content: prevalence of hepatic steatosis in the general population. Am J Physiol Endocrinol Metab 288:E462–468

Tabata T, Tani T, Endo Y, Hanasawa K (2002) Bacterial translocation and peptidoglycan translocation by acute ethanol administration. J Gastroenterol 37:726–731

Tamai H, Kato S, Horie Y, Ohki E, Yokoyama H, Ishii H (2000) Effect of acute ethanol administration on the intestinal absorption of endotoxin in rats. Alcohol Clin Exp Res 24:390–394

Turnbaugh PJ, Hamady M, Yatsunenko T, Cantarel BL, Duncan A, Ley RE, Sogin ML, Jones WJ, Roe BA, Affourtit JP, Egholm M, Henrissat B, Heath AC, Knight R, Gordon JI (2009) A core gut microbiome in obese and lean twins. Nature 457:480–484

Turnbaugh PJ, Ley RE, Mahowald MA, Magrini V, Mardis ER, Gordon JI (2006) An obesity-associated gut microbiome with increased capacity for energy harvest. Nature 444: 1027–1031

Uesugi T, Froh M, Arteel GE, Bradford BU, Thurman RG (2001) Toll-like receptor 4 is involved in the mechanism of early alcohol-induced liver injury in mice. Hepatology 34:101–108

Velayudham A, Dolganiuc A, Ellis M, Petrasek J, Kodys K, Mandrekar P, Szabo G (2009) VSL#3 probiotic treatment attenuates fibrosis without changes in steatohepatitis in a diet-induced nonalcoholic steatohepatitis model in mice. Hepatology 49:989–997

Wiest R, Chen F, Cadelina G, Groszmann RJ, Garcia-Tsao G (2003) Effect of *Lactobacillus*-fermented diets on bacterial translocation and intestinal flora in experimental prehepatic portal hypertension. Dig Dis Sci 48:1136–1141

Wiest R, Garcia-Tsao G (2005) Bacterial translocation (BT) in cirrhosis. Hepatology 41:422–433

Wigg AJ, Roberts-Thomson IC, Dymock RB, McCarthy PJ, Grose RH, Cummins AG (2001) The role of small intestinal bacterial overgrowth, intestinal permeability, endotoxaemia, and tumour necrosis factor alpha in the pathogenesis of non-alcoholic steatohepatitis. Gut 48:206–211

Wong F, Bernardi M, Balk R, Christman B, Moreau R, Garcia-Tsao G, Patch D, Soriano G, Hoefs J, Navasa M (2005) Sepsis in cirrhosis: report on the 7th meeting of the International Ascites Club. Gut 54:718–725

Yin M, Bradford BU, Wheeler MD, Uesugi T, Froh M, Goyert SM, Thurman RG (2001) Reduced early alcohol-induced liver injury in CD14-deficient mice. J Immunol 166:4737–4742

Yin M, Wheeler MD, Kono H, Bradford BU, Gallucci RM, Luster MI, Thurman RG (1999) Essential role of tumor necrosis factor alpha in alcohol-induced liver injury in mice. Gastroenterology 117:942–952

Zareie M, Johnson-Henry K, Jury J, Yang PC, Ngan BY, McKay DM, Soderholm JD, Perdue MH, Sherman PM (2006) Probiotics prevent bacterial translocation and improve intestinal barrier function in rats following chronic psychological stress. Gut 55:1553-1560

Zhao HY, Wang HJ, Lu Z, Xu SZ (2004) Intestinal microflora in patients with liver cirrhosis. Chin J Dig Dis 5:64–67

Chapter 14
Symbiotic Gut Microbiota and the Modulation of Human Metabolic Phenotypes

Lanjuan Li

Gut Microbiota and the Symbiotic Relationship with Human Beings

The human gut harbors an immensely diverse microbial ecological system. The gut microbial community is one of the most diverse and complex micro-ecological systems including trillions of microbes (approximately 1×10^{13} to 10^{14}) and 500–1000 species of microorganisms. The total number of microbial cells in the gut is estimated at 10 times that of our own cells, and its collective genome contains at least 100 times as many genes as our own genome. This composition of microorganisms has been thought of as the "microbiome" or "the second genome of human beings." Lederberg has emphasized the importance of having a broad macro-ecological view of our symbiotic microbiota as and such, human beings should be considered "superorganisms," composed of both human cells and the microbiome (Lederberg, 2000).

The Composition and Diversity of the Gut Microbiota

The human gastrointestinal tract consists mainly of three kinds of organisms: bacteria, archaea, and fungi (Ley et al., 2006), the majority of which are the bacteria. The human gut microbiota begins to take shape at birth and "matures" at about 2 years of age. The large intestine, particularly the colon, contains the highest density of microorganisms with up to 10^{12} 10^{14} bacteria, weighing about 1.2 kg. Most of the colon flora is composed of anaerobic bacteria and this accounts for 98% of the microorganisms. At the species and sub-species level, the intestinal flora shows a great diversity. However, at the level of the division, when compared with

L. Li (✉)
Dean of State Key Laboratory for Infectious Diseases Diagnosis and Treatment, The First Affiliated Hospital, College of Medicine, Zhejiang University, 79 QingChun Road, Hangzhou, Zhejiang 310003, People's Republic of China
e-mail: ljli@zju.edu.cn

K.E. Nelson (ed.), *Metagenomics of the Human Body*,
DOI 10.1007/978-1-4419-7089-3_14,

the Earth's "gastrointestinal tract" – the soil, diversity is lower in the human gastrointestinal tract. Regardless of race, whether in Orientals (Li et al., 2008) or the West (Eckburg et al., 2005), the human gut microbiota is dominated by the following divisions of bacteria: Firmicutes, Bacteroidetes, Proteobacteria, Actinobacteria, Verrucomicrobia, and Fusobacteria, with Bacteroidetes, Firmicutes, and one member of the archaea – *Methanobrevibacter smithii*, being most dominant (Li et al., 2008; Eckburg et al., 2005; Backhed et al., 2005). Whereas Proteobacteria are commonly found in the gut microbiota, they are usually not essential. The structure and composition of the intestinal microbiota reflect natural selection on two levels: (1) bacterial level, the proliferation speed and energy-utilizing patterns during the competition phase of initial colonization driving microbial cells to become functionally specialized; and (2) host level, the adaptability of the host to the gut microbiota and favoring the stable communities with a high degree of functional redundancy (Ley et al., 2006; Backhed et al., 2005).

The Function of Gut Microbiota

In recent years, with the availability of new and emerging techniques, the gut microbiota has been shown to influence many aspects of host biology, including dietary calorific bioavailability, epithelial development, and immune status, as well as drug metabolism and toxicity (Ley et al., 2006; Guarner and Malagelada, 2003; O'Hara and Shanahan, 2006) and is considered an important metabolic "organ" (O'Hara and Shanahan, 2006). The gut microbiota, which over certain periods of time is relatively stable, cannot only display individual characteristics, but also evolve throughout life from birth to old age. Therefore, human intestinal microbial groups are considered to be a human's "second fingerprint." In a healthy state, the gut microbiota and genetics (Zoetendal et al., 2001), age (Mariat et al., 2009), gender, and living environment (Ley et al., 2005) have significant correlations (Mueller et al., 2006). This has gained the attention of scientists to find if there are direct links between the gut microbiota and the living environment, age, health, and disease of the host; to study whether the microbiota can serve as a biomarker of disease; and to find effective intervention methods, directed at the microbiota to decrease disease susceptibility and to prevent and possibly treat disease.

Gut Microbiota in Health and Disease

Currently, most research indicates that our gut microbiota plays an intricate and crucial role in our health and disease.

Gut Microbiota with Infectious Disease

Accumulating evidence has demonstrated that the gut microbiota has a crucial role in the development of conditions such as spontaneous bacterial infections, hyperdynamic circulatory state of cirrhosis, and liver failure. First, bacterial translocation or imbalance of the gut microbiota often causes infection (Guarner and Malagelada, 2003; Xing et al., 2005; Xing et al., 2005). For example, *Bacteroides fragilis* is a virulent anaerobic pathogen that can cause considerable mortality. *Bacteroides fragilis* often causes a variety of diseases such as diarrhea, endogenous abdominal abscess, wound infection, and bacteremia (Duerden, 1994; Myers et al., 1987; Sack et al., 1994). Second, the dysbiosis of gut microbiota is often observed in patients with viral diseases, such as Hepatitis B, and could affect recovery or worsen the disease. For example, manipulation of the gut microbiota in Sprague-Dawley (Li et al., 2010) rats by different bacteria could have an effect on the prevention or exacerbation of acute liver injury, which has been shown to cause a disruption of the gut microbiota. The concept that gut microbiota matters in the pathogenesis of liver disease is not novel (see previous chapter). It has been highlighted that gut-derived bacteria and their products play a pivotal role in many aspects of liver disease, including the development of spontaneous bacterial infections and the worsening of the hyperdynamic circulatory state of cirrhosis, which leads to complications (Sheth and Garcia-Tsao, 2008). Therefore, we propose the hypothesis that disruption of the normal microbiota increases the risk for gut infections, disrupts colonization resistance, increases the permeability of the enteric mucosa, causes gut-derived endotoxemia, damages liver function, and eventually leads to liver failure. Therefore, the gut microbiota not only plays an important direct role in gut-derived bacterial infections (Solga and Diehl, 2004), but also indirectly influences the outcome or development of infectious diseases of viral origin (Garcia-Tsao et al., 1995; Lin et al., 1995).

With this in mind, scientists anticipate that many infectious diseases such as spontaneous bacterial peritonitis in cirrhotic patients, which might be closely associated with endotoxemia and bacterial translocation, can be prevented or treated using the endogenous gut microbiota. Several studies have provided evidence that manipulation of the intestinal microbiota could significantly reduce the incidence of infection, alleviate patient symptoms, improve quality of life, and finally increase the survival time of some patients, especially those with hepatic encephalopathy (Solga and Diehl, 2004; Qing Liu et al., 2004).

Gut Microbiota in Other Chronic Diseases

Much attention is currently being paid to the role that gut microbiota has in modifying the human metabolic phenotype, the metabolism, and toxicity of drugs, and the predisposition of humans to health and disease (Ley et al., 2006). Over the past decade, research has indicated that the microbiota may not only be exerting effects beyond infectious disease, but may also be involved in the development of numerous conditions, including Crohn's disease, obesity, type-2 diabetes, and nonalcohol fatty liver diseases. Patients with these diseases display an altered gut microbiota, which may be involved in the pathogenesis of these diseases and could be partly prevented and treated by antibiotics, probiotics, and prebiotics. Although it is well known that heredity is an important deciding factor of weight (Farooqi and O'Rahilly, 2006), the latest results from Jeff Gordon's group at Washington

University suggest that the gut microbiota is also an important environmental factor that affects energy balance in the host (Backhed et al., 2004). Soon after this initial study, they observed that obesity also affects the diversity of the gut microbiota (Ley et al., 2005). These findings raise scientists' hope of manipulating the microbiome to treat or prevent obesity and other associated diseases. In addition, Kelly and others found that gut microbes affect the transcription and regulation of the peroxisome proliferator-activated receptor (peroxisome proliferator-activated receptor-γ, PPAR-γ) in mammals (Kelly et al., 2004), which might have a close association with the occurrence of insulin resistance and the development of type-2 diabetes. In addition, studies show that there is a strong correlation between the gut microbiota and a fatty liver phenotype based on metabonomics investigation in insulin-resistant mice (Dumas et al., 2006b). Additionally, the excessive growth of certain intestinal *Clostridium* could produce neurotoxins, which may be the main reason for autism (Parracho et al., 2005).

Research Platforms and Technologies for Studying Gut Microbiota

Molecular Ecological Technologies Based on the 16S rRNA Gene

Characterization of the gut microbiota has been hindered by the refractory cultivation of most species. Modern molecular analyses based on 16S ribosomal (r)DNA sequences of human microbiota, such as fluorescence in situ hybridization (FISH), denaturing gradient gel electrophoresis (DGGE) (Zoetendal et al., 2001; Muyzer and Smalla, 1998), and terminal restriction fragment length polymorphism (T-RFLP) (Hayashi et al., 2003), allow us to quickly profile gut microbiota. In addition, comprehensive coverage of 16S rDNA clone libraries and random sequencing has been successfully used in the analysis of gut microbiota and variations associated with time, diet, race, and sex, which might contribute to differences in physiology between individuals or environments that predispose individuals to disease (Li et al., 2008; Eckburg et al., 2005). For example, using DNA sequencing-based analysis, Eckburg et al. (2005) and Li et al. (2008) revealed significant differences in the gut bacterial community between healthy adults. Additionally, Li et al. observed significant clustering of gut microbiota in a family cohort compared with the three American individuals included in the study by Eckburg et al. Such studies have, however, been limited by little or no functional correlation with the sequenced gene fragments. In addition, the molecular mechanisms of interaction between microbiota (microbiome) variation and human physiology have not been adequately described, nor have the key microbiota contributors been identified.

Metagenomics, which refers to the genomic analysis of microorganisms by direct extraction and cloning of DNA from an assemblage of microorganisms, provides insight into the genetic potential of complex microbial communities, including uncultured species, using culture-independent approaches (Handelsman, 2004).

Through metagenomic analysis of the human distal gut microbiome, Gill et al. were the first to show that our microbiome is significantly enriched for genes related to the metabolism of glycans, amino acids, and xenobiotics, methanogenesis, and the 2-methyl-D-erythritol 4-phosphate pathway-mediated biosynthesis of vitamins and isoprenoids. That study provided evidence that humans are actually superorganisms whose metabolism represents an amalgamation of microbial and human attributes (Gill et al., 2006).

Currently, established metabonomics techniques provide a powerful tool to detect the gut microbiota-derived metabolites, which are pertaining to pathophysiology and predisposition for health and disease (Nicholson et al., 2005; Nicholson and Lindon, 2008; Nicholson and Wilson, 2003). The development and utilization of metagenomics and metabolomics platforms that allow a high-throughput and robust analysis of the symbiotic microbiota will continue to improve our understanding of the co-evolution of human beings and their gut microbiota.

Metabonomics Study of "Co-metabolism" of Host and Microbiota

Metabolic products are the result of living activities regulated by genetic and environmental factors and are a much better representation of microbial/human activity than the information derived from genomics and proteomics. Emerging after genomics, transcriptomics, and proteomics, metabonomics is not only a complementary method to genomics, but is also an important component of systems biology. Metabonomics aims to measure the global, dynamic metabolic response of living systems to biological stimuli or genetic manipulation and focuses on understanding systemic change over time in complex muticellular systems (Nicholson and Wilson, 2003). With analytical techniques such as nuclear magnetic resonance (NMR) spectroscopy, gas or liquid chromatography-mass spectrometry (GC/LC-MS), and capillary electrochromatography (CEC), which are "mature" enough to be used, metabonomics has been widely applied in investigating human metabolic phenotypes and co-metabolism with gut microbiota. Because of the high throughput, high sensitivity, and high accuracy of the techniques used, metabonomics plays an important role in the understanding of molecular interactions and their mechanisms between the gut microbiota and host. For example, by employing (1)H NMR spectroscopy and multivariate pattern recognition techniques, Wang et al. found reduced levels of tricarboxylic acid cycle intermediates and a disturbance of amino acid metabolism in the urine of *Schistosoma mansoni*-infected mice, indicating disturbances in the gut microbiota (Wang et al., 2004). Using plasma and urine metabotyping, Dumas et al. revealed an intricate relationship between gut microbiota and host co-metabolic phenotypes associated with dietary-induced impaired glucose homeostasis and NAFLD in a mouse strain (129S6) and found that gut microbiota may play an active role in the development of insulin resistance (Dumas et al., 2006b). The gut microbiome has also been confirmed to contribute significantly to different metabolic profiles under identical treatments in experimental rats with identical genetic backgrounds (Ebbels et al., 2003). These studies also revealed that gut microbiota plays an important role in host metabolism, and that the changing of gut microbiota could modulate the metabolic phenotype of the host.

Recently, the "marriage" of metagenomics and metabonomics, profiling urinary, fecal, or other body fluids, has been proven to be a very effective approach. This strategy can provide information on holistic and dynamic biochemical changes, link phenotypic variation to different compositional structures of the gut microbiota, reveal the molecular mechanisms for the interaction between gut microbiota and the host, and help to recognize the comprehensive disease process by combining the DNA (RNA) information with the metabolite information by the help of multivariate statistical analyses. Many significant results have been acquired by this strategy (Li et al., 2008; Martin et al., 2009). Such a top-down strategy is crucial to resolve the contribution of gut microbiota to health and disease (Martin et al., 2007).

The Modulation of Gut Microbiota on the Host Metabolic Phenotype

Metabolism is the most essential process of living organisms. The unique composition of gut microbiota in each mammal may thus play an important role in the host's metabolism because these biotas add new functional diversity to the host at the whole organism level, which includes participation in the development of pathophysiology (Dumas et al., 2006b) and providing complimentary metabolic pathways for drugs and dietary nutrients (Nicholson et al., 2005). Evidence indicates that the gut microbiota is significantly involved in host metabolism. On the one hand, the composition of gut microbiota could directly determine host nutrition, drug metabolism, and absorption, including caloric extraction efficiency (Turnbaugh et al., 2006), vitamin synthesis (Bentley and Meganathan, 1982; Bhattacharyya et al., 1997), and calcium and magnesium absorption (Roberfroid et al., 1995; Younes et al., 2001).

The gut microbiota is composed of various enzymes and biochemical pathways, which our own genomes have not evolved. The major metabolic function of gut microbiota is fermentation of carbohydrates supplying energy for the host, especially for the gut epithelium (Guarner and Malagelada, 2003). Endpoint metabolites, such as acetate and propionate, show a modulating effect on glucose metabolism and insulin resistance (Brighenti et al., 1995; Venter et al., 1990). For example, gut microbiota could contribute extensively to inter-individual variation in isoflavone metabolism (Rowland et al., 2000). Dumas' large-scale population phenotyping in Aito Town, Japan ($n = 259$), Chicago, IL ($n = 315$), and Guangxi, China ($n = 278$) also shows that gut microbial factors could be related to cross-population differences in urinary metabolites (Dumas et al., 2006a).

On the contrary, modulating or intervention of gut microbiota could change not only the pathophysiology characteristics and metabolic phenotype of the host, but also the predisposition to some diseases. Symbiotic gut microorganisms interact closely with the mammalian host's metabolism and are important determinants of the host metabolic phenotype. The metabolic phenotype could also be considered a result of co-regulation of the host genome and the microbiome (Dumas et al., 2006b). Meanwhile, there are active metabolic exchanges and "co-metabolism" processes between our gut microbiota and ourselves (Nicholson et al., 2004). The diversity of gut microbiota contributes a lot to the complex metabolic characteristics of the mammalian host and the microbiome modulations correlate with the intestinal biochemistry (Martin et al., 2009). For example, genetically identical animals housed together can show inter-animal metabolic variation, which might be attributed to variations in their microbiota and therefore their co-metabolism (Nicholson et al., 2005). Recently, a top-down systems biology study of microbiome–mammalian metabolic interactions was performed in a mouse model and revealed that the microbiome modulates absorption, storage, and the energy harvest from the diet at the systems level (Martin et al., 2007). Additionally, the specific components of the gut microbiota might also be a determinant for the ability of some subjects to produce equol. In Bowey et al., the colonization of germ-free (GF) rats with a fecal flora from a human subject with the capacity to convert daidzein to equol, resulted in the rats excreting substantial amounts of the metabolite. By contrast, equol was undetectable in urine of human flora associated (HFA) rats associated with a fecal flora from a low equol-producing subject (Bowey et al., 2003).

The gut microbiota was recently proposed as an environmental factor closely correlating with the control of body weight and energy metabolism. The symbiotic gut microbiota can have a significant influence on host health, disease etiology, and predisposition to certain diseases, especially metabolic diseases (Nicholson et al., 2005; Rolando et al., 1996). Findings observed from the pharmacometabonomic analysis of predose metabolic profiles shows that the inter-subject variation of gut microbiota, to a great extent, predisposes different pathophysiological outcomes upon diet alteration or chemical stimulus (Nicholson et al., 2005; Rolando et al., 1996). A vast array of microbiota and mammalian co-metabolites constantly exchange through the enterohepatic circulation between the intestine and liver. The liver is an essential organ in mammals and is the central organ for metabolic processes. Previous data have shown that gut microbiota may play an active role in the development of insulin resistance, liver metabolism function, and even liver diseases such as NAFLD (Dumas et al., 2006b) and liver failure (Rolando et al., 1996; Li et al., 2004).

In addition to conferring active metabolite exchange, gut microbiota has an important metabolic capability of supplying complementary metabolic pathways for xenobiotics and drugs (Nicholson et al., 2005; Nicholson, 2006) and activating the mammalian liver enzyme systems (Nicholson and Wilson, 2003), which could also significantly influence personalized health treatment and care. With the accumulating evidence that gut microbiota has a crucial role in health and disease, scientists have begun exploring gut-microbiota-targeted modulation and therapies. Recent research into probiotics or prebiotics indicates that they could be used as dietary regulators to modulate the composition of gut microbiota, confer beneficial effects on the host, and finally modulate the mammalian phenotype. For example, transgenomic metabolic effects of exposure to either *Lactobacillus paracasei* or

Lactobacillus rhamnosus probiotics in humanized extended genome mice (GF mice colonized with human baby flora) indicate that probiotic exposure exerted microbiome modification and resulted in altered hepatic lipid metabolism coupled with lowered plasma lipoprotein levels and apparent stimulated glycolysis. Meanwhile, probiotic treatments also altered a diverse range of pathways and outcomes, including amino-acid metabolism, methylamines, and short-chain fatty acids (SCFAs) (Martin et al., 2008). In addition, various studies have reported many probiotic effects, including the management of minimal hepatic encephalopathy in patients with cirrhosis (Solga and Diehl, 2004), resistance to infection, improvement of allergic disease, anti-inflammatory properties, and NAFLD.

Perspectives for Future Gut Microbiota-Oriented Studies

Accumulating evidence shows that our gut microbiota plays intricate and key roles in our health and disease. The new emerging "omics" techniques, such as metagenomics and metabonomics, are now supplying us with new platforms for generating detailed insights into the gut microbial ecology, revealing the potential interaction mechanisms between the microbiome and our own genome, and supplying us with new clues for prevention and treatment of host diseases.

Although controversial, it is widely believed that the use of probiotics, prebiotics, or antibiotics needs to be explored in more detail as the composition of the commensal gut microbiota is readily changeable, and would allow for disruption in manipulating the microbiome to improve or correct microbiota-associated host disease. Therefore, we envisage that the combined use of probiotics, prebiotics, and antibiotics could be applied to manipulate the gut microbiota to restore the balance of gut micro-ecology in the host, modulate the symbiotic gut microbial–host metabolic interactions, and effectively prevent and treat disease.

References

Backhed F et al (2005) Host–bacterial mutualism in the human intestine. Science 307(5717):1915–1920

Backhed F et al (2004) The gut microbiota as an environmental factor that regulates fat storage. Proc Natl Acad Sci USA 101(44):15718–15723

Bentley R, Meganathan R (1982) Biosynthesis of vitamin K (menaquinone) in bacteria. Microbiol Rev 46(3):241–280

Bhattacharyya DK, Kwon O, Meganathan R (1997) Vitamin K2 (menaquinone) biosynthesis in *Escherichia coli*: evidence for the presence of an essential histidine residue in o-succinylbenzoyl coenzyme A synthetase. J Bacteriol 179(19):6061–6065

Bowey E, Adlercreutz H, Rowland I (2003) Metabolism of isoflavones and lignans by the gut microflora: a study in germ-free and human flora associated rats. Food Chem Toxicol 41(5):631–636

Brighenti F et al (1995) Effect of neutralized and native vinegar on blood glucose and acetate responses to a mixed meal in healthy subjects. Eur J Clin Nutr 49(4):242–247

Duerden BI (1994) Virulence factors in anaerobes. Clin Infect Dis 18(Suppl 4):S253–S259

Dumas ME et al (2006a) Assessment of analytical reproducibility of 1H NMR spectroscopy based metabonomics for large-scale epidemiological research: the INTERMAP study. Anal Chem 78(7):2199–2208

Dumas ME et al (2006b) Metabolic profiling reveals a contribution of gut microbiota to fatty liver phenotype in insulin-resistant mice. Proc Natl Acad Sci USA 103(33):12511–12516

Ebbels T, Beckonert O, Antti H, Bollard M, Holmes E, Lindon J, Nicholson J (2003) Toxicity classification from metabonomic data using a density superposition approach: 'CLOUDS'. Anal Chim Acta 490:109–122

Eckburg PB et al (2005) Diversity of the human intestinal microbial flora. Science 308(5728):1635–1638

Farooqi S, O'Rahilly S (2006) Genetics of obesity in humans. Endocr Rev 27(7):710–718

Garcia-Tsao G et al (1995) Bacterial translocation to mesenteric lymph nodes is increased in cirrhotic rats with ascites. Gastroenterology 108(6):1835–1841

Gill SR et al (2006) Metagenomic analysis of the human distal gut microbiome. Science 312(5778):1355–1359

Guarner F, Malagelada JR (2003) Gut flora in health and disease. Lancet 361(9356):512–519

Handelsman J (2004) Metagenomics: application of genomics to uncultured microorganisms. Microbiol Mol Biol Rev 68(4):669–685

Hayashi H et al (2003) Molecular analysis of fecal microbiota in elderly individuals using 16S rDNA library and T-RFLP. Microbiol Immunol 47(8):557–570

Kelly D et al (2004) Commensal anaerobic gut bacteria attenuate inflammation by regulating nuclear-cytoplasmic shuttling of PPAR-gamma and RelA. Nat Immunol 5(1):104–112

Lederberg J (2000) Infectious history. Science 288(5464):287–293

Ley RE, Peterson DA, Gordon JI (2006) Ecological and evolutionary forces shaping microbial diversity in the human intestine. Cell 124(4):837–848

Ley RE et al (2005) Obesity alters gut microbial ecology. Proc Natl Acad Sci USA 102(31): 11070–11075

Li H et al (2007) Pharmacometabonomic phenotyping reveals different responses to xenobiotic intervention in rats. J Proteome Res 6(4):1364–1370

Li LJ et al (2004) Changes of gut flora and endotoxin in rats with D-galactosamine-induced acute liver failure. World J Gastroenterol 10(14):2087–2090

Li M et al (2008) Symbiotic gut microbes modulate human metabolic phenotypes. Proc Natl Acad Sci USA 105(6):2117–2122

Li YT et al (2010) Effects of gut microflora on hepatic damage after acute liver injury in rats. J Trauma 68(1):76–83

Lin RS et al (1995) Endotoxemia in patients with chronic liver diseases: relationship to severity of liver diseases, presence of esophageal varices, and hyperdynamic circulation. J Hepatol 22(2):165–172

Mariat D et al (2009) The firmicutes/bacteroidetes ratio of the human microbiota changes with age. BMC Microbiol 9:123

Martin FP et al (2007) A top-down systems biology view of microbiome mammalian metabolic interactions in a mouse model. Mol Syst Biol 3:112

Martin FP et al (2008) Probiotic modulation of symbiotic gut microbial–host metabolic interactions in a humanized microbiome mouse model. Mol Syst Biol 4:157

Martin FP et al (2009) Topographical variation in murine intestinal metabolic profiles in relation to microbiome speciation and functional ecological activity. J Proteome Res 8(7):3464–3474

Mueller S et al (2006) Differences in fecal microbiota in different European study populations in relation to age, gender, and country: a cross-sectional study. Appl Environ Microbiol 72(2):1027–1033

Muyzer G, Smalla K (1998) Application of denaturing gradient gel electrophoresis (DGGE) and temperature gradient gel electrophoresis (TGGE) in microbial ecology. Antonie Van Leeuwenhoek 73(1):127–141

Myers LL et al (1987) Isolation of enterotoxigenic *Bacteroides fragilis* from humans with diarrhea. J Clin Microbiol 25(12):2330–2333
Nicholson JK, Wilson ID (2003) Opinion: understanding 'global' systems biology: metabonomics and the continuum of metabolism. Nat Rev Drug Discov 2(8):668–676
Nicholson JK, Lindon JC (2008) Systems biology: metabonomics. Nature 455(7216):1054–1056
Nicholson JK, Holmes E, Wilson ID (2005) Gut microorganisms, mammalian metabolism and personalized health care. Nat Rev Microbiol 3(5):431–438
Nicholson JK et al (2004) The challenges of modeling mammalian biocomplexity. Nat Biotechnol 22(10):1268–1274
Nicholson JK (2006) Global systems biology, personalized medicine and molecular epidemiology. Mol Syst Biol 2:52
O'Hara AM, Shanahan F (2006) The gut flora as a forgotten organ. EMBO Rep 7(7):688–693
Parracho HMRT et al (2005) Differences between the gut microflora of children with autistic spectrum disorders and that of healthy children. J Med Microbiol 54(10):987–991
Liu Q, Duan ZP, Ha DK, Bengmark S, Kurtovic J, Riordan SM (2004) Symbiotic modulation of gut flora improves minimal hepatic encephalopathy in cirrhotic patients. Hepatology 39:5
Roberfroid MB et al (1995) Colonic microflora: nutrition and health. Summary and conclusions of an International Life Sciences Institute (ILSI) [Europe] workshop held in Barcelona, Spain. Nutr Rev 53(5):127–130
Rolando N, Philpott-Howard J, Williams R (1996) Bacterial and fungal infection in acute liver failure. Semin Liver Dis 16(4):389–402
Rowland IR et al (2000) Interindividual variation in metabolism of soy isoflavones and lignans: influence of habitual diet on equol production by the gut microflora. Nutr Cancer 36(1):27–32
Sack RB, Albert MJ, Alam K, Neogi PKB, Akbar MS (1994) Isolation of entertoxigenic *Bacteroides fragilis* from Bangladeshi children with diarrhea: a controlled study. J clin Microbiol 32:9600–9963
Sheth AA, Garcia-Tsao G (2008) Probiotics and liver disease. J Clin Gastroenterol 42(Suppl 2):S80–84
Solga SF, Diehl AM (2004) Gut flora-based therapy in liver disease? The liver cares about the gut. Hepatology 39(5):1197–1200
Turnbaugh PJ et al (2006) An obesity-associated gut microbiome with increased capacity for energy harvest. Nature 444(7122):1027–1131
Venter CS, Vorster HH, Cummings JH (1990) Effects of dietary propionate on carbohydrate and lipid metabolism in healthy volunteers. Am J Gastroenterol 85(5):549–553
Wang Y et al (2004) Metabonomic investigations in mice infected with *Schistosoma mansoni*: an approach for biomarker identification. Proc Natl Acad Sci USA 101(34):12676–12681
Xing HC et al (2005) Effects of *Salvia miltiorrhiza* on intestinal microflora in rats with ischemia/reperfusion liver injury. Hepatobiliary Pancreat Dis Int 4(2):274–280
Xing HC et al (2005) Intestinal microflora in rats with ischemia/reperfusion liver injury. J Zhejiang Univ Sci B 6(1):14–21
Younes H et al (2001) Effects of two fermentable carbohydrates (inulin and resistant starch) and their combination on calcium and magnesium balance in rats. Br J Nutr 86(4):479–485
Zoetendal EG, Akkermans-van Vilet WM, de Visser JAGM, de Vos WM (2001) The host genotype affects the bacterial community in the human gastrointestinal tract. Microb Ecol Health Dis 13:129–134

Chapter 15
MetaHIT: The European Union Project on Metagenomics of the Human Intestinal Tract

S. Dusko Ehrlich and The MetaHIT Consortium

Introduction

There is widespread belief that we will require knowledge not only of the human genome but also of the human microbial metagenome in order to achieve detailed understanding of human biology. We live in intimate association with microbes that are present on all surfaces and in all cavities of the human body. The number of these microbes exceeds by at least ten-fold those of cells of our own body and the number of unique genes they encode was predicted to be at least 100-fold greater than the number of genes in our own genome (Ley et al., 2006a). These facts lead to a view that our body can be termed a "supra-organism" (Lederberg, 2000), composed of our own and of microbial cells. The ensemble of the genomes of human-associated microorganisms represents the human metagenome.

The complex and dynamic microbiota associated with us has a profound influence on our physiology, nutrition, and immunity. Disruptions of human-associated microbial communities are a significant factor in many diseases. Understanding the dynamic and variable nature of human microbial communities is a critical aspect of the challenge before us. To progress toward this ambitious goal, MetaHIT focuses on the microbiota of the intestinal tract, which is by far the most numerous and plays a particularly important role in human health and well-being.

Human intestinal microbiota is composed of up to ten trillion microbial cells, which can represent a weight of up to 2 kg. The intestinal microbes not only help digest food and synthesize vitamins that are needed by our body, but they also protect us by educating the immune system and controlling production of various drug-like metabolites (Backhed et al., 2005). Disorders of the gut microbial community have been associated with many different diseases, not only infectious but also noninfectious, such as obesity (Ley et al., 2005, 2006b; Turnbaugh et al., 2006) or inflammatory bowel diseases (IBDs) (Manichanh et al., 2006).

S.D. Ehrlich (✉)
Microbiology and the Food Chain Division, INRA, 78350, Jouy en Josas, France
e-mail: dusko.ehrlich@jouy.inra.fr

The list of Consortium members is appended to the Chapter.

K.E. Nelson (ed.), *Metagenomics of the Human Body*,
DOI 10.1007/978-1-4419-7089-3_15,

MetaHIT Objectives

The central objective of MetaHIT is to establish associations between the genes of the human intestinal microbiota and our health and disease. We focus on two disorders of increasing importance in our societies, IBDs and obesity.

To reach our central objective we integrate a number of activities.

(i) Establishing an extensive reference catalog of microbial genes and genomes present in the human intestine.
(ii) Developing profiling tools to determine which genes and genomes of the reference catalog are present in different individuals and at what frequency.
(iii) Gathering cohorts of individuals, some sick and some healthy, and determining which genes and genomes they carry.
(iv) Developing bioinformatic tools to store, organize, and interpret this information, and thus establish associations between intestinal microbiota and health and disease.
(v) Screening for and studying genes involved in host–microbe interactions.

MetaHIT seeks to be integrated in the world we live in. For this purpose, we actively participate in the international cooperation and coordination within the human metagenome field, via the International Human Microbiome Consortium (IHMC). We also seek to promote, on the one hand, the transfer of technology to industry, via an appropriate stakeholder platform, and on the other, the transfer of information about the project to the general public, by willingly accepting and even actively seeking contacts with the appropriate media.

MetaHIT Structure

MetaHIT is structured in six different pillars, each composed of two work packages, that address the different activities integrated within the project (Fig. 15.1).

The Sequencing Pillar encompasses two main aspects, sequencing of genes of all microorganisms and the genomes of individual microorganisms. The primary aim of the pillar is to produce the data for a reference set corresponding to the genetic repertoire of the intestinal microorganisms, both from metagenomic and whole-genome sequencing.

The Tool Pillar is dedicated to development of the tools that will enable profiling of an individual's microbial genes. The tools include DNA microarrays and high-throughput DNA sequencing technology.

→

Fig. 15.1 MetaHIT structure. The project is organized in six pillars, each composed of two work packages. An additional work package is devoted to the overall project management. The main activities of each pillar are indicated in Bold. The *arrows* highlight interactions between the different pillars

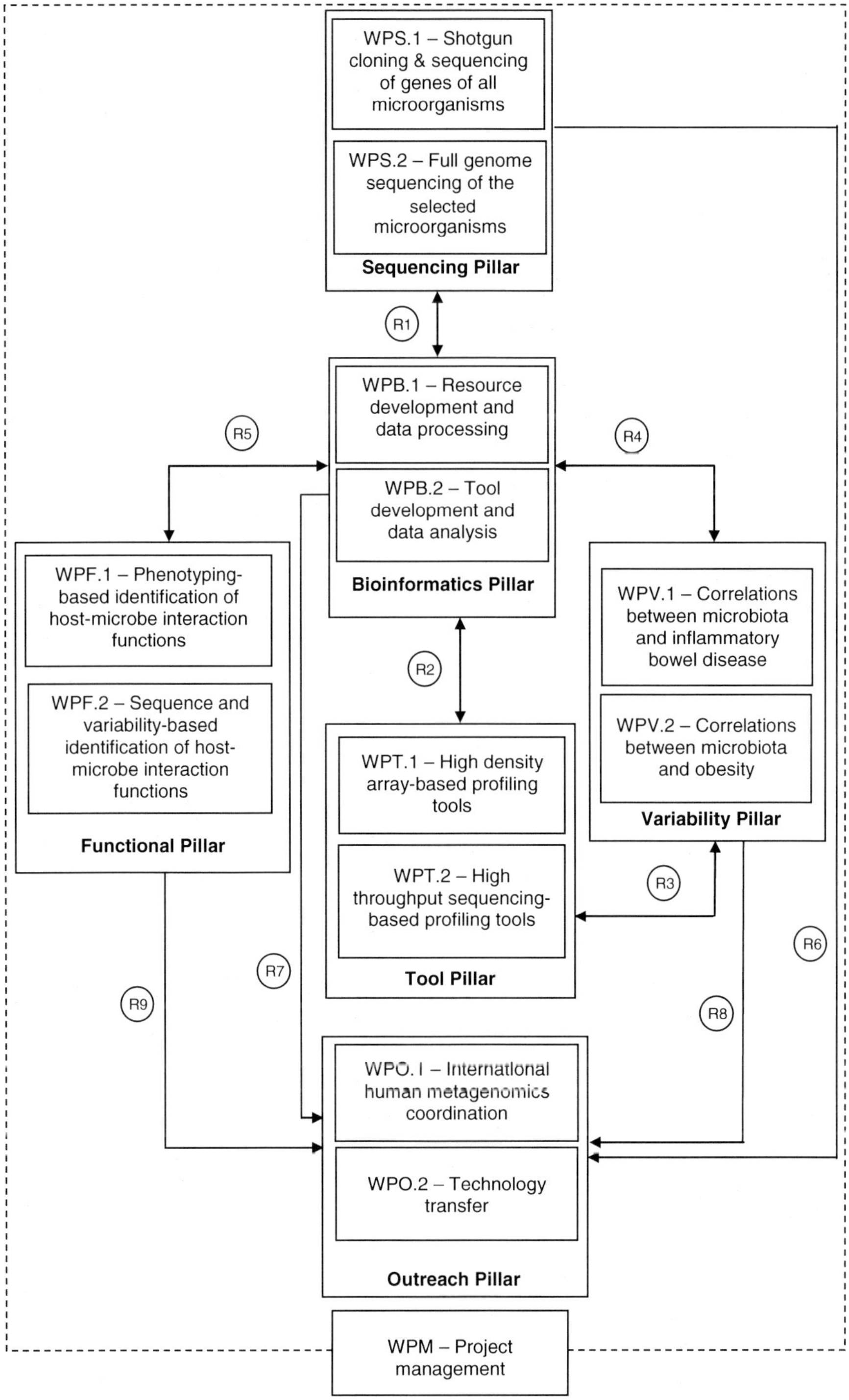
WPS.1 – Shotgun cloning & sequencing of genes of all microorganisms
WPS.2 – Full genome sequencing of the selected microorganisms
Sequencing Pillar
R1
WPB.1 – Resource development and data processing
WPB.2 – Tool development and data analysis
Bioinformatics Pillar
R5
R4
WPF.1 – Phenotyping-based identification of host-microbe interaction functions
WPF.2 – Sequence and variability-based identification of host-microbe interaction functions
Functional Pillar
R2
WPT.1 – High density array-based profiling tools
WPT.2 – High throughput sequencing-based profiling tools
Tool Pillar
WPV.1 – Correlations between microbiota and inflammatory bowel disease
WPV.2 – Correlations between microbiota and obesity
Variability Pillar
R3
R6
R7
R8
R9
WPO.1 – International human metagenomics coordination
WPO.2 – Technology transfer
Outreach Pillar
WPM – Project management

Fig. 15.1 (continued)

The Bioinformatics Pillar is devoted to database development, on the one hand, and data processing and analysis, on the other. One of the primary objectives is to construct a "pipeline" leading from the raw sequence to database that integrates all that is known about the homologs of a given putative gene product. Another is to allow easy analysis and interpretation of the massive data generated by the profiling tools. A final objective is the ability to carry out a high-level meta-analysis of different types of information generated in the project.

The Variability Pillar uses the profiling tools to identify genes correlated with health and disease states in terms of both presence and expression, targeting IBDs and obesity.

The Function Pillar seeks to explore host–microbe interactions relevant to health. It includes phenotyping-based identification of host–microbe interaction functions as well as the sequence- and variability-based identification of host–microbe interaction functions.

The Outreach Pillar is aimed, on the one hand, at coordination of research in human metagenomics and dissemination of knowledge to stakeholders and the general public and, on the other, to technology transfer of the knowledge generated in the project promoting health.

MetaHIT Partnership and Funding

MetaHIT gathers 13 partners from 8 countries (Fig. 15.2). Of these, nine are public and four are private institutions. The breadth of the complementary expertise that they collectively possess is commensurate with the complexity of our project.

Institution	Key personnel	Country
Institut National de la Recherche Agronomique	S.Dusko Ehrlich & Joël Doré	France
Danone Research	Raish Oozer	France
European Molecular Biology Laboratory	Peer Bork	Germany
Hospital Universitari Vall d 'Hebron	Francisco Guarner & Natalia Boruel	Spain
Istituto Europeo di Oncologia	Maria Rescigno	Italy
Novo Nordisk Hagedorn Research Institute	Oluf Pedersen	Denmark
UCB Pharma S.A	Miguel Forte	Spain
Wageningen Universiteit	Willem de Vos & Michiel Kleerebezem	Netherlands
Wellcome Trust Sanger Institute	Julian Parkhill	United Kingdom
Commissariat à l'Energie Atomique - Genoscope	Jean Weissenbach	France
Bejing Genomics Institute Shenzhen	Wang Jun	China
Mérieux Alliance	Christian Brechot	France
Kopenhagen University	Karsten Kristiansen	Denmark

Fig. 15.2 MetaHIT partners. Participating institutions and key scientists

The total cost of MetaHIT has been estimated to be over 20 million €, with a contribution from the European Commission of 11.4 million €. The remainder is contributed not only by the participating institutions, mostly in kind, but also by a direct cash contribution of the private companies, approaching 1 million €. The project started on January 1, 2008 and is scheduled to last for 4 years.

Gene Catalog of Gut Microbes

We follow a two-pronged approach to generate the catalog of intestinal genes and genomes. One is to determine the complete sequence of genomes from bacteria that we know how to cultivate. As they represent only a minority, about 20%, of all intestinal bacteria, we will also sequence the total bacterial DNA prepared from stool samples. This provides information about genes of bacteria that we cannot cultivate.

Our objective is to sequence full genomes of 100 bacterial species and the total stool DNA of a cumulative length corresponding to some 1,000 bacterial species. The genomes of the first 30 cultivable species have been sequenced and 15 more are under way. Their choice has been coordinated within the IHMC, and in particular the USA National Institutes of Health (NIH), which funds sequencing of intestinal bacterial species carried out within the Human Microbiome Project (HMP; http://nihroadmap.nih.gov/hmp/). We are now developing a methodology to sequence genomes of single bacterial cells and therefore accede to genomes of cells that cannot be cultivated at present.

We plan to sequence total stool DNA from different individuals, in order to maximize the reference gene catalog, as different individuals carry different microbiota. Bacterial diversity of 35 individuals from the IBD and obesity cohorts was examined and the eight individuals with the most divergent microbiota were identified. Two technologies that allow us to obtain long sequencing reads (di-deoxy- and pyrosequencing) were used to sequence the total stool DNA of these individuals. About 6 Gb of the sequence has been obtained, the length corresponding to 1,500–2,000 bacterial genomes.

In parallel, we have explored the use of the short read Illumina technology for metagenomic sequencing. Fecal DNA of 124 individuals was sequenced, and a total of almost 0.6 Tb was generated. The sequence was successfully assembled into contigs and a catalog of 3.3 million nonredundant genes was constituted (Qin et al., 2010), which is 150-fold more than the human gene complement. This catalog contains over 85% of the prevalent (that is, present at a frequency above 7×10^{-7}) genes from the cohort studied and includes 85 and 70% of the microbial sequences from the human gut identified in two previous studies, carried out in the United States (Turnbaugh et al., 2009) and Japan (Kurokawa et al., 2007), respectively. Over 99% of the genes are of bacterial origin, suggesting that the cohort harbors between 1,000 and 1,150 abundant bacterial species, encoding an average of 3,400 genes. Each of the individuals from the cohort harbors, on average, 540,000 genes, or at least 170 bacterial species. Most of the genes and bacterial species must therefore be shared between the individuals of the cohort and likely among the entire humankind. We conclude that the intestinal metagenome may well be of a finite and not an overly large size, at least with respect to the abundant bacterial species.

Microbial Gene Profiling

Two technologies are being used to determine the presence, frequency, and expression of microbial genes in different individuals, which we term gene profiles. One is based on high-density micro-arrays and the other on very high-throughput DNA sequencing. Both use information from the reference gene catalog.

A procedure to create high-density arrays from a complex gene catalog has been developed. The present arrays allow for the detection of over 3.4 million microbial genes, revealed by metagenomic and full genome sequencing, and are being used to generate DNA- and RNA-based gene profiles.

High-throughput DNA sequencing has been calibrated on DNA samples of known composition and has been used to analyze samples from 370 individuals. Procedures to compute and compare gene profiles for different individuals are being tested at present.

IBD and Obesity Patient Cohorts

IBD includes two diseases, Crohn's disease (CD) and ulcerative colitis (UC). Two studies are planned for each. One compares patients in remission with healthy relatives (30 for each disease), the other intends to follow the 30 patients over a 2-year period, during which about 50% are expected to relapse. The first aims to identify bacterial genes associated with the disease, whereas the second aims to identify genes associated with relapse. In addition, one study targets a particular CD phenotype, resulting in perianal fistulae, and will compare two groups of 20 patients having the fistulizing or nonfistulizing disease form with 20 healthy individuals. This study is supported by one of the industrial partners of MetaHIT, Mérieux Alliance. Currently, 87 individuals have been enrolled in the IBD studies and sampled. Of these, 39 have been profiled, using the high-throughput sequencing approach.

Obesity is associated with a number of co-morbidities, such as diabetes, hypertension, and cardiovascular disease. These co-morbidities are found typically in individuals with visceral rather than subcutaneous obesity. We are comparing 60 individuals of each type with 60 healthy individuals, aiming to detect bacterial genes associated with obesity that could be used as risk predictors. All the individuals have been sampled and analyzed by the high-throughput sequencing technology.

Information Organization and Analysis

Sequence information must be interpreted (annotated) in terms of genes, proteins, and the functions they perform. An automated sequence assembly and annotation pipeline was developed to analyze sequence data from the human intestinal metagenome. The function of up to 75% of the genes can be deduced in this way, which is similar to the proportion of genes in individual genomes that it is possible to annotate automatically. The automated annotation is very consistent, which is of great importance for comparing different metagenomes. The pipeline was used to process sequences generated within MetaHIT, as well as those from other worldwide projects targeting the human intestinal metagenome, published previously or conducted currently. In total, bacterial sequences equivalent to some 1,000 full genomes have been integrated successfully.

Function of Bacterial Genes Associated with Disease

Two approaches are planned to uncover the function of bacterial genes associated with disease. One addresses the role of genes in bacteria whereas the other focuses on the effect of gene products on the host. The former involves inactivation of the genes in appropriate model bacteria and studying the effects of inactivation on the bacterial phenotypes. The latter is centered on overexpression of proteins in standard bacterial hosts and determining the effects of the proteins on appropriate human cell lines. The focus will be on two types of effects, cell proliferation and activation of inflammatory pathways. Sixteen different screening procedures have been established for this purpose. A pilot study with some 5,000 *Escherichia coli* clones that harbor fosmids with 40-kb-long metagenomic inserts revealed clones that stimulate or inhibit the NF-κB signaling in a human colonic epithelial cell line.

Studies with Industry – Transfer of Technology

Two clinical studies of IBD are being carried out in association with MetaHIT industrial partners, in addition to the study of the fistulizing phenotype of the CD, presented above. The first concerns the effects of a fermented milk product on the stability of microbiota in UC patients and involves Danone Research, whereas the second compares the microbiota of patients who do or do not respond to a particular drug treatment and involves UCB Pharma. Both are based on the profiling methodology developed in MetaHIT. The clinical part of the former, with 50 patients and 12 healthy controls, has been finished and the samples are being profiled by sequencing. The second has been approved by the institutional ethical committee and is ready to begin.

MetaHIT Communication and Coordination

To promote the necessary international cooperation and coordination in the human metagenome field, MetaHIT takes an active role in the IHMC. IHMC was first presented at the kick-off meeting of MetaHIT (Jouy en Josas, France, April 2008), and its constitution was publicly announced at the second MetaHIT meeting (Heidelberg, Germany, October 2008). The MetaHIT coordinator was the first co-chairman of the IHMC.

To support and accompany the technology transfer policy of MetaHIT, and communicate with patients' associations, a Stakeholder Platform was established. The first meeting of the platform was organized in conjunction with the 3rd MetaHIT meeting (Barcelona, Spain, March 2009). Participants from various industries, including human health, food, and biotechnology, took part in the meeting, as did the European Federation of Crohn′s & Ulcerative Colitis Associations.

To inform the general public about the field, the project and its achievements, a MetaHIT website (http://www.metahit.eu) has been created. In conjunction with MetaHIT meetings, press releases have been issued, resulting in a number of articles in the national and international press, as well as radio and TV coverage. In addition, two short human metagenomic documentary films oriented toward general public have been produced and are accessible on the MetaHIT web site. MetaHIT can be followed on twitter http://twitter.com/MetaHIT, netwibes http://www.netvibes.com/metahit#Live_News and facebook http://www.facebook.com/pages/MetaHIT/195087858830?v=info#!/pages/MetaHIT/195087858830?v=info.

Expected Achievements

The expected MetaHIT achievements are the establishment of the methodology to characterize individual intestinal metagenomes and the discovery of associations between bacterial genes and human disease. This should open the way for the development of novel diagnostic and prognostic tools. Furthermore, the description of the human intestinal microbiota, its dynamics, and its interaction with the human host will lead to a much more complete understanding of global human biology.

In the longer term, we expect to open avenues for reasoned modulation of our microbiota. The ability to characterize individual intestinal microbiota and follow its evolution with time, coupled to the understanding of the microbe–host interactions, should allow to explore the effects of factors such as food, environment, and age on the dynamics of our microbial populations and to develop interventions to optimize these populations. This should open novel possibilities to improve human health and well-being.

Acknowledgments The research has received funding from the European Community′s Seventh Framework Program (FP7/2007-2013): MetaHIT, grant agreement HEALTH-F4-2007-201052

The MetaHIT Consortium Members

Maria Antolin (Hospital Universitari Val d′Hebron, Ciberehd, Barcelona, Spain), François Artiguenave (Commissariat à l′Energie Atomique, Genoscope, Evry, France), Manimozhiyan Arumugam (European Molecular Biology Laboratory, Heidelberg, Germany), Jean-Michel Batto (Institut National de la Recherche Agronomique), Marcelo Bertalan (Technical University of Denmark, Lyngby, Denmark), Hervé Blottiere (Institut National de la Recherche Agronomique, Jouy en Josas, France), Peer Bork (EMBL), Natalia Borruel (HUVH), Christian Brechot

(Mérieux Alliance, Lyon, France), Søren Brunak (DTU), Thomas Bruls (CEA Genoscope), Kristoffer Solvsten Burgdorf (Novo Nordisk Hagedorn Research Institute, Copenhagen, Denmark), Jianjun Cao (Beijing Genomics Institute Shenzhen), Francesc Casellas (HUVH), Christian Chervaux (Danone Research, Palaiseau, France), Jean-Frederic Colombel (Mériex Alliance), Antonella Cultrone (INRA), Joel Doré (INRA), Christine Delorme (INRA), Gérard Denariaz (Danone Research), Rozenn Dervyn (INRA), Miguel Forte (UCB Pharma SA, Madrid, Spain), Carsten Friss (DTU), Francisco Guarner (HUVH), Maarten van de Guchte (INRA), Eric Guedon (INRA), Florence Haimet (INRA), Torben Hansen (NN HRI), Alexandre Jamet (INRA), Min Jian (BGI Shenzhen), Catherine Juste (INRA), Ghalia Kaci (INRA), Michiel Kleerebezem (Wageningen Unviersiteit, the Netherlands), Jan Knol (Danone Research), Karsten Kristiansen (Kopenhagen University, Copenhagen, Denmark), Severine Layec (INRA), Karine Le Roux (INRA), Marion Leclerc (INRA), Denis Le Paslier (CEA Genoscope), Florence Levenez (INRA), Dongfang Li (BGI Shenzhen), Junhua Li (BGI Shenzhen), Shaochuan Li (BGI Shenzhen), Shengting Li (BGI Shenzhen), Songgang Li (BGI Shenzhen), Yingrui Li (BGI Shenzhen), Huiqing Liang (BGI Shenzhen), Daniel Mende (EMBL), Emmanuelle Maguin (INRA), Chaysavanh Manichanh (HUVH), Raquel Melo Minardi (CEA Genoscope), Christine Mrini (Mériex Alliance), H. Bjørn Nielsen (DTU), Trine Nielsen (NN HRI), Raish Oozeer (Danone Research), Julian Parkhill (Wellcome Trust Sanger Institute, Cambridge, UK), Nicolas Pons (INRA), Oluf Pedersen (NN HRI), Eric Pelletier (CEA Genoscope), Junjie Qin (BGI Shenzhen), Jeroen Raes (EMBL, presently at Vrije Universiteit Brussel, 1050 Brussels, Belgium), Pierre Renault (INRA), Maria Rescigno (Istituto Europeo di Oncologia, Milan, Italy), Ruiqiang Li (BGI), Nicolas Sanchez (INRA), Thomas Sicheritz-Ponten (DTU), Sebastian Tims (WU), Toni Torrejon (HUVH), Keith Turner (WTSI), Encarna Varela (HUVH), Willem de Vos (WU), Bo Wang (BGI Shenzhen), Jian Wang (BGI Shenzhen), Jun Wang (BGI Shenzhen & KU), Jean Weissenbach (CEA Genoscope), Yohanan Winogradsky (INRA), Yinlong Xie (BGI Shenzhen), Junming Xu (BGI Shenzhen), Takuji Yamada (EMBL), Huanming Yang (BGI Shenzhen), Chang Yu (BGI Shenzhen), Xiuqing Zhang (BGI Shenzhen), Huisong Zheng (BGI Shenzhen), Hongmei Zhu (BGI Shenzhen), Yan Zhou (BGI Shenzhen), Erwin Zoetendal (WU).

References

Backhed F, Ley RE, Sonnenburg JL, Peterson DA, Gordon JI (2005) Host–bacterial mutualism in the human intestine. Science 307:1915–1920. doi:307/5717/1915 [pii] 10.1126/science.1104816

Kurokawa K et al (2007) Comparative metagenomics revealed commonly enriched gene sets in human gut microbiomes. DNA Res 14:169–181. doi:dsm018 [pii] 10.1093/dnares/dsm018

Lederberg J (2000) Infectious history. Science 288:287–293

Ley RE, Peterson DA, Gordon JI (2006a) Ecological and evolutionary forces shaping microbial diversity in the human intestine. Cell 124:837–848. doi:S0092-8674(06)00192-9 [pii] 10.1016/j.cell.2006.02.017

Ley RE, Turnbaugh PJ, Klein S, Gordon JI (2006b) Microbial ecology: human gut microbes associated with obesity. Nature 444:1022–1023. doi:4441022a [pii] 10.1038/4441022a

Ley RE et al (2005) Obesity alters gut microbial ecology. Proc Natl Acad Sci USA 102:11070–11075. doi:0504978102 [pii]

Manichanh C et al (2006) Reduced diversity of faecal microbiota in Crohn's disease revealed by a metagenomic approach. Gut 55:205–211. doi:gut.2005.073817 [pii] 10.1136/gut.2005.073817

Qin J et al (2010) A human gut microbial gene catalog established by deep metagenomic sequencing, Nature (in press)

Turnbaugh PJ et al (2006) An obesity-associated gut microbiome with increased capacity for energy harvest. Nature 444:1027–1031. doi:nature05414 [pii] 10.1038/nature05414

Turnbaugh PJ et al (2009) A core gut microbiome in obese and lean twins. Nature 457:480–484. doi:nature07540 [pii] 10.1038/nature07540

Chapter 16
Implications of Human Microbiome Research for the Developing World

Appolinaire Djikeng, Barbara Jones Nelson, and Karen E. Nelson

Introduction

New high-throughput sequencing and data analysis approaches (Costello et al., 2009; Turnbaugh et al., 2009), along with novel diversity screens and even more intrinsic single cell approaches to isolating new species (Lasken, 2009), have presented the sciences with a unique opportunity to investigate and interrogate the microorganisms that are associated with the human body, all at a greater depth than previously appreciated.

From the earliest studies, the greater scientific community has recognized a high level of microbial diversity in nature beyond imaginations. This includes observations on the oceans, soils, and on animals. With respect to humans, it became increasingly apparent that the species on and in our bodies make significant contributions to our health and development. These species maintain normal cell function in the gastrointestinal tract (for example, see Eckburg et al., 2005; Bik et al., 2006; Gill et al., 2006). We are also dependent on these species for the efficient digestion of food components, such as plant material and xenobiotics (Gill et al., 2006), and to ward off certain diseases. In parallel, these microbes have been associated with, and can result in, many common diseases such as cavities, stomach ulcers, bacterial vaginosis (BV), and irritable bowel syndrome (Foster et al., 2008; Dorer et al., 2009).

Most of the initial studies on the microbial diversity associated with the human body focused either on traditional culturing approaches or on sequencing and phylogenetic analysis of the 16S ribosomal (r)RNA genes derived from microbial samples taken from the human body (Eckburg et al., 2005; Bik et al., 2006). The limit to culturing or 16S rRNA sequencing was primarily a reflection of the availability of molecular tools and approaches, and the cost associated with earlier versions of available sequencing technologies. The 16S rRNA sequencing and analysis invariably revealed a higher level of microbial diversity than that seen with conventional

K.E. Nelson (✉)
Department of Human Genomic Medicine, The J. Craig Venter Institute, 9704 Medical Center Drive, Rockville, MD 20850, USA
e-mail: kenelson@jcvi.org

K.E. Nelson (ed.), *Metagenomics of the Human Body*,
DOI 10.1007/978-1-4419-7089-3_16,

culture techniques (Gao et al., 2007; Gao et al., 2008). From the human stomach, for example, although the highly acidic environment was thought to only contain *Helicobacter* types, Bik et al. (2006) used 1,833 16S rRNA sequences obtained from 23 gastric endoscopic biopsy samples to identify 128 phylotypes of bacteria that potentially reside in the human stomach. The majority of these phylotypes was shown to belong to the Proteobacteria, Firmicutes, Actinobacteria, Bacteroidetes, and Fusobacteria. This work also described that 10% of the clones represented organisms that were previously uncharacterized.

Subsequent ongoing studies continue to reveal high levels of diversity from the microbial species that inhabit the human body, with high levels of both intra- and interspecies diversity (Costello et al., 2009; Turnbaugh et al., 2009). Most of these studies that have investigated the diversity of the microbial species associated with the human body have however left important questions unanswered such as the identity of the nondominant community members and their biological roles, and which metabolic processes the populations that are present encode and carry out. In addition, the past 15 years of genomic research have made it clear that closely related species, and even species that are identical at the 16S rRNA level, can have wide variation in gene content (Perna et al., 2001; Kudva et al., 2002). Terms such as pan-genome and core-genome have been coined over the years to address the variations that are apparent in closely related species and have now been applied in a similar fashion to metagenomic populations (Callister et al., 2008; Bentley, 2009).

The initial publication of a shotgun sequencing of the human microbiome focused on the analysis of fecal samples from the human gastrointestinal tract (Gill et al., 2006). This study along with subsequent applications of shotgun techniques to the study of the human microbiome again highlighted the extent of microbial diversity associated with the human body (Grice et al., 2009). Gordon and co-workers, for example, investigated the interplay between the gastrointestinal ecology and energy balance of animals on a Western diet. Here they found that obesity that was induced by the diet resulted in an increased proportion of one class of the Firmicutes and that this same population could be reduced by manipulation of the diet. Transplantation of the microbial populations from the obese mice to lean germ-free mice resulted in a higher level of the deposition of fat than when these microbial populations were taken from lean donors (Turnbaugh et al., 2008). More recently, Gordon's group presented the results of a monozygotic and dizygotic twin pair study, where twins were concordant for leanness or obesity, and their mothers (Turnbaugh et al., 2009). The aim of this study was to address how host genotype, environmental exposure, and host adiposity influence the gut microbiome (Turnbaugh et al., 2009). The comparative analysis of fecal samples that were derived from 154 individuals yielded 9,920 near full-length and 1,937,461 partial bacterial 16S rRNA sequences. In addition, 2.14 Gb of metagenomic data was obtained from their microbiomes. The results from this analysis suggest that the gastrointestinal microbiome is shared among family members, but variations are present within each individual with respect to the lineages that can be observed. This variation was evident in both the monozygotic and dizygotic twin pairs. The results however suggest that there is a core functional microbiome that can vary depending on physiological states.

The Promise of the Human Microbiome

The genomic era created the possibility of studying the detailed genetics of many microbial species. These include pathogens and nonpathogens and species that have both positive and negative impacts on the environment. The developments in the genomics arena have taken advantage of emerging and improving sequencing technologies, as well as improved assembly algorithms and approaches coupled with reduced costs for generating genomic data. The developments also placed the greater scientific community in positions to ask in-depth gene-based questions and get real answers.

The advent of metagenomics was a natural progression of the genomics field, and in particular took advantage of the ability to sequence DNA that was derived from a mixed community and assemble this genetic information to reconstitute metabolic and physiological information of any community of choice. Metagenomic approaches have by now been successfully applied to environments as diverse as soils, the oceans (Nealson and Venter, 2007; Rusch et al., 2007; Yooseph et al., 2007; Yutin et al., 2007), and the human body (Gill et al., 2006; Costello et al., 2009; Turnbaugh et al., 2009) in an attempt to describe and decipher the microbial species that are present in these niches. Entire systems can now be studied with respect to viral, microbial, and fungal diversity, over varying time courses, and before and after perturbation (Costello et al., 2009). On the human side, when coupled with 16S rRNA analyses for detailed estimates of microbial diversity, metagenomics approaches present the perfect opportunity to address questions related to human health and associated problems. This is particularly relevant in the developing world, which presents its own series of challenges, some of which are presented below.

One of the most valuable examples to date of the potential benefits from knowledge gained with human microbiome studies comes from a series of studies performed by Dore and colleagues at INRA. The significance of the studies conducted by this group relates to how the evolution of initial studies focusing on the microbiome can result in recommendations to improve human health conditions. Their results evolved from initial metagenomic studies on human gastric samples (Manichanh et al., 2006) using fosmid libraries from six healthy patients, and six patients with Crohn's disease (CD). The group was able to identify 125 nonredundant ribotypes mainly represented by the phyla Bacteroidetes and Firmicutes, of which 43 distinct ribotypes were identified in the healthy microbiota, and only 13 in CD. This metagenomic approach that was initially published gave the first insight into the reduced microbial diversity in patients with CD.

Their work continued to focus on microbiology of CD (Seksik et al., 2006; Sokol et al., 2006; Sokol et al., 2007; Seksik et al., 2009). They most recently compared fecal samples of 22 patients active for CD (A-CD) patients, 10 CD patients in remission (R-CD), 13 active ulcerative colitis (A-UC) patients, four UC patients in remission (R-UC), eight infectious colitis (IC) patients, and 27 by 16S PCR and found that members of Firmicutes (*Clostridium leptum* and *C. coccoides* groups in particular) were less represented in A-IBD patients compared to healthy subjects (HS) with *Faecalibacterium prausnitzii* species (a major representative of the

C. leptum group) in lower abundance in A-IBD and IC patients compared with HS. As a result of the initial work, the group proposed that *F. prausnitzii* was important for gut homeostasis.

In 2008 members of the same group had published on the composition of the mucosa-associated microbiota of CD patients at the time of surgical resection and 6 months following using FISH analysis (Sokol et al., 2008), and again found reduced abundance of *F. prausnitzii* being correlated with an increased risk of postoperative recurrence of ileal CD. They further studied the anti-inflammatory effects of *F. prausnitzii* and showed that the bacterium exhibited anti-inflammatory effects on cellular and TNBS colitis models.

NIH-Funded HMP and International Components of the HMP

The US National Institutes of Health (NIH) initiated a roadmap program focused on the human microbiome (Peterson et al., 2009). The project has been described as a community resource, with overarching aims to help determine the core human microbiome, to understand the changes in the human microbiome that can be correlated with changes in human health, to develop new technological and bioinformatics tools to support these goals, and to address the ethical, legal, and social issues raised by human microbiome research (http://nihroadmap.nih.gov/hmp/). The project has a heavy sequencing and data analysis component that currently is underway at the four large NHGRI/NIAID-funded sequencing centers (J. Craig Venter Institute (JCVI), Baylor College of Medicine, The Broad Institute and Washington University). The current sequencing focus includes generating at least 3000 reference genomes at various levels of finishing (Chain et al., 2009), coupled with significant 16S rRNA sequencing and metagenomic sequencing from 15 to 18 body sites on 300 individuals some of which would be repeat sample donors (http://nihroadmap.nih.gov/hmp/). A number of "Investigator"-driven demonstration projects have also been awarded. These demonstration projects aim to understand the implications of a number of diseases including CD, BV, psoriasis, and esophageal cancer to name a few (Peterson et al., 2009). It is anticipated that the results from these demonstration projects will give additional insights into the relationship between human health and disease and changes in the human microbiome. Finally, the Human Microbiome Project (HMP) roadmap initiative has awarded funds to investigate new technologies for improving knowledge of the human microbiome, as well as for the development of computational tools that will increase the value of metagenomic data (http://nihroadmap.nih.gov/hmp/fundedresearch.asp), and the ethical, legal, and social implications of this work.

In summary, and as captured in the recent publication by Peterson et al. (2009), the goals of the HMP are to demonstrate the characterization of the human microbiome with population, genotype, disease, age, etc., and also catalog the influence of disease. The aim also is to present a standardized data resource and technological benefits. The project will go over 5 years at a funding level of close to 157 million US dollars.

Since the launch of this roadmap initiative in 2007, a number of other HMPs have been described. An overview of available projects as of mid-2008 was presented in an editorial (Mullard, 2008). Projects beyond the large NIH USA based human microbiome efforts include work in Europe, China, Australia, and Canada. In 2007, the European Commission committed close to 31 million US dollars to a 4-year initiative called the Metagenomics of the Human Intestinal Tract (METAHIT) where the primary focus is the microorganisms that inhabit the gut, and how they contribute to obesity and inflammatory bowel disease (Mullard, 2008). A review of this effort is presented in another chapter written by Ehrlich and colleagues.

We are all cognizant of the fact that age, diet, and geographical location contribute to variations in the human microbiome. Consequently, the more initiatives that we have in diverse parts of the world, the better positioned we will be to fully understand the key components of the microbiome and their interactions with the host under various environmental and physiological cues.

Implications of the HMP for the Developing World

Because of a slow rate of progress in the areas of scientific research, along with low levels of available funding and investment in sciences in most developing countries, there has been very little scientific contribution toward solving major problems that hinder their global development. As Coloma and Harris (2009) nicely put it, "researchers in many developing countries will not be participating in genomics research, mainly because of their technological isolation and their limited resources and capacity for genomics research combined with the urgency of many other health priorities." Areas such as public health, emerging infectious diseases, and agricultural development, which are key to long-lasting and sustainable national development, still lack the funding and innovation required to mitigate their inability to contribute to global development. The global health sector is of particular importance given the increasing number of diseases that plague the developing world (some of which are making a comeback after several years under effective control). Examples of some of these are detailed below. Consequently, in most developing countries throughout the world, and specifically in sub-Saharan Africa, South America, and Asia, there is a serious need to improve public health. In these countries, communicable diseases caused by known and even unknown pathogens (see below) remain a leading cause of mortality. Emerging infectious diseases are a major cause for alarm, and malnutrition and associated effects are also major issues that need to be effectively addressed.

If one takes emerging infectious disease as an example, this captures many viral and bacterial agents. Severe acute respiratory syndrome (SARS) was the first infectious disease to emerge in the 21st century, and other emerging viral infectious diseases according to The World Health Organisation (WHO) include but are not limited to Ebola and Marburg hemorrhagic fevers. In addition, in an earlier publication by WHO ("New WHO office to help developing countries control emerging diseases"; J Environ Health 63, 2001) it was stated that in 1998 alone, communicable diseases caused the death of over 13 million people worldwide, mainly among the poorest populations of developing countries. Since then, more than 30 new communicable diseases have been identified, and this list includes several diseases that were thought to be almost extinct that apparently have come back into the human population. Certain food-borne diseases are also considered to be emerging as they now occur more often, and that list includes outbreaks of salmonellosis, which have increased significantly in the past 25 years. *Listeria monocytogenes* also falls into this category as its major role in food-borne diseases has become more recently appreciated, and some food-borne trematodes are also emerging as a serious public health concern. Although food-borne infections with *Escherichia coli* serotype O157:H7 were first described in 1982, it has rapidly emerged to be a leading cause of infections, which in turn result as a major cause of bloody diarrhea and acute renal failure, with an infection that is sometimes fatal. Finally, while cholera devastated much of Asia and Africa for years, its introduction for the first time in almost a century on the South American continent in 1991 makes it another example of an emerging infectious disease.

In addition, very little to none of the successful metagenomics stories in understanding the human microbiome and its role in aspects of human health have been conducted or duplicated in developing countries. Notwithstanding ongoing efforts focusing on vaccines, better diagnostics, and improved treatment of many of these diseases, it is becoming increasingly essential to complement such approaches with an investigation of the role of the human microbiome on human health. Several areas of anticipated important contribution of the human microbiome include zoonotic diseases and other emerging and re-emerging infectious diseases, sexually transmitted diseases, diarrheal diseases, respiratory diseases, eukaryotic diseases, malnutrition, and the integration of probiotics for improving human health. In addition to the translation of findings into practical approaches for improving human health, other opportunities offered by the human microbiome initiative are related to the transfer of technology to developing countries and the associated long-term benefits to training local populations in these developing sciences so that nations can retain the benefits. This will further strengthen capacity in genomics and bioinformatics and all the associated downstream applications that come with these areas of research.

It is now appreciated that there is a resident (which constitutes the core microbiome) and a nonresident microbiota. The nonresident microbiota contains known and unknown microbes that cause a wide range of human diseases, most of which remain to be effectively controlled in both the developed and the developing world. Human losses in the developing world in terms of mortality and contributions to economic development appear to be greater, however. Currently, for example, communicable diseases caused by eukaryotic parasites such as *Plasmodium* spp., *Leishmania* spp., *Trypanosoma* spp., and various viruses, among others, remain serious public health concerns in the developing world and affect more than 1.2 billion people (Mahmoud and Zerhouni, 2009). In this context, scientific challenges that include genomics, metagenomics, proteomics, and metabolomics-related activities need to be expanded to ultimately encompass system and ecological understanding of communities of microbes and their interactions with humans. It has in fact been anticipated that the control of these diseases may be accelerated by the complete understanding of the genomes of these species, coupled with an understanding of the changes of the human microbiome that favor or reduce/eliminate their virulence and/or transmissibility. The adaptation processes, for example, by which zoonotic microorganisms that enter the human population adapt to become pathogens over-time can also be accelerated by longitudinal studies that focus on the populations on the bodies of both healthy and diseased individuals.

The Promise for Technology Development in the Developing World

As with most advanced technological and scientific approaches, and as is evident from the developing countries reports presented above, the development and applications of technological advances probably will take a significant amount of time to trickle to the developing world. In the realm of genomics and metagenomics as applied to human health, there is limited evidence that this will happen soon enough to allow developing countries to actively participate and shape research in these new fields. However, a recent award from the Bill & Melinda Gates Foundation (BMGF) to Dr. Jeffrey Gordon at The Washington State University in St. Louis to study childhood malnutrition in developing countries (http://www.gatesfoundation.org/press-releases/Pages/child-malnutrition-microbial-cells-study-090331.aspx) suggests that there will be more movement in the direction of applying these technologies to problems in the developing world. For the above-mentioned study, Gordon's group will investigate the microbes in severe malnutrition with a major focus on severely malnourished infants living in Malawi and Bangladesh, and whether their microbial flora varies from that found in healthy infants who live in the same environment. It is anticipated that as a result of these studies, we will be better positioned to develop effective treatment regimes for these disease conditions. This award is part of an initiative by the BMGF to fund research on malnutrition (http://mednews.wustl.edu/news/page/normal/13840.html?emailID=23653).

In addition to that award and the anticipated outcome, there have been a significant number of plant and microbial genome projects initiated and conducted in the developing world. The range of microbial species that have been sequenced includes human, plant, and animal pathogens, as well as organisms that have potential benefit to the environment. Some of these species include *Actinobacillus pleuropneumoniae* serovar 3 str. JL03 that causes fibrinous and necrotizing pleuropneumonia in swine, and that was sequenced by the Huazhong Agricultural University in China and *Haemophilus parasuis* SH0165 also sequenced by the Huazhong Agricultural University/Institute of Pathogen Biology/Chinese Academy of Medical Sciences/Peking Union Medical College. *Chromobacterium violaceum* ATCC 12472 was sequenced by the LNCC (National Laboratory of Scientific Computing in Rio de Janiero, Brazil); this bacterium carries the bacteriocidal pigment violacein and can also cause diarrhea and septicemia in humans. *Ehrlichia ruminantium* str. Welgevonden was sequenced at the University of Pretoria, South Africa. *Leifsonia xyli* subsp. xyli str. CTCB07, the causative agent of ratoon stunting disease in sugar cane, was sequenced by the Sao Paulo state (Brazil) Consortium and *Leptospira interrogans* serovar Copenhageni str. Fiocruz L1-130 and *Xylella fastidiosa* were also sequenced by the same group. *Leptospira interrogans* serovar Lai str. 56601 sequenced by the Chinese National HGC, Shanghai, and *Lysinibacillus sphaericus* C3-41 sequenced by the Chinese Academy of Sciences/Beijing Genomics Institute, Chinese Academy of Sciences.

The developing world has also been involved in the sequencing and analysis of some of the major eukaryotic parasites such as *Trypanosoma brucei*, *Trypanosoma cruzi*, *Leishmania major*, and *Theileria parva* (Nene et al., 2000; Berriman et al., 2005; Bishop et al., 2005; El-Sayed et al., 2005; Gardner et al., 2005). There have also been several initiatives that have looked at host genotyping in many developing countries. According to Coloma and Harris (2009), Thailand, South Africa, Indonesia, Brazil, Mexico, and India have all devoted resources to studies on human genetics and variation in human populations.

As a result of many of these initiatives in developing countries, a limited capacity of tool development for genomics and bioinformatics approaches has occurred. However, much more remains to be achieved in technology and knowledge transfer, particularly in sub-Saharan Africa and Latin America. The main focus should be on developing genomics platforms leveraging on the next-generation sequencing approaches that remain to be established in much of the developing world.

Monitoring of Zoonotic Infections and Global Surveillance of Emerging and Reemerging Infectious Diseases

Events of emerging and reemerging infectious diseases in the human population remain constant occurrences in sub-Saharan Africa, Southeast Asia, and South America. Emerging infectious disease events such as SARS (Field, 2009) and the most recent pandemic of H1N1 (Gibbs et al., 2009) illustrate and confirm the constant flow of pathogens from wild and domesticated animals, and other reservoirs into the human population. Chikungunya fever, which is an arboviral infection, reemerged in Asia in 2005–2006 after a long period of quiescence (Bhatia and Narain, 2009). It is thought that factors including microbial, climatic, social, and economic aspects influenced the reemergence of the disease and the pace at which it spread, eventually resulting in high death rates (Bhatia and Narain, 2009).

Indeed, there are many microorganisms (viral, bacterial, and eukaryotes) that have moved into the human population and remain part of the human microbiota, which, when able to effectively survive, can cause either new diseases or disease with a much higher severity. Such cases of unknown and potentially pathogenic microorganisms in circulation in the human population are usually favored by factors related to (1) the ability of microorganisms to adapt in new hosts, (2) human actions (interactions with wild animals, hunting and effects on the environment leading to ecological disturbances), and (3) human movements as a result of global world travel (Field, 2009). Consequently, at any particular time, there could be a set of known and unknown microorganisms present in a given individual or a population as a result of their presence in and interaction with a specific environment or organisms therein such as animal reservoirs (i.e., bats, mice, and rats) of known and unknown microorganisms. The main challenge in the context of forecasting, and better yet preventing emerging and reemerging infectious diseases has been early detection and genetic identification of such known and unknown microorganisms. Global human microbiome studies using metagenomics analysis of known and unknown microorganisms provide unique but powerful opportunities to uncover the near-complete composition of the microbial content of an individual or a population at any given time, thus setting the stage for a comprehensive inventory of the genetic characteristics of potential human pathogens.

Studies of the human microbiome in the developing world will likely contribute significantly to the discovery of emerging pathogens (viruses, bacteria, and others) in circulation in humans. In fact, in both developed and developing countries the issues of early identification of emerging pathogens have been an impediment for the prevention of emerging and reemerging infections. Based on recent studies, human metagenomics coupled with the next-generation DNA sequencing provides an opportunity for early detection of microbial organisms even when there is significantly low abundance (Relman, 2002).

Because of the extreme importance of monitoring zoonotic infections, metagenomics studies should in principle be extended from humans to domesticated and nondomesticated animals. In fact, based on the technologies already available for human metagenomics studies, there has been increasing interest in launching animal metagenomics initiatives. Such initiatives will not only provide insights into the resident and transient microbial populations but also, in the case of natural reservoirs of given microorganisms, provide an opportunity to pinpoint the genetic changes that must occur for their adaptation to a new host – the human body. This is applicable in particular to invertebrate vectors and bats that are known to be host to a number of highly pathogenic viruses that pose significant public health problems to humans.

The Case of Selected Emerging Infectious Diseases: Tuberculosis (TB) and Leptospirosis

Developing countries are particularly affected by the impact of *Mycobacterium tuberculosis* virulence and TB drug resistance. This has been primarily because of genome plasticity in the causative agent. Unfortunately, available microarray-based platforms to identify strain diversity have not been fully implemented with the greatest TB incidence largely due the HIV/AIDS epidemic. The renewed interest and funding for top infectious diseases has recently revamped efforts to accelerate TB research, with a particular focus on the use of integrated approaches to find better control measures. In this context, it is proposed and highly anticipated that key aspects such as the integration of large-scale "omics," datasets focusing on parasite genetic determinants, host genetics, and host–parasite interactions will be crucial for this quest for better control measures. In addition, and given recent reports, the human microbiome would be a great addition to this integration of data in the context of systems biology. To this end, the evaluation of the human microbiome in cases of latent, nonlatent TB, and drug-resistant TB infections will provide insights into the role of the human (resident and nonresident) flora in various aspects of TB infections. Such information would most likely contribute to improving diagnosis, control, and spread of TB infection.

Another example of the potential to come from using human metagenomic research and approaches in the developing world relates to another emerging infectious pathogen that causes *Leptospirosis*. The Leptospires cause an infection that is associated with very high levels of mortality annually, but have received relatively little attention, probably because the infection is concentrated in the tropical regions and in the developing world. More than half a million cases are reported annually, and majority of these cases are associated with human exposure to pathogenic *Leptospira* species in the environment. Mortality rates as high as 25% have been recorded.

The genus *Leptospira* is serologically divided into two species: *L. interrogans*, which is pathogenic to humans and animals, and *L. biflexa*, a free-living nonpathogenic species found in water and wet soil. More than 16 pathogenic and saprophytic species have been recognized. Many animals including rodents and dogs are known to be reservoirs of *Leptospira*, and humans are considered to be the accidental hosts of this pathogen. Transmission of the pathogen is primarily from soil and water to mammalian tissues (often noticed following on large-scale flooding), with the infection occurring via penetrating leptospires through mucosa or open skin. Symptoms of leptospirosis include meningitis, pneumonitis, hepatitis, nephritis, pancreatitis, erythema nodosum, and death. No human vaccine against leptospirosis is available, and mild leptospirosis is treated with doxycycline, ampicillin, or amoxicillin. For severe leptospirosis, the primary therapy is penicillin G. The molecular diagnosis of Leptospirosis has been with traditional approaches such as restriction enzyme analysis, nucleic acid probes and hybridization, pulse field gel electrophoresis (PFGE), and varying ribotyping approaches.

Genome sequences from at least six *Leptospiras* have become available in the past few years, and these genomes are providing insight on the diversity of these species. In addition, the availability of these genomes is allowing for the identification of novel virulence factors, and ultimately will facilitate vaccine development. Recently, the genome sequence of the free-living *L. biflexa* was completed (Picardeau et al., 2008) and shown to contain 3,590 protein-coding genes distributed across three circular replicons. In the current study, it has been estimated that 2,052 genes (61%) represent a progenitor genome that existed before divergence of pathogenic and saprophytic *Leptospira* species. Basically, nearly one-third of the *L. biflexa* genes are absent in pathogenic *Leptospira*. In addition, 1,431 pathogen specific genes that are found in the pathogenic Leptospires are not present in *L. biflexa*. Of these, 893 genes have no known function suggesting that there are mechanisms that are unique to *Leptospira* and that the pathogenic specific genes need further study. The resulting genome studies suggest that there is still a significant amount of information that is not understood about the *Leptospiras*, particularly as it relates to how the species adapts to new environments and how the genomes mutate. Metagenomic studies of samples derived from infected populations will present an opportunity to study the pathogen without repeated passage where it has been shown to have genome rearrangements. In parallel, the pathogen can be studied directly in the environment when it is in transition from its natural host to humans (the accidental host). Interestingly, there is a large NIAID-funded project underway at the JCVI to sequence the genomes of an additional 400 *Leptospira* isolates (Joe Vinetz, personal communication; http://gsc.jcvi.org/).

Leptospirosis is another example of an emerging infectious disease that is prevalent in tropical environments and has not received as much attention as the major diseases in the developed world although the causative organisms result in a high mortality rate. Genomics and metagenomics approaches have the potential to increase the understanding of these species and their impact on human health.

Potential for Understanding and Control of Diarrheal Diseases

Diarrheal diseases remain one of the leading causes of deaths worldwide (Culligan et al., 2009). Specifically, diarrheal diseases are the second most common cause of child deaths worldwide, and more than 80% of child deaths due to diarrhea occur in Africa and south Asia. Worldwide, 88% of deaths from diarrhea are due to unsafe water and poor sanitation or hygiene. Three-quarters of all deaths from diarrhea in children younger than 5 years occur in 15 countries. There are about 2.5 billion cases of diarrhea among children each year, in addition to those who die from the disease. The UN reports that vaccines and better hygiene could decrease the number of deaths from diarrhea among children.

Since the 1970s, oral rehydration therapy has been the cornerstone of treatment programs. This therapy prevents dehydration that is associated with diarrhea. Giving zinc supplements with oral rehydration salts has also been shown to reduce the length of the illness and also the risk of more diarrhea episodes. Sixty percent (60%) of children in developing countries do not get the recommended treatment for diarrhea, which is vaccination against rotavirus, the leading cause of the disease. In fact, rotavirus causes about 40% of hospital admissions of children below 5 years suffering from diarrhea. Current therapies are focused on rehydration therapies but the studies from a human microbiome approach, coupled with the development of novel antibiotics and/or probiotics holds significant potential (Culligan et al., 2009).

Many diarrhoeal diseases have been associated with viruses (Ramani and Kang, 2009). Recent results suggest that viruses are present in as much as 43% of diarrheal samples in the developing world (Ramani and Kang, 2009). There are however a significant number of cases of diarrhea without obvious causes, thus making it difficult to control them. In addition and specifically in the case of rotaviruses, because of their high genetic diversity, the emergence of new genotypes, and the reassortment between different genotypes (Matthijnssens et al., 2009), there is constant need for surveillance of circulating strains. Human metagenomics studies hold the promise for increasing our understanding of the diversity of rotavirus and other etiological agents of diarrheal diseases. Based on previous studies, gastrointestinal tract metagenomics studies in both healthy and diarrheal patients in developing countries may lead to the identification and association of additional microorganisms (bacteria, viruses, and eukaryotes) with various cases of diarrheal diseases (Finkbeiner et al., 2008).

As an example, recent human microbiome studies have led to the discovery of a novel virus of the Cosavirus genus and its association with acute diarrhea in a child in Australia (Holtz et al., 2008). Regular and comprehensive metagenomics analyses focusing on acute and difficult-to-cure cases of diarrhea and diarrhea cases with known and unknown causes primarily in developing countries may provide opportunities for (1) a constant assessment of the diversity of known causative agents of diarrhea and (2) identification of new microorganisms as they relate to cases of diarrheal diseases.

Potential for Understanding Sexually Transmitted Diseases

Sexually transmitted diseases (STDs) are common infections throughout the developed and the developing world. STDs can result in premature birth, stillbirth, and neonatal infections (De Schryver and Meheus, 1990).

Many ongoing studies on BV aim to understanding the microbial populations that are present in the vaginal ecosystem and how they vary under health and disease conditions. Recent studies that are focused on 16S rRNA gene analysis have suggested that the extent of microbial diversity in the vaginal tract is not fully understood, which in turn has implications for current treatment regimes. This has potentially significant implications for asymptomatic disease conditions for example. Additional results show a lack of homogeneity within the vaginal tract, highlighting a complex ecosystem (Kim et al., 2009). Metagenomic approaches to studying this environment promise to give additional insights into the extent of diversity within this niche.

Ongoing studies in several parts in sub-Saharan Africa reveal that there is some relationship between the population of microbes that exists in the vaginal tract and STDs. Recently, van de Wijgert et al. (2008) described a study in which they investigated the relationships among BV, vaginal yeast, and vaginal practices, mucosal inflammation, and HIV acquisition. From a cohort of 4,531 HIV-negative women, they observed that women who were positive for BV or vaginal yeast had a higher likelihood to acquire HIV, and they concluded that BV and yeast may contribute more to the HIV epidemic than previously appreciated (van de Wijgert et al., 2008). Similar observations have been made in a review of all available literature on the extent to which BV may increase the risk of HIV acquisition (Atashili et al., 2008).

Earlier, in 2000, van De Wijgert et al. (2000) studied 169 Zimbabwean women to determine if intravaginal practices could be associated with changes in the vaginal flora and acquisition of STDs. In this study, they found that some disturbances of the flora could be associated with increased likelihood of STDs and HIV; the absence of *Lactobacilli* from the vaginal flora was associated with being positive for HIV (van De Wijgert et al., 2000). Martin et al. (1999) had similarly looked at a cohort of sex workers in Kenya and demonstrated that although only 26% of these women were colonized with *Lactobacillus* species at baseline, follow-up studies showed that the absence of culturable vaginal lactobacilli could be associated with the increased likelihood of acquiring HIV-1. Abnormal vaginal flora on Grams-stain was associated with increased risk of both HIV-1 acquisitions. This group proposed that the treatment of BV and the use of lactobacilli to colonize the vaginal cavity should be evaluated for reduce risk of acquiring HIV-1, gonorrhea, and trichomoniasis (Martin et al., 1999). How the microbial populations in the vaginal cavity can contribute to reduce chances of HIV infection is one of those major areas that need attention, and that will undoubtedly benefit from human microbiome research.

Potential for Enhancement of Malaria Treatment Regimens

According to The World Health Organisation (WHO, http://www.who.int/mediacentre/factsheets/fs094/en/), every 30 seconds a child dies of malaria, a disease that can be prevented and cured. In 2006 there were 247 million cases of malaria, and these resulted in nearly 1 million deaths mostly among African children. In fact, 90% of all malaria deaths occur in sub-Saharan Africa. People who live in lower-income communities, i.e., approximately half of the world's population, are at risk of the disease. The WHO reports that in 2006 malaria was present in 109 countries and territories.

The disease, however, can be eradicated, says Bill Gates. In an interview with the BBC World Services World Today program in January 2010, Gates said "*we have a vaccine that's in the last trial phase – called phase three.*" He added that "a partially effective vaccine could be available within 3 years." A vaccine that is fully effective against malaria would take 5–10 years to develop.

Although most cases of malaria are found in sub-Saharan Africa, there are other countries, including in Asia, Latin America, the Middle East, and parts of Europe, that are also affected. Key interventions include prompt and effective treatment with artemisin-based combination therapies; people at risk using insecticide nets; and indoor residual spraying with insecticide to control the vector mosquitoes. Genomics approaches have already been used to elucidate the genomes of several of the *Plasmodium* species (Gardner et al., 2002; Carlton et al., 2008; Pain et al., 2008; Mitsui et al., 2009), but new metagenomics approaches present opportunities to monitor the impact of the parasite of the microbial communities that reside on and in the human body, with a longer-term potential to develop novel probiotic approaches to supplement nutrition of infected individuals while the parasite runs its course.

In countries that have a high rate of malaria, economic growth rates may be cut by as much as 1.3%. In addition, genomic studies on the environments, in which the mosquitoes reside and breed, are being and will continue to allow for an increased understanding of the communities that they require for their survival (Ponnusamy et al., 2008a, b, 2009). This is particularly relevant since mosquitoes breed in areas where there are wet conditions, and the transmission of the disease can differ according to local factors such as rainfall, proximity of mosquito breeding sites to people, and the mosquito species in the area. A November 2009 report from Susan Anyangu in Nairobi, Kenya, carried by Inter Press Service (IPS) states that the RTS.S vaccine being developed is to be used specifically in Africa. It will be for infants and children aged less than 5 years (the most vulnerable to malaria). The vaccine could be ready for use in 5 years time.

Supplementing nutrition of people with malaria with probiotic solutions that have been derived from a metagenomic approach to understand the human microbiome holds significant promise. The FAO/WHO defines probiotics as "Live microorganisms which, when administered in adequate amounts confer a health benefit on the host." Probiotics have become more and more valuable over the past few years and are available in a number of food sources, including yogurts and other milk products, fermented and unfermented milk, and some juices. These live microorganisms are in most cases bacteria that are similar to beneficial microorganisms found in the gastrointestinal tract. Each species that is present in the gut environment would seem to hold some potential for use as a probiotic and therefore in human health.

Probiotics have been shown to be effective in treating irritable bowel syndrome (IBS), childhood and traveler's diarrhea, prevention and treatment of vaginal yeast infection and urinary tract infection, preventing and treating inflammation of the colon after surgery, reduction of the recurrence of bladder cancer, shortening the time of intestinal infections, and preventing eczema in children. Although the benefits of probiotics are evident, they have yet to be adapted extensively in the developing world (Reid and Devillard, 2004). Other ideas on the use of probiotics for reducing the morbidity and mortality associated with HIV/AIDS have been explored and proposed (Reid et al., 2005) where it has been proposed that lactic acid bacteria could play a role in maintaining the health of the human gut. We can only hope that as a result of the initiatives of the human microbiome project, new probiotics for a range of human health conditions may be developed based on baselines for people in different geographic locales.

Challenges: Funding and Technology Transfer

The efficient implementation of human microbiome research relies on the advanced instrumentation necessary for the processing of collected clinical samples, preparation and amplification of nucleic acid, and DNA sequencing. In addition, DNA sequence analysis also requires advanced bioinformatics resources. All genomics-related technologies developed over the past 10 years remain very expensive to be acquired by developing countries. This is usually justified by low-use volume and high costs of equipment and maintenance (Coloma and Harris, 2009). Therefore, as suggested by these authors, involvement of laboratories and institutions in developing countries should take advantage of "North–South" and "South–South" collaborations. Previous examples of successful "North–South" collaborations could be leveraged to initiate new ones in the context of human microbiome studies.

For the past several years, there have been numerous initiatives in developing countries to reduce the technological divide and hence begin to actively contribute to genomics research. In this context, activities have included training and capacity building in genomics and bioinformatics. In addition, there has also been an emphasis on the development of "Centers for Excellence" to provide resources and a critical scientific mass at regional levels. Four such regional "Centers of Excellence" are currently being established in eastern and central Africa, southern Africa, west Africa, and north Africa.

One of the most advanced "Centers for Excellence," Biosciences for Eastern and Central Africa (BecA), located at the International Livestock Research Institute (ILRI) in Nairobi, Kenya, has established facilities (with advanced genomics and bioinformatics resources) to support and accelerate research in a wide range of disciplines, including plant/crop sciences and animal sciences. Such infrastructure would ideally be poised for use as a focal point for the implementation of a regional initiative on the human microbiome. The existence of such facilities would normally be used to engage various African institutions in South–South collaborations. The "South–South" collaborations indeed provide opportunities to strengthen the scientific capacity of institutions in developing countries, which would be translated into their effective participation in North–South initiatives.

Genomics and metagenomics initiatives are usually quite expensive, and obviously, most institutions in the developing world would not be able to fund such activities independently. However, given the existence of several human microbiome projects in the United States, Canada, Europe, China, Japan, Singapore, and Australia, components in developing countries could easily be justified. For example, an African component of the human microbiome would provide elements to answering important outstanding microbiome questions, among which are included: (1) Is there a core human microbiome? (2) Does the composition of the human microbiome vary from one geographical region to another?

Given the anticipation of such interesting outcomes, existing initiatives could further provide seeds to launch other initiatives in the developing world. Furthermore, in the context of the use of biosciences for Africa's development, a strong case should be made to various stakeholders such as The African Union and other regional organizations to fund the African component of the human microbiome. This next wave of genomics research will not be without its own set of challenges. Recent studies, for example, show that many diseases present with similar observations, and as such initial surveys into the human microbiome under health and disease may give unexpected outcomes (Yazdanbakhsh and Kremsner, 2009).

Acknowledgement The authors wish to acknowledge the invaluable information found on the World Health Organisation (WHO) website and on the Mayo Clinic website.

References

Atashili J, Poole C, Ndumbe PM, Adimora AA and Smith JS (2008) Bacterial vaginosis and HIV acquisition: a meta-analysis of published studies. AIDS 22:1493–1501

Bentley S (2009) Sequencing the species pan-genome. Nat Rev Microbiol 7:258–259

Berriman M, Ghedin E, Hertz-Fowler C, Blandin G, Renauld H, Bartholomeu DC, Lennard NJ, Caler E, Hamlin NE, Haas B, Bohme U, Hannick L, Aslett MA, Shallom J, Marcello L, Hou L, Wickstead B, Alsmark UC, Arrowsmith C, Atkin RJ, Barron AJ, Bringaud F, Brooks K, Carrington M, Cherevach I, Chillingworth TJ, Churcher C, Clark LN, Corton CH, Cronin A, Davies RM, Doggett J, Djikeng A, Feldblyum T, Field MC, Fraser A, Goodhead I, Hance Z, Harper D, Harris BR, Hauser H, Hostetler J, Ivens A, Jagels K, Johnson D, Johnson J, Jones K, Kerhornou AX, Koo H, Larke N, Landfear S, Larkin C, Leech V, Line A, Lord A, Macleod A, Mooney PJ, Moule S, Martin DM, Morgan GW, Mungall K, Norbertczak H, Ormond D, Pai G, Peacock CS, Peterson J, Quail MA, Rabbinowitsch E, Rajandream MA, Reitter C, Salzberg SL, Sanders M, Schobel S, Sharp S, Simmonds M, Simpson AJ, Tallon L, Turner CM, Tait A, Tivey AR, Van Aken S, Walker D, Wanless D, Wang S, White B, White O, Whitehead S, Woodward J, Wortman J, Adams MD, Embley TM, Gull K, Ullu E, Barry JD, Fairlamb AH, Opperdoes F, Barrell BG, Donelson JE, Hall N, Fraser CM, Melville SE and El-Sayed NM (2005) The genome of the African trypanosome *Trypanosoma brucei*. Science 309:416–422

Bhatia R, Narain JP (2009) Re-emerging chikungunya fever: some lessons from Asia. Trop Med Int Health 14:940–946

Bik EM, Eckburg PB, Gill SR, Nelson KE, Purdom EA, Francois F, Perez-Perez G, Blaser MJ, Relman DA (2006) Molecular analysis of the bacterial microbiota in the human stomach. Proc Natl Acad Sci USA 103:732–737

Bishop R, Shah T, Pelle R, Hoyle D, Pearson T, Haines L, Brass A, Hulme H, Graham SP, Taracha EL, Kanga S, Lu C, Hass B, Wortman J, White O, Gardner MJ, Nene V, de Villiers EP (2005) Analysis of the transcriptome of the protozoan *Theileria parva* using MPSS reveals that the majority of genes are transcriptionally active in the schizont stage. Nucleic Acids Res 33: 5503–5511

Callister SJ, McCue LA, Turse JE, Monroe ME, Auberry KJ, Smith RD, Adkins JN, Lipton MS (2008) Comparative bacterial proteomics: analysis of the core genome concept. PLoS One 3:e1542

Carlton JM, Adams JH, Silva JC, Bidwell SL, Lorenzi H, Caler E, Crabtree J, Angiuoli SV, Merino EF, Amedeo P, Cheng Q, Coulson RM, Crabb BS, Del Portillo HA, Essien K, Feldblyum TV, Fernandez-Becerra C, Gilson PR, Gueye AH, Guo X, Kang'a S, Kooij TW, Korsinczky M, Meyer EV, Nene V, Paulsen I, White O, Ralph SA, Ren Q, Sargeant TJ, Salzberg SL, Stoeckert CJ, Sullivan SA, Yamamoto MM, Hoffman SL, Wortman JR, Gardner MJ, Galinski MR, Barnwell JW, Fraser-Liggett CM (2008) Comparative genomics of the neglected human malaria parasite *Plasmodium vivax*. Nature 455:757–763

Chain PS, Grafham DV, Fulton RS, Fitzgerald MG, Hostetler J, Muzny D, Ali J, Birren B, Bruce DC, Buhay C, Cole JR, Ding Y, Dugan S, Field D, Garrity GM, Gibbs R, Graves T, Han CS, Harrison SH, Highlander S, Hugenholtz P, Khouri HM, Kodira CD, Kolker E, Kyrpides NC, Lang D, Lapidus A, Malfatti SA, Markowitz V, Metha T, Nelson KE, Parkhill J, Pitluck S, Qin X, Read TD, Schmutz J, Sozhamannan S, Sterk P, Strausberg RL, Sutton G, Thomson NR, Tiedje JM, Weinstock G, Wollam A, Detter JC (2009) Genomics. Genome project standards in a new era of sequencing. Science 326:236–237

Coloma J, Harris E (2009) Molecular genomic approaches to infectious diseases in resource-limited settings. PLoS Med 6:e1000142

Costello EK, Lauber CL, Hamady M, Fierer N, Gordon JI and Knight R (2009) Bacterial community variation in human body habitats across space and time. Science 326: 1694–1697

Culligan EP, Hill C, Sleator RD (2009) Probiotics and gastrointestinal disease: successes, problems and future prospects. Gut Pathog 1:19

De Schryver A, Meheus A (1990) Epidemiology of sexually transmitted diseases: the global picture. Bull World Health Organ 68:639–654
Dorer MS, Talarico S, Salama NR (2009) *Helicobacter pylori's* unconventional role in health and disease. PLoS Pathog 5:e1000544
Eckburg PB, Bik EM, Bernstein CN, Purdom E, Dethlefsen L, Sargent M, Gill SR, Nelson KE, Relman DA (2005) Diversity of the human intestinal microbial flora. Science 308:1635–1638
El-Sayed NM, Myler PJ, Blandin G, Berriman M, Crabtree J, Aggarwal G, Caler E, Renauld H, Worthey EA, Hertz-Fowler C, Ghedin E, Peacock C, Bartholomeu DC, Haas BJ, Tran AN, Wortman JR, Alsmark UC, Angiuoli S, Anupama A, Badger J, Bringaud F, Cadag E, Carlton JM, Cerqueira GC, Creasy T, Delcher AL, Djikeng A, Embley TM, Hauser C, Ivens AC, Kummerfeld SK, Pereira-Leal JB, Nilsson D, Peterson J, Salzberg SL, Shallom J, Silva JC, Sundaram J, Westenberger S, White O, Melville SE, Donelson JE, Andersson B, Stuart KD, Hall N (2005) Comparative genomics of trypanosomatid parasitic protozoa. Science 309:404–409
Field HE (2009) Bats and emerging zoonoses: henipaviruses and SARS. Zoonoses Public Health 2009-May 28th
Finkbeiner SR, Allred AF, Tarr PI, Klein EJ, Kirkwood CD, Wang D (2008) Metagenomic analysis of human diarrhea: viral detection and discovery. PLoS Pathog 4:e1000011
Foster JA, Krone SM, Forney LJ (2008) Application of ecological network theory to the human microbiome. Interdiscip Perspect Infect Dis 2008:839501
Gao Z, Tseng CH, Pei Z, Blaser MJ (2007) Molecular analysis of human forearm superficial skin bacterial biota. Proc Natl Acad Sci USA 104:2927–2932
Gao Z, Tseng CH, Strober BE, Pei Z, Blaser MJ (2008) Substantial alterations of the cutaneous bacterial biota in psoriatic lesions. PLoS One 3:e2719
Gardner MJ, Bishop R, Shah T, de Villiers EP, Carlton JM, Hall N, Ren Q, Paulsen IT, Pain A, Berriman M, Wilson RJ, Sato S, Ralph SA, Mann DJ, Xiong Z, Shallom SJ, Weidman J, Jiang L, Lynn J, Weaver B, Shoaibi A, Domingo AR, Wasawo D, Crabtree J, Wortman JR, Haas B, Angiuoli SV, Creasy TH, Lu C, Suh B, Silva JC, Utterback TR, Feldblyum TV, Pertea M, Allen J, Nierman WC, Taracha EL, Salzberg SL, White OR, Fitzhugh HA, Morzaria S, Venter JC, Fraser CM, Nene V (2005) Genome sequence of *Theileria parva*, a bovine pathogen that transforms lymphocytes. Science 309:134–137
Gardner MJ, Hall N, Fung E, White O, Berriman M, Hyman RW, Carlton JM, Pain A, Nelson KE, Bowman S, Paulsen IT, James K, Eisen JA, Rutherford K, Salzberg SL, Craig A, Kyes S, Chan MS, Nene V, Shallom SJ, Suh B, Peterson J, Angiuoli S, Pertea M, Allen J, Selengut J, Haft D, Mather MW, Vaidya AB, Martin DM, Fairlamb AH, Fraunholz MJ, Roos DS, Ralph SA, McFadden GI, Cummings LM, Subramanian GM, Mungall C, Venter JC, Carucci DJ, Hoffman SL, Newbold C, Davis RW, Fraser CM, Barrell B (2002) Genome sequence of the human malaria parasite *Plasmodium falciparum*. Nature 419:498–511
Gibbs AJ, Armstrong JS, Downie JC (2009) From where did the 2009 'swine-origin' influenza A virus (H1N1) emerge? Virol J 6:207
Gill SR, Pop M, Deboy RT, Eckburg PB, Turnbaugh PJ, Samuel BS, Gordon JI, Relman DA, Fraser-Liggett CM, Nelson KE (2006) Metagenomic analysis of the human distal gut microbiome. Science 312:1355–1359
Grice EA, Kong HH, Conlan S, Deming CB, Davis J, Young AC, Bouffard GG, Blakesley RW, Murray PR, Green ED, Turner ML, Segre JA (2009) Topographical and temporal diversity of the human skin microbiome. Science 324:1190–1192
Holtz LR, Finkbeiner SR, Kirkwood CD, Wang D (2008) Identification of a novel picornavirus related to cosaviruses in a child with acute diarrhea. Virol J 5:159
Kim TK, Thomas SM, Ho M, Sharma S, Reich CI, Frank JA, Yeater KM, Biggs DR, Nakamura N, Stumpf R, Leigh SR, Tapping RI, Blanke SR, Slauch JM, Gaskins HR, Weisbaum JS, Olsen GJ, Hoyer LL, Wilson BA (2009) Heterogeneity of vaginal microbial communities within individuals. J Clin Microbiol 47:1181–1189

Kudva IT, Evans PS, Perna NT, Barrett TJ, Ausubel FM, Blattner FR, Calderwood SB (2002) Strains of *Escherichia coli* O157:H7 differ primarily by insertions or deletions, not single-nucleotide polymorphisms. J Bacteriol 184:1873–1879

Lasken RS (2009) Genomic DNA amplification by the multiple displacement amplification (MDA) method. Biochem Soc Trans 37:450–453

Mahmoud A, Zerhouni E (2009) Neglected tropical diseases: moving beyond mass drug treatment to understanding the science. Health Aff (Millwood) 28:1726–1733

Manichanh C, Rigottier-Gois L, Bonnaud E, Gloux K, Pelletier E, Frangeul L, Nalin R, Jarrin C, Chardon P, Marteau P, Roca J, Dore J (2006) Reduced diversity of faecal microbiota in Crohn's disease revealed by a metagenomic approach. Gut 55:205–211

Martin HL, Richardson BA, Nyange PM, Lavreys L, Hillier SL, Chohan B, Mandaliya K, Ndinya-Achola JO, Bwayo J, Kreiss J (1999) Vaginal lactobacilli, microbial flora, and risk of human immunodeficiency virus type 1 and sexually transmitted disease acquisition. J Infect Dis 180:1863–1868

Matthijnssens J, Bilcke J, Ciarlet M, Martella V, Banyai K, Rahman M, Zeller M, Beutels P, Van Damme P, Van Ranst M (2009) Rotavirus disease and vaccination: impact on genotype diversity. Future Microbiol 4:1303–1316

Mitsui H, Arisue N, Sakihama N, Inagaki Y, Horii T, Hasegawa M, Tanabe K, Hashimoto T (2009) Phylogeny of Asian primate malaria parasites inferred from apicoplast genome-encoded genes with special emphasis on the positions of *Plasmodium vivax* and *P. fragile*. Gene 2010, Gene 450:32–38

Mullard A (2008) Microbiology: the inside story. Nature 453:578–580

Nealson KH, Venter JC (2007) Metagenomics and the global ocean survey: what's in it for us, and why should we care? ISME J 1:185–187

Nene V, Bishop R, Morzaria S, Gardner MJ, Sugimoto C, ole-MoiYoi OK, Fraser CM, Irvin A (2000) *Theileria parva* genomics reveals an atypical apicomplexan genome. Int J Parasitol 30:465–474

Pain A, Bohme U, Berry AE, Mungall K, Finn RD, Jackson AP, Mourier T, Mistry J, Pasini EM, Aslett MA, Balasubrammaniam S, Borgwardt K, Brooks K, Carret C, Carver TJ, Cherevach I, Chillingworth T, Clark TG, Galinski MR, Hall N, Harper D, Harris D, Hauser H, Ivens A, Janssen CS, Keane T, Larke N, Lapp S, Marti M, Moule S, Meyer IM, Ormond D, Peters N, Sanders M, Sanders S, Sargeant TJ, Simmonds M, Smith F, Squares R, Thurston S, Tivey AR, Walker D, White B, Zuiderwijk E, Churcher C, Quail MA, Cowman AF, Turner CM, Rajandream MA, Kocken CH, Thomas AW, Newbold CI, Barrell BG, Berriman M (2008) The genome of the simian and human malaria parasite *Plasmodium knowlesi*. Nature 455: 799–803

Perna NT, Plunkett G 3rd, Burland V, Mau B, Glasner JD, Rose DJ, Mayhew GF, Evans PS, Gregor J, Kirkpatrick HA, Posfai G, Hackett J, Klink S, Boutin A, Shao Y, Miller L, Grotbeck EJ, Davis NW, Lim A, Dimalanta ET, Potamousis KD, Apodaca J, Anantharaman TS, Lin J, Yen G, Schwartz DC, Welch RA, Blattner FR (2001) Genome sequence of enterohaemorrhagic *Escherichia coli* O157:H7. Nature 409:529–533

Peterson J, Garges S, Giovanni M, McInnes P, Wang L, Schloss JA, Bonazzi V, McEwen JE, Wetterstrand KA, Deal C, Baker CC, Di Francesco V, Howcroft TK, Karp RW, Lunsford RD, Wellington CR, Belachew T, Wright M, Giblin C, David H, Mills M, Salomon R, Mullins C, Akolkar B, Begg L, Davis C, Grandison L, Humble M, Khalsa J, Little AR, Peavy H, Pontzer C, Portnoy M, Sayre MH, Starke-Reed P, Zakhari S, Read J, Watson B, Guyer M (2009) The NIH human microbiome project. Genome Res 19(12):2317–2323

Picardeau M, Bulach DM, Bouchier C, Zuerner RL, Zidane N, Wilson PJ, Creno S, Kuczek ES, Bommezzadri S, Davis JC, McGrath A, Johnson MJ, Boursaux-Eude C, Seemann T, Rouy Z, Coppel RL, Rood JI, Lajus A, Davies JK, Medigue C, Adler B (2008) Genome sequence of the saprophyte *Leptospira biflexa* provides insights into the evolution of Leptospira and the pathogenesis of leptospirosis. PLoS One 3:e1607

Ponnusamy L, Wesson DM, Arellano C, Schal C, Apperson CS (2009) Species composition of bacterial communities influences attraction of mosquitoes to experimental plant infusions. Microb Ecol 2010 59:158–73

Ponnusamy L, Xu N, Nojima S, Wesson DM, Schal C, Apperson CS (2008a) Identification of bacteria and bacteria-associated chemical cues that mediate oviposition site preferences by *Aedes aegypti*. Proc Natl Acad Sci USA 105:9262–9267

Ponnusamy L, Xu N, Stav G, Wesson DM, Schal C, Apperson CS (2008b) Diversity of bacterial communities in container habitats of mosquitoes. Microb Ecol 56:593–603

Ramani S, Kang G (2009) Viruses causing childhood diarrhoea in the developing world. Curr Opin Infect Dis 22:477–482

Reid G, Devillard E (2004) Probiotics for mother and child. J Clin Gastroenterol 38:S94–101

Reid G, Anand S, Bingham MO, Mbugua G, Wadstrom T, Fuller R, Anukam K, Katsivo M (2005) Probiotics for the developing world. J Clin Gastroenterol 39:485–488

Relman DA (2002) New technologies, human–microbe interactions, and the search for previously unrecognized pathogens. J Infect Dis 186(Suppl 2):S254–258

Rusch DB, Halpern AL, Sutton G, Heidelberg KB, Williamson S, Yooseph S, Wu D, Eisen JA, Hoffman JM, Remington K, Beeson K, Tran B, Smith H, Baden-Tillson H, Stewart C, Thorpe J, Freeman J, Andrews-Pfannkoch C, Venter JE, Li K, Kravitz S, Heidelberg JF, Utterback T, Rogers YH, Falcon LI, Souza V, Bonilla-Rosso G, Eguiarte LE, Karl DM, Sathyendranath S, Platt T, Bermingham E, Gallardo V, Tamayo-Castillo G, Ferrari MR, Strausberg RL, Nealson K, Friedman R, Frazier M, Venter JC (2007) The sorcerer II global ocean sampling expedition: northwest Atlantic through eastern tropical Pacific. PLoS Biol 5:e77

Seksik P, Sokol H, Lepage P, Vasquez N, Manichanh C, Mangin I, Pochart P, Doré J, Marteau P (2006) The role of bacteria in onset and perpetuation of inflammatory bowel disease. Aliment Pharmacol Ther 24 Suppl 3:11–18

Seksik P, Cosnes J, Sokol H, Nion-Larmurier I, Gendre JP, Beaugerie L. (2009) Incidence of benign upper respiratory tract infections, HSV and HPV cutaneous infections in inflammatory bowel disease patients treated with azathioprine. Aliment Pharmacol Ther 29:1106–1113

Sokol H, Lepage P, Seksik P, Doré J, Marteau P (2006) Temperature gradient gel electrophoresis of fecal 16S rRNA reveals active Escherichia coli in the microbiota of patients with ulcerative colitis. J Clin Microbiol 44: 3172–3177

Sokol H, Lepage P, Seksik P, Doré J, Marteau P (2007) Molecular comparison of dominant microbiota associated with injured versus healthy mucosa in ulcerative colitis. Gut 56:152–154

Sokol H, Pigneur B, Watterlot L, Lakhdari O, Bermudez-Humaran LG, Gratadoux JJ, Blugeon S, Bridonneau C, Furet JP, Corthier G, Grangette C, Vasquez N, Pochart P, Trugnan G, Thomas G, Blottiere HM, Dore J, Marteau P, Seksik P, Langella P (2008) *Faecalibacterium prausnitzii* is an anti-inflammatory commensal bacterium identified by gut microbiota analysis of Crohn disease patients. Proc Natl Acad Sci USA 105:16731–16736

Turnbaugh PJ, Backhed F, Fulton L, Gordon JI (2008) Diet-induced obesity is linked to marked but reversible alterations in the mouse distal gut microbiome. Cell Host Microbe 3:213–223

Turnbaugh PJ, Hamady M, Yatsunenko T, Cantarel BL, Duncan A, Ley RE, Sogin ML, Jones WJ, Roe BA, Affourtit JP, Egholm M, Henrissat B, Heath AC, Knight R, Gordon JI (2009) A core gut microbiome in obese and lean twins. Nature 457:480–484

van De Wijgert JH, Mason PR, Gwanzura L, Mbizvo MT, Chirenje ZM, Iliff V, Shiboski S, Padian NS (2000) Intravaginal practices, vaginal flora disturbances, and acquisition of sexually transmitted diseases in Zimbabwean women. J Infect Dis 181:587–594

van de Wijgert JH, Morrison CS, Cornelisse PG, Munjoma M, Moncada J, Awio P, Wang J, Van der Pol B, Chipato T, Salata RA, Padian NS (2008) Bacterial vaginosis and vaginal yeast, but not vaginal cleansing, increase HIV-1 acquisition in African women. J Acquir Immune Defic Syndr 48:203–210

Yazdanbakhsh M, Kremsner PG (2009) Influenza in Africa. PLoS Med 6:e1000182

Yooseph S, Sutton G, Rusch DB, Halpern AL, Williamson SJ, Remington K, Eisen JA, Heidelberg KB, Manning G, Li W, Jaroszewski L, Cieplak P, Miller CS, Li H, Mashiyama ST, Joachimiak

MP, van Belle C, Chandonia JM, Soergel DA, Zhai Y, Natarajan K, Lee S, Raphael BJ, Bafna V, Friedman R, Brenner SE, Godzik A, Eisenberg D, Dixon JE, Taylor SS, Strausberg RL, Frazier M, Venter JC (2007) The sorcerer II global ocean sampling expedition: expanding the universe of protein families. PLoS Biol 5:e16

Yutin N, Suzuki MT, Teeling H, Weber M, Venter JC, Rusch DB, Beja O (2007) Assessing diversity and biogeography of aerobic anoxygenic phototrophic bacteria in surface waters of the Atlantic and Pacific Oceans using the global ocean sampling expedition metagenomes. Environ Microbiol 9:1464–1475

Index

Note: The letters 'f' and 't' following locators refer to figures and tables respectively.

K.E. Nelson (ed.), *Metagenomics of the Human Body*,
DOI 10.1007/978-1-4419-7089-3,

B

Q

R

S

T

U

MIX
Papier aus verantwortungsvollen Quellen
Paper from responsible sources
FSC® C105338

Printed by Books on Demand, Germany